ORIGINS

ORIGINS

The Search for Our Prehistoric Past

FRANK H. T. RHODES

Comstock Publishing Associates
a division of
Cornell University Press
Ithaca and London

First published 2016 by Cornell University Press

Printed in the United States of America

Library of Congress Cataloging-in-Publication Data

Names: Rhodes, Frank Harold Trevor, author.
Title: Origins : the search for our prehistoric past / Frank H.T. Rhodes.
Description: Ithaca : Comstock Publishing Associates, a division of Cornell University Press, 2016. | Includes bibliographical references and index.
Identifiers: LCCN 2015046455 | ISBN 9781501702440 (cloth : alk. paper)
Subjects: LCSH: Evolution (Biology)—Popular works. | Life—Origin—Popular works. | Phylogeny—Popular works.
Classification: LCC QH367 .R495 2016 | DDC 570—dc23
LC record available at http://lccn.loc.gov/2015046455

Cornell University Press strives to use environmentally responsible suppliers and materials to the fullest extent possible in the publishing of its books. Such materials include vegetable-based, low-VOC inks and acid-free papers that are recycled, totally chlorine-free, or partly composed of nonwood fibers. For further information, visit our website at www.cornellpress.cornell.edu.

Cloth printing 10 9 8 7 6 5 4 3 2 1

CONTENTS

PREFACE

Scores of books have been written on origins, ranging from the most comprehensive to the very specific: from Ron Redfern's superbly illustrated book on the *Origins of Continents and Oceans,* to Richard Leakey and Roger Lewin's *Origins,* from Neil de Grasse Tyson and Donald Goldsmith's comprehensive *Origins: Fourteen Billion Years of Cosmic Evolution*, for example, to Sabir Abdu Samee's *Origins of the Bangladesh Army,* and Shidao Xu's *Origins of Chinese Cuisine*. The list goes on: *The Origin of Wealth, The Origin of Consciousness, The Origin of Stories, The Origin of Languages,* The *Origin of the Soul, The Origin of Geometry, The Origin of Ball Games, The Origin of the Bhutanese Mask Dances, The Origin of Satan, The Origin of Concepts, The Origin of Buddhist Meditation, The Origin of Phyla, The Origin of Mormonism, The Origin of Basketball.* And then, of course, there's the best known of all: *The Origin of Species,* by Charles Darwin, as well as the fervently opposed and mildly disrespectful *The Origin of Specious Nonsense* by John J. May.

And television series, movies, and e-books on origins are scarcely less common, from the factual—Neil de Grasse Tyson's PBS Nova Series *Origins*, for example—to the fictional: X-Men *Origins: Wolverine*. There are several university centers devoted to origins. NASA has an Origins of Life program. And there is at least one scholarly journal, *Origins of Life*, published by the International Society for the Study of the Origins of Life. The Leakey Foundation is devoted to research related to human origins.

This plethora of publications and associations devoted to the study of origins reflects our profound interest in understanding how all things present have come to be what they now are. But our particular preoccupation with our own origins and relationships reflects, perhaps, something more: our distinctive need to understand our own place in nature and so secure a baseline, an essential starting point, from which to view everything else. That, I suppose, is why the first book of the Bible, Genesis, begins with the words "In the beginning." Every subsequent relationship and activity is influenced, to some degree, by our origins.

One of the several hazards confronting any writer on the origin and development of living things is to so burden the account with the marvelous particularities of ancient forms—the ostracoderms, the labyrinthodonts, the stegocephalians, the titanotheres, and a hundred other fascinating and long-vanished groups—as to overwhelm and obscure the extent and wonder of this dazzling diversity over an almost inconceivable span of time. Any general account of this evolutionary cavalcade requires a framework of fact and particular example, but it also requires some deliberate attempt to stand back and gaze at what Darwin called the "grandeur in this view of life."

This becomes a particular challenge, because a wholesale loss of the more familiar traditional group names and classification has occurred with the broad adoption by the biological community over the last decade or so of a system of classification known as "cladistics." Cladistics is based on supposed inherited characteristics, rather than observable structural similarities. Cladistics uses these inferred relationships, based on lines of descent, to arrange biological groups (taxa) into a branching hierarchy in which all members of a given category share a presumed common ancestor. These groups, known as "clades," are branches, each consisting of an ancestral organism and all its supposed descendants, both living and fossil. Thus birds, dinosaurs, crocodiles, and all their descendants constitute a single clade. Clades, in short, emphasize supposed ancestry at the expense of readily observable characteristics.

This system poses particular problems for the general reader in dealing with generally recognized groups such as reptiles, for example. The familiar definition of reptiles as a distinctive group of cold-blooded vertebrates, usually egg-laying, with scaly skin, is replaced because it is presumed the reptiles inherited these characteristics from their common ancestor with birds, though birds themselves are not cold-blooded. Thus dinosaurs, a category once universally recognized and understood, has now been redefined as a group that includes birds such that "dinosaurs" now become "nonavian dinosaurs." An additional complication is that many paleontologists have declined to use the new cladistic method of classification, making the situation increasingly confusing.[1]

I have therefore thought it best to illustrate the new cladistic clarification based on presumed evolutionary relationships by use of several general cladograms, (e.g., figs. 8.4, 12.1, 12.3), showing broadly accepted patterns of ancestry, while retaining the use of familiar and traditional categories such as reptiles, amphibians, and so on, in the descriptions in the text. My hope is that this system will render the text accessible to the general reader without doing violence to the current direction of taxonomic practice.

Sprinkled throughout the various chapters, I have included thumbnail sketches of a few of the many individuals involved in unraveling the long history of life. It is, after all, far too easy to accept our present version of that history and the various events and processes behind it, as "given": something we can assume and take for granted. But the unraveling of this ancient story involved competing viewpoints, challenges to cherished orthodoxies, tensions with long-held convictions—even, in some cases, the contradiction of "common sense." Understanding something of those who led those distant debates provides a useful background against which to understand both present orthodoxies and continuing contention. For no history is ever complete, no explanation infallible. The search continues to be open ended, subject to the next discovery in the field or the latest experiment in the lab.

ACKNOWLEDGMENTS

I am greatly indebted to a number of colleagues who have very kindly reviewed sections of my manuscript and given me the benefit of their comments and advice. These include Warren Allmon (chapters 1–5, 15), John Cisne (chapters 9, 10, 12, 13), William Crepet (chapters 7 and 11), the late Kenneth Kennedy (chapter 14), Amy McCune (chapter 6), William Schopf (chapter 3), and Keith Thomson (chapters 6, 8, and 16).

They are not responsible, of course, for whatever mistakes may remain.

I am also deeply grateful for the help and support of two other people. My executive assistant, M. Joy Wagner, has helped me at every turn, given me endless technical help and support, and encouraged and helped me to complete the manuscript.

Rachel Parks Rochefort, lawyer, dedicated environmentalist, and public policy exponent, typed and retyped the manuscript with endless patience, skill, and goodwill. She unearthed references, checked dates, organized figures, clarified my descriptions, and, at every stage, played the role of constructive critic and patient enabler. My debt to her is great, and I am profoundly grateful for the countless ways she has helped, encouraged, and supported my writing.

At Cornell University Press I warmly thank Kitty Liu, Emily Powers, and Dina Dineva for their contributions. I am especially grateful for the generous help of Ange Romeo-Hall, managing editor, and for her consummate editorial skills.

A small portion of chapter 17 is drawn from a paper I read before the Geologists' Association, and I offer my kind thanks to the association for its appearance here.

ORIGINS

1

DEFROSTING THE MAMMOTH

In spring 2007, in the remote Yamal Peninsula of Arctic Russia, Yuri Khudi, a nomadic reindeer breeder and hunter, and his three sons discovered a frozen carcass, lying partly exposed in the melting snow. Familiar with the mammoth tusks and bones that are frequently found in the region, Khudi recognized the remains as those of a baby mammoth. Later measurements showed it to be a female, some 3 feet tall, weighing 110 pounds, perhaps only a month old, perfectly preserved in the ice after its death, some 40,000 years ago. Lyuba, as the baby subsequently came to be called, after Khudi's wife, has proved a treasure trove of information.[1] Lyuba is not the only frozen mammoth known, but she is by far the best preserved. Already she is providing new information on the diet, soft parts, and early development of these creatures.

Lyuba is a member of the elephant-like species *Mammuthus primigenius*. Like all species of mammoths, this one is now extinct. Mammoths emerged in Africa about 4 million years ago and later expanded into Eurasia during the closing years of the last Ice Age, a period extending from some 3 million to 10,000 years ago. Their migration to North America some 1.8 million years ago involved crossing the land bridge over the Bering Strait. Woolly mammoths were about the size of living elephants, but they had a distinctive appearance, with a humped and sloping back, huge curved tusks, small ears, and a long shaggy coat. Their appearance is confirmed by prehistoric cave paintings found in human rock shelters in several parts of Europe, showing them to be long-haired, humpbacked creatures.

Dwarf versions of the woolly mammoth have been found in the Channel Islands off California and on Wrangle Island off the coast of northern Russia, where they survived until about 4,700 years ago.

Fossil remains of woolly mammoths have long been a source of fascination. Early legends described gigantic creatures that lived below the ice of the Arctic; others thought them the remains of the war elephants used by Hannibal, the great Carthaginian general, as part of his army that crossed the Pyrenees and the Alps as it invaded northern Italy in 218 BC.

One person fascinated by mammoths was Thomas Jefferson, the third president of the United States. Jefferson, at his own expense, commissioned William Clark (of Lewis and Clark fame) to search for mammoth remains at Big Bone Lick, Kentucky. Jefferson also led a fund-raising campaign for a thousand guineas to support a transcontinental search for mammoth remains. Such was his enthusiasm that the unfinished East Room of the White House was filled with fossils and became known as the Mastodon Room.[2] Mastodons ("breast teeth") had molar teeth whose knobbly structure is a reflection of their browsing habits and diet of twigs and leaves, in contrast to the flat, ridged platelike teeth of grazing mammoths. Jefferson's enthusiasm reflected two distinctive beliefs. He was convinced that the American mammoth was larger than the European, thus refuting the claims of the French naturalist Georges-Louis Leclerc, Comte de Buffon, that New World animals were smaller than, and thus inferior to, their Old World counterparts. Jefferson was also convinced that no species could become extinct, and so his western expedition was based in part on the search for living mammoths, which, he thought, might still be surviving in the unexplored west.

The person who provided clear evidence that the mammoth was, indeed, extinct, was the great French naturalist Georges Cuvier (1769–1832). Cuvier, who was largely self-taught, was appointed by the revolutionary government as professor of anatomy at the newly established National Museum of Natural History in Paris. Cuvier studied the fossil remains of elephants (what we call mammoths and mastodons today) found in rocks of the Paris basin and concluded that they were distinct from both the living African and the Asian elephants. In this belief he echoed the conclusion of Jefferson, but, unlike Jefferson, he concluded that the mastodon fossils must represent creatures now extinct. The idea of extinction was unpalatable to many of his contemporaries, who argued that it implied a lack of providential concern for creatures forming essential links in the Great Chain of Being, within which every creature and object had its ordained place in a divine hierarchical order.

Cuvier countered this objection by suggesting that the extinction had taken place during a global catastrophe that, like the flood of Noah, led to the destruction of all living things, to be followed by a new creation. Cuvier's recognition of the fossil mastodon as a separate species, distinct from both living elephants and fossil mammoths, was of major significance. Cuvier was one of the greatest anatomists of all time, and this clear demonstration of the reality of extinction was an important contribution to the broader understanding of the history of life, even though the Catastrophe Theory he suggested to explain it is no longer supported. By this concept, the Earth had experienced a number of worldwide catastrophes—of which the flood of Noah was the most recent—each of which had destroyed the existing creation, which was replaced by one that was newly created.

In embracing this theory of catastrophism, Cuvier was opposed by his contemporary, Jean Baptiste Chevalier de Lamarck, who, with others, pioneered the concept of evolution.

Mammoths and living African and Asian elephants, though distinct, are closely related, having all evolved from a common ancestral stock about 5 million years ago. They differ from mastodons, whose teeth—as we have seen—had pointed cusps, adapted to crushing the coarse plants and bushes on which they fed. Mammoths and living elephants, in contrast, had a quite different diet from the mastodons. Their flat-ridged teeth were adapted to grazing on harsh prairie grasses. The number of plates and the amount of wear of mammoth molars can provide an indication of the age of individuals.

But if mammoths and mastodons were so abundant and so widespread, what caused them to become extinct? They had survived for some 4 million years and had spread to four continents. What brought about their extinction? Various suggestions have been made, but the most plausible seems to be the changing climate at the end of the Ice Age, which led to drastic changes in vegetation and food supplies, as well as human competition and predation. About 10,000 years ago, warming temperatures produced the melting of glaciers and created profound changes in climate and vegetation across the Northern Hemisphere. Recent studies suggest the decline of grasses, willows, and drier herbaceous vegetation on which these creatures thrived, and their replacement by birch shrubland and wet tundra with extensive peat bogs, were significant factors. But earlier warming episodes had occurred that mammoths and other large mammals had somehow survived without the mass decimation that took place 10,000 years ago. What could explain this difference? It seems likely that growing predation by human hunters in this changing landscape may have compounded the effects of these climatic changes and hastened the decline of mammoths.

The problem becomes more complex, however, when we note that, in spite of comparably significant changes in climate in the Southern

Hemisphere, the Asian and African elephants were able to survive. Extinction is rarely a simple process in which a single factor leads to the demise of a group. Though fossils shed some light on the process, they are generally tantalizingly inadequate to tell us the whole story.

THE VALUE OF FOSSILS

The mammoth story is a useful introduction to a broader exploration and discussion of the history of life. Let me suggest half a dozen reasons.

The "parable" of the mammoth provides a classic example of the *recognition and the reconstruction of ancient creatures*. Fossils have long been objects of curiosity and speculation. The Roman emperor Augustus is said to have decorated his villa in Capri with fossil bones, while for centuries Chinese people ground up fossil teeth ("dragon teeth") for medicinal use. In some areas of England, fossil oysters were once used in the preparation of medication for cattle, and fossil cephalopod belemnites were used for the treatment of eye infections in horses.

But these uses of fossils as curios and charms imply no particular views of or even interest in, the nature or value of fossils, although a few early writers recognized fossils as the remains or direct indication of prehistoric life. A number of Greek writers, including Xenophanes of Colophon (c. 614 BC) and Herodotus (484–425 BC), for example, did realize the true nature of fossils and grasped their significance in indicating former patterns of land and sea. During the Dark and Middle Ages fossils were variously regarded as sports of nature, or as unsuccessful attempts of a life force (the *vis plastica* of Avicenna [980–1280], or the *virtus formativa* of other writers) to create living creatures out of the rocks. Still others regarded them as the products of living seeds, carried inland by winds from the sea.

It was not until the fifteenth century that the revival of learning, the invention of printing, the rise of universities, and the great age of exploration gave impetus and significance to almost every aspect of human interest. Leonardo da Vinci (1452–1519), for example, during his younger years, worked in canal construction in northern Italy, where he identified fossils as organic remains. Robert Hooke (1635–1703) in England not only recognized what we now regard as the true nature of fossils but also speculated that earthquakes might explain changes in the distribution of land and sea implied by the presence of marine fossils in inland rocks.

But recognizing the *nature of fossils* was one thing; *reconstructing* the creatures they represented was quite another. Sometimes that was easy, as with the well-preserved Lyuba, but even in fossils lacking such preservation, many fossil clams look broadly similar to living clams, for example. So do fossil elephants and fossil sea urchins. But in the case of other extinct forms, with no living representatives, recognizing the affinities and building the reconstruction of the animal or plant may be much more difficult. Consider, for example, a group of microfossils known as conodonts. These toothlike microfossils, ranging up to 0.2 inch (0.5 mm) in size, are made of calcium phosphate, with characteristics much like those of the mineral apatite. Now long extinct, they existed throughout the world from Late Cambrian to Late Triassic times, a span of some 300 million years. Because they underwent rapid evolutionary change, they have proved to be among the best and most refined "index fossils" for rocks of that age, providing the basis for a series of time biozones, the most precise of which represents a time interval of an extraordinarily short duration of 0.5 million years. That's a blink of an eye against the 4.5-billion-year span of Earth. Conodonts have also been used to determine the environmental nature and relationships, as well as the thermal history, of the rocks that contain them.

Although they have been known since the 1850s and intensively studied since the 1920s, there was, until relatively recently, no agreement about the biological affinities of conodonts. They were variously interpreted as the remains of algae, higher plants, mollusks, arthropods, various wormlike groups, lophophorates, chaetognaths, and chordates. Only in 1983 did convincing evidence of their nature emerge from the study of specimens discovered in Scotland. Well-preserved features of soft parts showed the conodont elements to be parts of elongated, segmented, large-eyed eel-like marine animals, about 1.5 to 2.5 inches long, with evidence of chevron-shaped structures along the body and an asymmetrical finlike tail. The conodonts, arranged near the head, probably played some role in food capture, perhaps in grasping and "chewing" prey (fig. 1.1).

But what were they, these enigmatic animals? It now seems very probable that they were chordates, perhaps related to the living jawless hagfish, or lampreys, or perhaps even to *Amphioxus*, the lancelet, a cephalochordate.[3] To have gone from plants to chordates is an indication of just how difficult it may sometimes be to recognize the affinities of creatures now long extinct.

In many cases, however, fossils *can* provide us with indications of the *origin, ancestry, relationships, and evolution* of both living and extinct groups. We saw in the case of mammoths how fossils can provide such evidence,

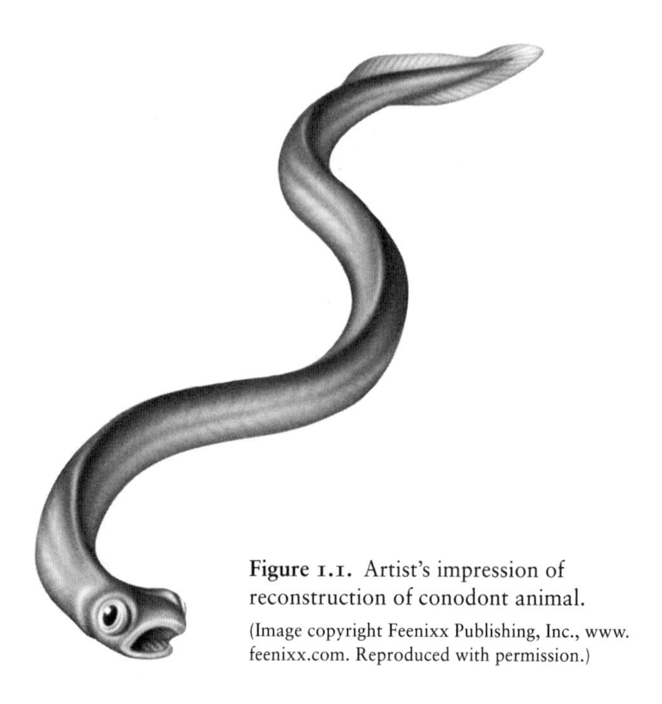

Figure 1.1. Artist's impression of reconstruction of conodont animal. (Image copyright Feenixx Publishing, Inc., www.feenixx.com. Reproduced with permission.)

and it is no less available for many other groups, from horses, camels, and dinosaurs, for example, to sea urchins and foraminifera. Consider the camel, that "ship of the desert," that has played, and still plays, such a conspicuous part in the life and economy of the Middle Eastern nations. Like horses, camels originated in North America, where the major part of their evolutionary development took place. They originated in Eocene times some 56 million years ago and underwent marked increases in size, loss of toes from four to two, changes in relative lengths of neck and limbs, and changes in both overall form and size (fig. 1.2). They became extinct in North America only at the end of Pleistocene times, 10,000 years ago.

The evidence provided by fossils on the affinity and relationships between various biological groups is also of major importance. Living and fossil elephants, for example, are broadly related to two very different groups: the ponderous, finned, herbivorous Sirenia or sea cows of the seas and rivers, on the one hand, and the rodentlike conies or hyracoids that inhabit the Middle East and Africa on the other. Dissimilar as all these animals now are, their fossil remains show that they arose from a common ancestor in Eocene times, some 56 million years ago (fig. 1.3).

Fossils also provide information on the *geography and climate of prehistoric times.* For example, South America, Africa, India, Australia, and

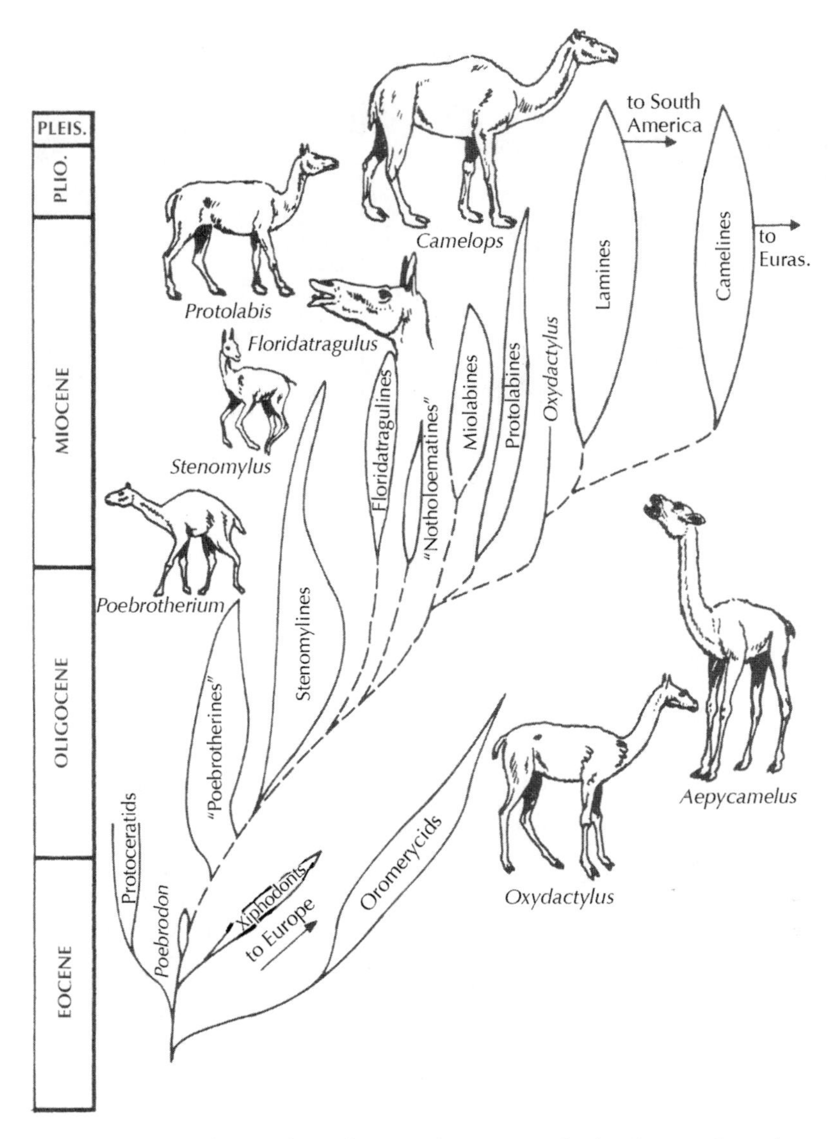

Figure 1.2. The evolution of camels in North America. The family tree of camels, showing the great diversity of forms, from small primitive deerlike creatures to the gazellelike stenomylines, the short-legged protolabines and miolabines, the long-legged, long-necked "giraffe camels," and the modern humpless South American camels (alpaca, llama, vicuña, guanaco), which are more typical of the whole family. Only the living African dromedary and the two-humped Asian Bactrian camels have humps.

(Drawing by C. R. Prothero; after Prothero 1994. From Donald R. Prothero, *Evolution: What the Fossils Say and Why It Matters* [New York: Columbia University Press, 2007], 314. Reproduced with permission of the publisher.)

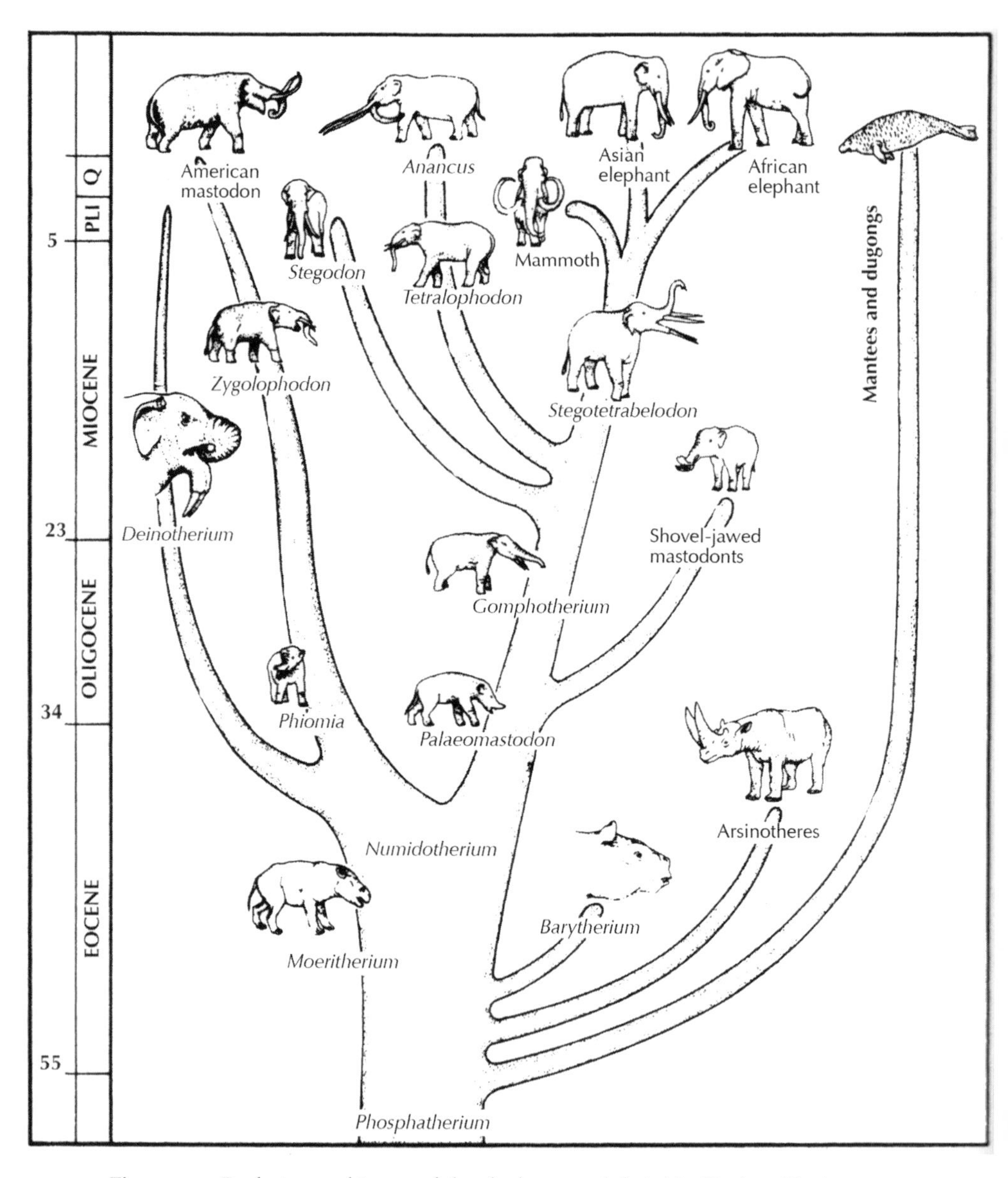

Figure 1.3. Evolutionary history of the elephants and their kin (Proboscidea), starting with pygmy hippolike forms like *Moeritherium* with no trunk or tusks through mastodonts with short trunks and tusks, and concluding with the huge mammoths and the two living species. Early in their history, the other tethytheres branched off from the Proboscidea. These include the manatees, order Sirenia, the extinct desmostylians, and the extinct horned arsinotheres.

(Drawing by C.R. Prothero; from Prothero 1994. From Donald R. Prothero, *Evolution: What the Fossils Say and Why It Matters* [New York: Columbia University Press, 2007], 324. Reproduced with permission of the publisher.)

Antarctica are now widely separated, but fossils reveal that for an extended period of earlier Permian and Triassic times (about 299–237 million years ago), they were so closely situated that they shared certain distinctive animals and plants, including two lumbering land reptiles, *Cynognathus* and *Lystrosaurus*, a freshwater reptile, *Mesosaurus*, that was 3 feet or so long, with a long deep tail and paddlelike feet, and two genera of seed fern plants, *Dicroidium* and *Glossopteris*. The distribution of these fossils confirmed earlier suggestions, based on the jigsaw-like "fit" of these now isolated continents, that they had once formed a continuous southern landmass, the "Supercontinent" Gondwana, or Gondwanaland (fig. 1.4)

Fossils can also provide information on the *general ecology*, *living habits*, and detailed *environments* of animals and plants of distant times. Chemical methods of measuring the oxygen isotope composition (O^{16}:O^{18} ratio) of marine fossils can provide a record of the temperatures of ancient oceans, for example. Changing distribution of fossil trees in rocks of Pleistocene age can be used to mark the fluctuations in the extent of continental

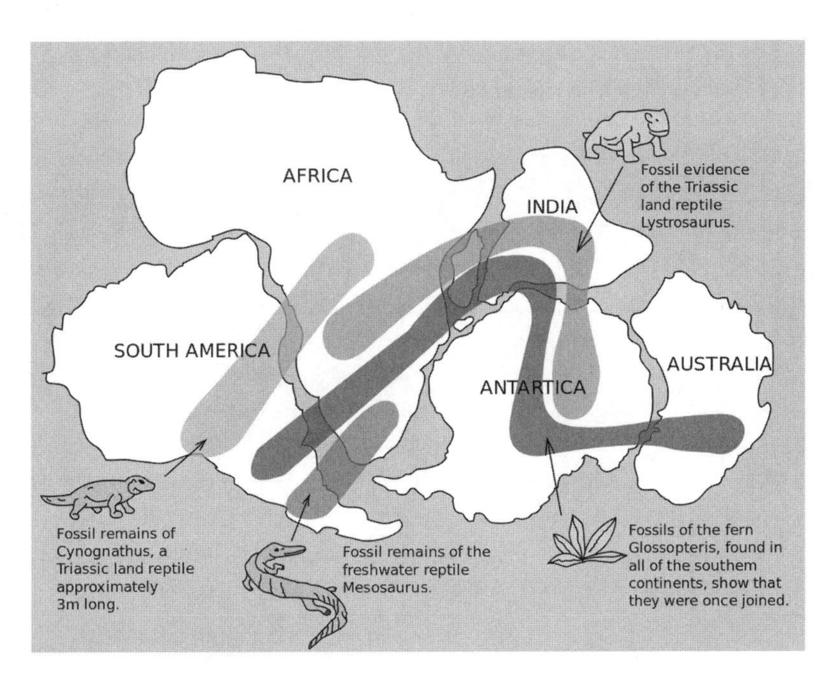

Figure 1.4. Continental drift fossil evidence. As noted by Snider-Pellegrini and Alfred Wegener, the locations of certain fossil plants and animals on present-day, widely separated continents would form definite patterns (shown by the bands of colors) if the continents were rejoined.

(Image courtesy of the U.S. Geological Survey.)

glaciers, while the rise and evolutionary change of Tertiary fossil horses can be shown to reflect their changing diet, as widespread forests on which they browsed were replaced by harsh prairie grasses (fig. 1.5)

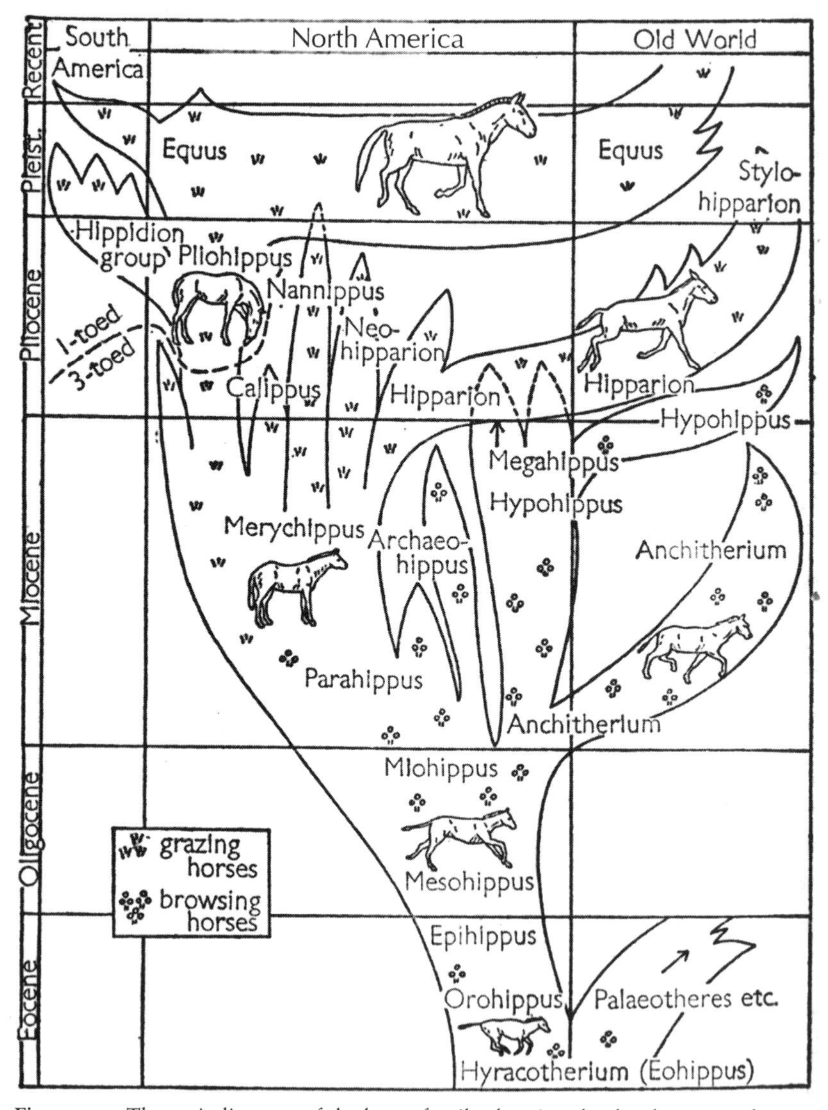

Figure 1.5. The main lineages of the horse family showing the development of one-toed grazers and the geological and geographical distribution of the family. The restorations are to scale.

(After Simpson. From Frank H. T. Rhodes, *Evolution of Life* [New York: Penguin, 1976], fig. 40. Reproduced with permission.)

Fossils can, as we have already suggested, also provide clues to the broad and puzzling questions of *survival* and *extinction*, though these are often matters of lively debate and controversy. This is especially true in episodes of mass extinction, when major groups of once dominant organisms disappeared within a geologically brief time span. The extinction of the dinosaurs and other reptiles, 65 million years ago, for example, at the end of the Mesozoic Era—the era of middle life—was one of several mass extinctions that have marked the long history of life. Dinosaurs had dominated life on the land for more than 100 million years, and they existed on all the continents, yet their demise 65 million years ago was worldwide, and, in a geological sense, sudden. But they were not alone. A host of large and powerful marine reptiles, (plesiosaurs and mosasaurs) as well as the flying reptiles (pterosaurs), also vanished.

Table 1.1. Data on the rates of extinction of vertebrates at the K-T boundary*

Group	Families extant	Families extinct	Extinction rate %
Chondrichthyes	44	8	18
Bony fishes	50	6	12
Amphibians	11	0	0
Reptiles	71	36	51
Turtles	15	4	27
Lizards and snakes	16	1	6
Crocodilians	14	5	36
Pterosaurs	2	2	100
Dinosaurs	21	21	100
Plesiosaurs	3	3	100
Birds	12	9	75
Mammals	22	5	23
Basal groups	11	1	9
Marsupials	4	3	75
Placentals	7	1	14
All vertebrates	210	64	30
Fishes	94	14	15
Tetrapods	116	50	43
Amniotes	105	50	48

* Figures are based on the numbers of families extant during the Maastrichtian Stage and the numbers that died out some time during that time interval.

Source: Michael Benton, *Vertebrate Paleontology*, 3rd ed. (Oxford: Wiley-Blackwell, 2005), table 8.1, p. 251. Reproduced with permission.

So, too, did some long-lived groups of marine invertebrates, which also became extinct at this time. These included such widespread and long-lived groups as ammonites, belemnites, and various bivalves. The cause of this mass extinction has been widely debated. At various times, volcanoes, climatic change, disease, predation by mammals, and several other agents have all been suggested. An asteroid impact is the latest favorite. Yet, even more puzzling, some other reptiles—including turtles, lizards, crocodiles, and snakes—survived. The debate continues (chapter 16).

Fossils have made one more very practical contribution to our contemporary society: they provide a widely used method of *"dating" and correlating the rocks* in which they occur. We will discuss this in more detail in chapter 2, but it has great value in everything from petroleum exploration to the reconstruction of geologic history. Oil, gas, and coal are, in fact, fossil fuels, formed from the remains of once-living organisms.

Valuable as are all these "practical" uses of fossils, they have, perhaps, one other intrinsic value, not as charms or curios, but as *medals of creation*, as they were once called. For they provide glimpses of the inhabitants of worlds long vanished, of fellow tenants who once shared this ancient planet, which we, for a brief time, call home.

BECOMING A FOSSIL: THE PROCESS OF FOSSILIZATION

Fossils, then, are the remains of, or the direct indication of, life of the geologic past. Exquisite as is the preservation of Lyuba, the baby mammoth, frozen in ice for 40,000 years, most known fossils are vastly older than that, and considerably less well preserved. There are, however, other rare instances in which some soft parts of organisms have been preserved. The soft parts of one Ice Age woolly rhinoceros (*Rhinoceras tichorhinus)* were recovered from asphalt deposits in Galicia, Poland. Some soft remains may be preserved by dehydration. Remains of the skin of extinct giant ground sloths have been preserved in asphalt deposits, and "mummified" dinosaur skin impressions have been found, together with their bones, in rocks of Cretaceous age. A different type of preservation is involved in the case of insects, leaves, and flowers preserved in exquisite external detail in amber. Amber is fossil resin from ancient pine trees and has long been known from Tertiary rocks of the Baltic and elsewhere. Insects and plant fragments become trapped in this and their outline is preserved in carbon, the more volatile constituents having been lost.

But instances of this kind require unusual conditions of burial and preservation. And they are rare. Of the vast hordes of animals and plants that once lived, relatively few have been preserved as fossils. That is because fossilization requires rapid burial of the organism under bacteria-free environmental conditions. In most cases, after organisms die, their soft remains tend to decompose rapidly, though their hard parts—shells, bones, teeth—may be preserved.

There are, however, a few sites in the world where rapid burial under oxygen-free conditions has preserved the outlines of even the soft structures of organisms: the Middle Cambrian Burgess Shale of British Columbia and the Solnhofen Limestone of Bavaria are notable examples.

These then are some of the ways in which the soft parts of both animals and plants may be preserved. Most fossils, however, consist only of hard and more resistant remains of organisms, although the soft parts are frequently represented on them by the various impressions by which they are attached. Here again it is somewhat unusual to find these hard parts in their unaltered original condition. In a few animal groups the chitinous or phosphatic skeletons, teeth, and shells, may be preserved, however, as also may the resistant cuticle of plants. Striking examples are provided by the microscopic fossil jaws of polychaete worms, many of them 500 million years old, which are so perfectly preserved, they are still flexible. Again, the teeth of many vertebrates of considerable age and scales of fish are often preserved as fossils with little or no apparent change in composition. More frequently these hard parts are more or less altered, however, and this alteration usually involves the leaching out of the more soluble components and their partial or total replacement by other substances.

The simplest example of this type of preservation is represented by shells and bones from relatively young deposits, which usually show evidence of slight but distinctive dissolution. These may have a rather bleached appearance, and all feel light in weight. Most fossils, however, are strikingly heavy, and this is the result of the infilling of the cavities left by decay and dissolution with other minerals carried in solution. Fossil wood and bone are both rather "spongy" and therefore unusually heavy when the pores are filled by this process of permineralization. The mineral deposited in the pores may sometimes be similar to that making up the original shell or bones but is usually different, the most common replacing minerals being calcite ($CaCO_3$), silica (SiO_2), and various iron salts. Sometimes this replacement gives a faithful preservation of the original microstructure, as, for example, in much petrified wood, where the cellular structure is often clearly visible in polished sections. Usually, however, although the exterior features of the shell may be unchanged, the detailed structure is lost, either by recrystallization of the original material (the calcite of shells, for example) or by its replacement by another mineral. A small number of fossil eggs have been recorded. The best known of these were found in wind-deposited Cretaceous sandstones of Mongolia and are those of the dinosaur *Protoceratops*. They are sometimes found in "nests." In a few eggs discovered in 1993 the remains of well-developed embryos have been preserved. These embryos have turned out to be the remains not of *Protoceratops*, but of *Oviraptor*, which had been identified as an "egg thief."

The color of the original hard parts is only very rarely preserved, but a few good examples exist even among very ancient fossils. A fossil brachiopod, at least 350 million years old, shows a series of distinct maroon bands, for example.

Now, all the types of fossils we have so far discussed have been the variously preserved remains of organisms. But other fossils have been found that are mere ghosts, traces of the former existence of an animal or plant of which nothing any longer remains. An obvious example of this type of fossil is the leaching away of *all* the original material, so that a hollow mold is left in the rock. This may later be filled in by some other mineral to produce a natural cast of the original organism. These casts and molds are common in permeable rocks such as sandstones.

Other traces of former life may be rather more indirect, such as fossil tracks, trails, and borings that are sometimes preserved. Some fossil shells display the borings of gastropods and sponges, worm burrows are found in many rocks, some bones of extinct vertebrates show marks made by gnawing rodents, and some tracks, such as those of dinosaurs, are often quite spectacular. Fossil excreta are relatively common and may provide information about the diet of the creatures they represent. Stomach stones are sometimes found in association with fossil skeletons. The fossil skeleton of a marine reptile (*Alzadasaurus pembertoni*) found in South Dakota contained within its abdomen 255 of these, together weighing more than 18 pounds.

Of particular interest are the fossil remains of our own species; of these the most common are the simple stone implements (artifacts), tools, and weapons made by early humans. These show a gradual increase in quality of workmanship over time and are of value in determining the relative ages of the more recent strata in which they are found.

Not everything that looks at first glance like a fossil turns out to be a fossil. Concretions and dendritic mineral deposits are often confused with fossils.

These then are the main types of fossils. We have discussed them at some length because they are the foundation of all that will follow, the sole source of our knowledge of the life of the past. But from what we have already said, it will be clear that the process of fossilization is a very hit or miss business, and indeed its hazards and vagaries are such that the overwhelming majority of organisms have left no fossil record of their former existence. In most cases the fossilization of an organism depends on the presence of hard parts (bone, shell, teeth, etc.) and rapid burial. Fossilization favors widely dispersed, long-lived groups whose members have hard parts, but, even in these cases, the original material is rarely preserved, but instead is replaced by other materials, carried in percolating groundwater solutions. The fossil record is therefore not only very incomplete ("only the skimmings of the pot of life," as T. H. Huxley described it in a letter to Darwin in May 1864),[4] but also biased and selective, lacking almost all trace of soft-bodied organisms and having very few representatives from those environments where rapid burial is unlikely (such as most of those on the land). Even this incompleteness is not all, however, for many organisms that survive all these vicissitudes of fossilization to become fossils are subsequently destroyed by metamorphism (the alteration of rocks by heat and pressure within the earth) or eroded away when the rocks that bear them are exposed to weathering on the surface of the Earth.

These are the fragments from which, piece by laborious piece, the great mosaic of the history of life has been constructed. Here and there—as we shall see in chapter 2—we can supplement these meager scraps by the use of biochemical markers or geochemical signatures that add useful information, but, even with such additional help, our reconstructions and our models of descent are often tentative. For the fossil record is, as we have seen, as biased as it is incomplete. But fragmentary, selective, and biased though it is, the fossil record, with all its imperfections, is still a treasure. Though whole chapters are missing, many pages lost, and the earliest pages so damaged as to be, as yet, virtually unreadable, this—the greatest biography of all—is one in whose closing pages we find ourselves.

2

TERRESTRIAL TIMEPIECES

THE SIGNIFICANCE OF TIME

We live in a world preoccupied with time: with clocks, with calendars, with deadlines. At every turn, from birthday parties to Medicare eligibility, from warranty limits to parking meter expiration, we encounter time. Geology, and paleontology, too, to a degree unknown in most other sciences, are rooted in time. If we are to comprehend our past, we need a handle on time. That's as true of human history, of course, as it is of Earth history. So we speak of the post–World War II period, the Roaring Twenties, the Prohibition era, pre-Revolutionary times, the Roosevelt presidency, the Depression years. And, if our history is sound and our memory good, we can put rough dates on these times.

We can also recognize periods characterized by the dominance of certain ruling nations: the Egyptian Era, the Greek Era, the Roman Era, and so on. In other countries, the United Kingdom, for example, ruling dynasties give a time framework. Thus we can piece together a succession of periods based on the ruling monarchs: Norman, Plantagenet, House of York, House of Lancaster, Tudor, Stewart, and so on. And we can place particular rulers within each division. Thus the Norman period is marked by the reigns of William the Conqueror, William Rufus, and so on. And events and legislation can be placed in these periods: the Domesday Book, Magna Carta, War of the Roses, and so on.

Now these periods are useful because they place events in sequence and we can use them this way. If someone says he grew up during the Depression, we have a rough idea of how old he is. We speak, too, of Colonial mansions, Revolutionary battlefields, Norman castles, Tudor cottages, and we conjure up images, not only of the objects but also of their approximate age. And we can do that because, with the advent of writing, we preserved a chronological record of these events.

But if we go back to prehistory, this association of events and dates is no longer possible. Though we can, in some cases, deduce the relative age of objects and events from the distant past, we lack the precise time scale provided by recorded human history.

It's much the same with rocks. Long before any such numerical time scale was developed for geology, it had become clear that the *relative* age of stratified sedimentary rocks could be determined from the order of their superposition. In an undisturbed rock sequence, the older rocks would lie below the younger. Tracing them laterally might reveal their relationship to younger or older rocks that might have been preserved. The Grand Canyon is a classic example of this layer-cake superposition. The first to enunciate this principle was Nicolas Steno (1638–1686), a Danish Catholic bishop, physician, anatomist, and geologist.

WILLIAM SMITH AND THE RECORD OF TIME

William Smith (1769–1839), an English surveyor and civil engineer, used this principle and developed it in a critical way. William Smith was born in Oxfordshire in 1769, the son of a blacksmith who died when Smith was only seven years old. Largely self-taught, he found work as a surveyor and in 1791 was employed to make a survey of a great estate in Somerset that contained workable coal deposits. Smith was hired to map and evaluate them, and later to plan and excavate a series of canals by which the coal could be transported to market. During these excavations, he became fascinated that the stratified rocks, "like so many slices of bread and butter," could be traced over great distances, each containing "fossils peculiar to itself." He worked at these tasks for eight years and then, for reasons that are not fully clear, was suddenly dismissed and became unemployed in 1799.

For much of the next five years, Smith went on to find part-time work over much of the country. He surveyed, enclosed, and drained great tracts of waterlogged land to prepare it for effective agriculture, taking jobs wherever he could find them from farmers and landowners, including several influential members of the aristocracy and nobility. Meanwhile, he continued developing a map of the rocks he encountered and painstakingly collecting fossils, which he used to identify the various strata. He determined that he would create a great map of the geology of the whole of England and Wales and in 1801 published a four-page prospectus of the map and book he intended to produce, inviting subscribers to reserve copies. But no sooner had he announced his intention, than his publisher went bankrupt. That particular publication never appeared, but Smith continued his travels and surveys all over Britain for the next ten years. Singlehandedly, he rode and tramped across Britain, back and forth, project by project, wherever his journeyman work took him, patiently piecing together the information needed to create a great geological map of the whole 50,000 square miles of England and Wales, as well as a part of Scotland. He built up a vast collection of fossils, some 700 different species of them, which he used to characterize and correlate the rocks he studied. To map so vast an area at a scale of 5 miles to the inch was achievement enough, but even that was not Smith's greatest achievement. He, himself, invented the concept of a geological map. He, himself, discovered the value of fossils in the identification and correlation of the rocks in which they occur.

By 1815 the great, hand-colored map—more than 8 feet high by 6 feet wide—was finished. On August 1, 1815, it was formally published, with a dedication to Smith's patron, Sir Joseph Banks, the greatest scientist of his day. Smith later published descriptions of his fossils. Four hundred copies of the map were printed. It was, wrote Simon Winchester, Smith's biographer, "the map that changed the world." So great was his passion that he became known as "Strata Smith" to his acquaintances. Not for nothing was he later celebrated as the Father of English Geology.[1]

The map was not only of scientific interest: it was also of immediate practical value. On it, Smith marked the positions of roads and tunnels and the various geological formations through which they passed as well as the locations of lead, copper, and tin mines and collieries. These items were of great significance in a nation rapidly developing a manufacturing economy. Even more important, by creating the first geological map, the first attempt to illustrate the succession of strata making up the landscape, he also provided an accurate picture of the subsurface structure, using fossils to identify the rocks that contain them. It marked, in fact, the establishment of a new field of study—geology—that was to provide the economic basis of the modern world.

But even at the moment of his triumph, Smith was undermined. His map was shamelessly plagiarized by a group led by the wealthy George Bellas Greenough, to whom Smith had shown his work. Greenough and his colleagues published and sold "their" map at a price lower than that

Figure 2.1. William Smith's geological map of England and Wales.

(CP15/040 British Geological Survey © NERC. All rights reserved.)

of Smith in a move that, whether intentional or not, effectively ruined him. So dire was his plight that he became bankrupt and was jailed in a common debtor's prison, and his London home and property were seized.

Released from jail in 1819, at the age of forty-four, homeless, penniless, and unemployed, with an ailing young wife and his orphaned nephew, John Phillips, to support, he abandoned London for Yorkshire. Arriving there, he set about rebuilding what remained of his life, finding work wherever he could. Slowly, job by job, assignment by assignment, he not only rebuilt his career, but also gained a growing group of admirers and influential sponsors.

It was from members of this group, as well as from some older friends, that recognition finally came. These people, understanding the true value and significance of his work, summoned the sixty-two-year-old Smith back from his Yorkshire exile to London and, before the assembled Geological Society, which had once denied him membership, recognized him as the first recipient of its highest award, the Wollaston Medal, for his research "into the mineral structure of the earth." The society's president, Adam Sedgwick, the Woodwardian Professor of Geology at Cambridge, praised Smith for his scientific contributions and hailed him as "the Father of English Geology."

And so he was: that, and perhaps more. His twenty long years of geological wandering, his single-handed unraveling of the geology and structure of a whole country, his unique method of correlating rocks and so contributing a vital part to the creation of a geologic time scale, surely entitle him to a larger paternity: he was, in a larger sense, The Father of Modern Geology. To his parentage, our modern society owes a unique and lasting debt. Adam Sedgwick added a fulsome tribute to his influence:

"If, in the pride of our present strength, we were disposed to forget our origin, our very speech betrays us: for we use the language which he taught us in the infancy of our science. If we, by our united efforts, are chiseling the ornaments and slowly raising up the pinnacles of one of the temples of nature, it was he that gave the plan, and laid the foundations, and erected a portion of the solid walls, by the unassisted labour of his hands."

William Smith led a contented life in his later years, respected and honored in Scarborough, Yorkshire, where he settled, supported by a royal pension. And, having himself once been shunned by the leaders of the scientific establishment, it seems somehow fitting that Smith's nephew, John Phillips, whom he had long supported, later became Reader in Geology at

Oxford, and, fourteen years after Smith, was also awarded the Wollaston Medal of the Geological Society, of which he also became president. And Phillips, too, contributed his terms to the scale of geologic time, coining the names of the Mesozoic and Kainozoic (Cenozoic) eras, which are now recognized worldwide to identify the two most recent eras of life.

Using this same William Smith approach—walking the outcrops and identifying rocks by their distinctive fossils—other geologists slowly pieced together a geological sequence of rocks. They named the subdivisions of their succession after regions where these rocks were well represented and studied. Thus they named successive periods of the time scale beginning with the Cambrian—the Roman name for Wales; Ordovician—after the Ordovices, an ancient warlike tribe that inhabited the area of northwest Wales; Silurian—after the Silures, an ancient tribe living in what is now southeastern Wales; Devonian—after Devon, an English county; Carboniferous—after the coal-bearing rocks of that system; Mississippian and Pennsylvanian, two component parts of the Carboniferous, named for the regions where they are well developed; Permian, after the region of Perm in Russia; Triassic, after its three-part development; Jurassic, after the Jura Mountains of Europe; Cretaceous, after its chalklike composition in many areas; and Tertiary and Quaternary, names from an older fourfold classification.

This sequence of rocks, first developed largely in England and Wales, later proved to be recognizable and applicable throughout Europe, North America, and other areas and was widely adopted. By the mid-nineteenth century it was in general use. With it, one can recognize the relative age of a rock or fossil and compare it with that of one found elsewhere.

Using this method of rock identification and classification, it is possible to build up a history of living things. The following outline shows a framework for the major divisions of geologic time.

The eras and systems are applied and agreed upon worldwide, but each of the various regions—North America, Asia, and so on—uses distinctive regional names for epochs and ages.

But if a geologist is asked, "How old is this fossil?" and he or she replies "It's Silurian," that's of little help except to another geologist. What most people generally want to know is how many years old an object is: how many years ago this fossil animal lived, what geologists generally describe as "absolute age," as opposed to the "relative age" provided by terms like "Silurian." How, then, can we estimate the age of the Earth and its various component rocks?

Table 2.1. Framework for the major divisions of geologic time using rock identification and classification

Rock succession chronostratigraphy	Time periods geochronology	Typical duration	Example
Eonothem	Eon	Half a billion or more years	Phanerozoic
Erathem	Era	Total: 4 Several hundred million years	Paleozoic
System	Period	Total: 10 Variable	Silurian
Series	Epoch	Few to tens of millions of years	Ludlow
Stage	Age	Few million years	Ludfordian

THE MEASUREMENT OF TIME

An interest in the age of the Earth is almost as old as humanity itself, as the creation records of many ancient civilizations bear witness. The Brahmins of ancient India believed the Earth to be eternal. Babylonian astrologers concluded that humans had appeared half a million years ago. As long ago as 450 BC Herodotus suggested that the rate of deposition of sediment by the River Nile indicated that the building of its delta must have required many thousands of years.

But not all lines of evidence suggested such great antiquity. Archbishop James Ussher of Armagh (1581–1656), primate of All Ireland, always gets a bad press in this regard. Let me explain why. By the seventeenth century, speculation had grown concerning the age of the Earth and evidence was sought from whatever source might be fruitful. The great astronomer Johannes Kepler (1571–1630) combined the study of science—using cycles of solar eclipses to date the darkness of the crucifixion—and the Genesis narrative of creation to calculate a date of 3993 BC for the origin of the Earth. Sir Isaac Newton, perhaps the greatest scientist of all time, refined that to 3998 BC. It is here that Ussher took up his calculation. Ussher was not only primate of All Ireland, but also provost of Trinity College, Dublin, and has been described as one of the greatest Hebrew scholars of his day. His extensive library formed the core of the great library of Trinity College. In his *Annals of the World*, published in 1650, he suggested a date for the origin of the Earth as "the entrance of the night preceding the twenty-third day of October, in the year of the Julian Calendar, 710" as the best date that could be deduced from all the substantial array of ancient documents he had collected. The year 710 of the Julian calendar is equivalent to 4004 BC. Ussher based his calculations on the lineage and ages of individuals from Adam to Jesus recorded in the Bible. The indulgence with which Ussher's calculation is now treated overlooks the serious scholarship behind his work. Ussher and Kepler based their chronology on a combination of historical and scientific records—surely a not unreasonable approach at the time—but, in addition to assuming the accuracy of the records, Ussher also made one major assumption: that the age of the Earth was the same as the age of humanity. That assumption, we now know, was wrong, but the very concept of time and the experience of history are profoundly influenced by one's view of life.

Two experiences seem always to have been intertwined: first, the sense of life's basis in cycles, of day and night, of sleeping and waking, of monthly cycles, of phases of the moon, of movements of the planets, of seedtime and harvest, of spring and fall and the regularity of the seasons. Perhaps, then, the Earth itself, perhaps life itself, was cyclical, seasonal, changing yet unchanging. That was (and still is) the Buddhist and the Hindu view. It was the view of Plato, who suggested a great cycle of 72,000 years, first of advance, then of decline. It was something like the view, far more recently, of a group of eminent astronomers—Fred Hoyle, Herman Bondi, and Tommie Gold—who in 1948 and later suggested a cosmic pattern of "continuous creation."

But against that view was the sharp reality of birth and death: of childhood, youth, maturity, and age: of parents, children, family, and friends, whose lives showed, not cyclicity, but directionality, linearity, from a beginning to an end. It was this view that was embedded in the great Abrahamic faiths of Judaism, Christianity, and Islam, based as they are on the history of divine creation and providential purpose. And that's why, one supposes, the first serious attempts to construct a chronology of Earth came from those who were not only well skilled in science, but also well versed in the scriptures. Johannes Kepler and Sir Isaac Newton were among the most respected biblical scholars of their day and are also counted among the greatest scientists of all time.

But no calculation based only on human history could calibrate the span of prehuman history, though even the Genesis narrative bore witness to such a period. Increasingly, the more one looked at Earth itself and the processes that shaped its surface, the more those observations implied its antiquity.

This same general approach to estimating geologic time based on geological processes was used in the nineteenth century, with attempts to calculate the total thickness of sedimentary rocks and then to divide that by the annual rate of deposition of sediment. But such estimates varied widely, largely because it was impossible to obtain a meaningful figure for either total rock thickness or rate of deposition. Both vary from place to place and from one geologic period to another, and rocks are both eroded and redeposited.

In 1898, Charles Darwin's second son, George Darwin, suggested an age of 56 million years for the Earth, basing his calculation on the increase in orbital period of the Moon. He assumed that the Moon had formed from a molten Earth and he calculated the time that would be required for tidal friction to slow Earth's rotation to its present period.

In 1899, John Joly, professor of geology at Trinity College, Dublin, suggested another quantitative approach to estimating the age of the Earth, based on the salinity of the oceans. Joly assumed that the oceans were originally fresh water and argued that if all the sodium chloride in the oceans came from the weathering of igneous rocks, and if one could calculate the annual rate of addition from erosion of rocks, the oceans must be 80 to 100 million years old. He later increased this upper limit to 150 million years. We now recognize that erosion of Earth's crust and recycling of salt greatly increase that figure.

But this estimate challenged the calculations of the greatest physicist of the age. In the mid-nineteenth century, William Thompson, who later became Lord Kelvin, professor of natural philosophy at the University of Glasgow, had calculated the Earth's age as between 20 million and 400 million years, basing his estimates on the time required by it to cool from an originally molten state. As he later refined his estimate it became closer to 20 million to 40 million years. Such low estimates proved troubling to geologists and paleontologists, including both Charles Lyell and Charles Darwin, who believed they provided inadequate time for the slow transformations of Earth and life processes.

Kelvin's estimates assumed that all Earth's heat came from its cooling from an originally molten condition, but in 1896 Henri Becquerel made a discovery for which Kelvin could have made no estimate. Becquerel discovered the existence of radioactivity in uranium; it was soon realized that this discovery introduced both a new source of heat, for which Kelvin had made no allowance, and a new means of measuring geologic time. Certain radioactive elements, such as uranium, break down to give stable end products, such as lead. The rate of this breakdown is measurable, slow, and constant, none of the physical and chemical factors that typically affect other reactions having any effect on it. If, therefore, the ratio of "disintegrated" lead to "undisintegrated" uranium is measured, it can provide an estimate of the age of the rock.

Sir Ernest Rutherford, who worked with others to develop and apply this concept, once delivered a lecture on the age of the Earth in which he described the experience as follows:

> I came into the room which was half dark and presently spotted Lord Kelvin in the audience, and realised that I was in for trouble at the last part of my speech dealing with the age of the Earth, where my views conflicted with his.
>
> To my relief, Kelvin fell fast asleep, but as I came to the important point, I saw the old bird sit up, open an eye and cock a baleful glance at me.
>
> Then a sudden inspiration came, and I said Lord Kelvin had limited the age of the Earth, provided no new source [of heat] was discovered. The prophetic utterance referred to what we are considering tonight, radium. Behold! The old boy beamed upon me.[2]

It is noteworthy that John Joly later also realized the potential value of radioactive elements in the Earth's crust and worked with Rutherford to suggest an age of at least 400 million years for the beginning of the Devonian Period. This is an estimate that is remarkably close to current estimates.

Important as were Joly's contributions to estimating the age of the Earth, he is also renowned as a pioneer in the extraction of uranium and its use in the treatment of cancer.

Developing and using these radioactive techniques, Arthur Holmes, who became professor of geology at Durham and later at Edinburgh, developed a geologic time scale that he published in an influential book, *The Age of the Earth,* which went through fourteen editions between 1913 and 1949.

Many, many other calculations of the ages of rocks from all parts of the world have now been made, and there is a remarkable consistency in their results. Although the uranium-238 to lead-706 breakdown is still widely used, other radioactive series have also proved very valuable, not least because they cover different time spans. Carbon-14, for example,

Table 2.2. The most commonly used isotopes in radiometric age dating

Isotopes		Half-life of parent (years)	Effective dating range (years)	Minerals and other materials that can be dated
Parent	Daughter			
Carbon-14	Nitrogen-14	5,730	100–70,000	Anything that was once alive: wood, other plant matter, bone, flesh, and shells; also carbon in carbon dioxide dissolved in ground water, deep layers of the ocean, or glacier ice
Potassium-40	Argon-40 Calcium-40	1.3 billion	50,000–4.6 billion	Muscovite Biotite Hornblende Whole volcanic rock
Uranium-238	Lead-206	4.5 billion	10 million–4.6 billion	Zircon
Uranium-235	Lead-207	710 billion		Uraninite and Pitchblende
Thorium-232	Lead-208	14 billion		
Rubidium-87	Strontium-87	47 billion	10 million–4.6 billion	Muscovite Biotite Potassium feldspar Whole metamorphic or igneous rock

Source: Republished with permission of Brooks/Cole, a division of Cengage Learning, from Graham R. Thomson and Jonathan Turk, Earth Science and the Environment, 3rd ed., 2005; permission conveyed through Copyright Clearance Center, Inc.

can be used to date materials, such as wood, charcoal, and bone, associated with prehistoric humans. The four principal radioactive elements used in radiometric dating are uranium-238, potassium-40, rubidium-87, and carbon-14.

These various methods now provide the basis for putting dates on our scale of relative time. The most widely used version is shown in figure 2.2. Now, instead of knowing only that Devonian fossils, say, are younger than Silurian fossils, we can date each one to specify how much younger they are. Hundreds of measurements of the ages of rocks from almost every part of the world have been made and they provide a remarkably consistent pattern.

How old are Earth's oldest rocks? The oldest rocks so far discovered come from the Acasta Gneiss in northwestern Canada, which has an age of 4.03 billion years. Slightly younger rocks are known from Michigan, Swaziland, and western Australia. The oldest minerals so far discovered on Earth are detrital zircon crystals found in sedimentary rocks in western Australia with an age of 4.3 billion years. Is this, then, the age of the Earth? No. These crystals must have been derived by erosion of still older rocks, of which we have no record. All Earth's oldest rocks are likely to have been recycled and destroyed by processes deep within the planet. What these ages do provide is a minimum age for Earth.

To get a more precise estimate for Earth's age, we need to go beyond the Earth. Because meteorites and the Moon are assumed to have formed at the same time as Earth but have been unaffected by the recycling and destruction that characterize terrestrial rocks, their ages provide a proxy for the age of the Earth. Iron meteorites have an age of 4.54 billion years, while lunar rocks have ages of 4.4 and 4.5 billion years. Other meteorites have ages ranging from 4.53 to 4.58 billion years.

So how old is the Earth? The figure generally accepted is 4.55 billion years. That's an almost impossible figure to comprehend, so let's try to make it a little more intelligible. Perhaps it's only by analogy that we can get any sense of how vast these numbers are. Suppose that a cosmic historian began to write at the rate of one line every century when Earth was first created. Suppose this same historian completed a set of volumes, each comprising 500 pages and each with fifty lines to a page. That would see the completion of a book every 2.5 million years. There would now be a series of more than 1,800 such volumes, and they would stretch along a library shelf more than half the length of a football field.

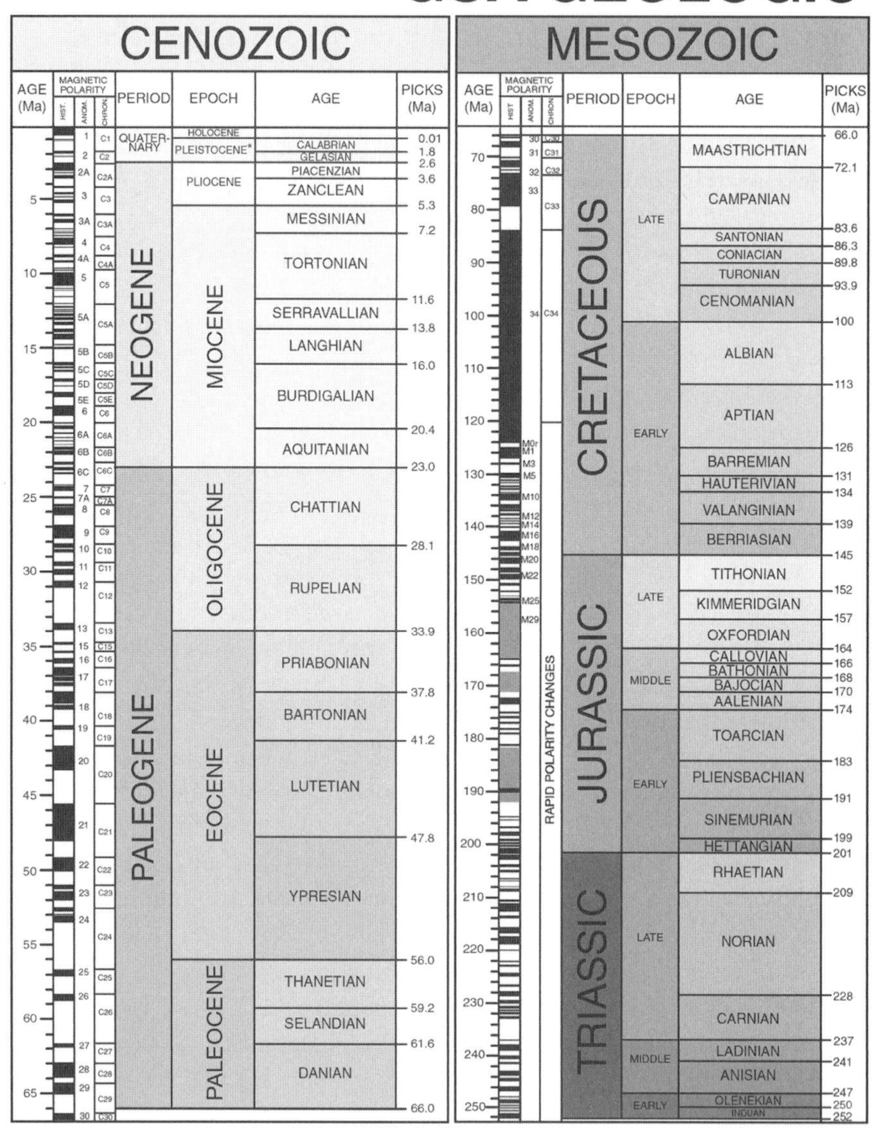

Figure 2.2. Geologic Time Scale (2012).

(Courtesy of the Geological Society of America.)

*The Pleistocene is divided into four ages, but only two are shown here. What is shown as Calabrian is actually three ages—Calabrian from 1.8 to 0.78 Ma, Middle from 0.78 to 0.13 Ma, and Late from 0.13 to 0.01 Ma.

Walker, J.D., Geissman, J.W., Bowring, S.A., and Babcock, L.E., compilers, 2012, Geologic Time Scale v. 4.0: Geological Society of America, doi: 10.1130/2012.CTS004R3C.

TIME SCALE v. 4.0

PALEOZOIC

PERIOD	EPOCH	AGE	PICKS (Ma)
			252
PERMIAN	Lopingian	CHANGHSINGIAN	254
		WUCHIAPINGIAN	260
	Guadalupian	CAPITANIAN	265
		WORDIAN	269
		ROADIAN	272
	Cisuralian	KUNGURIAN	279
		ARTINSKIAN	290
		SAKMARIAN	296
		ASSELIAN	299
CARBONIFEROUS – PENNSYLVANIAN	LATE	GZHELIAN	304
		KASIMOVIAN	307
	MIDDLE	MOSCOVIAN	315
	EARLY	BASHKIRIAN	323
CARBONIFEROUS – MISSISSIPPIAN	LATE	SERPUKHOVIAN	331
	MIDDLE	VISEAN	347
	EARLY	TOURNAISIAN	359
DEVONIAN	LATE	FAMENNIAN	372
		FRASNIAN	383
	MIDDLE	GIVETIAN	388
		EIFELIAN	393
	EARLY	EMSIAN	408
		PRAGIAN	411
		LOCHKOVIAN	419
SILURIAN	PRIDOLI		423
	LUDLOW	LUDFORDIAN	426
		GORSTIAN	427
	WENLOCK	HOMERIAN	430
		SHEINWOODIAN	433
	LLANDOVERY	TELYCHIAN	439
		AERONIAN	441
		RHUDDANIAN	444
ORDOVICIAN	LATE	HIRNANTIAN	445
		KATIAN	453
		SANDBIAN	458
	MIDDLE	DARRIWILIAN	467
		DAPINGIAN	470
	EARLY	FLOIAN	478
		TREMADOCIAN	485
CAMBRIAN	FURONGIAN	AGE 10	490
		JIANGSHANIAN	494
		PAIBIAN	497
	Epoch 3	GUZHANGIAN	501
		DRUMIAN	505
		AGE 5	509
	Epoch 2	AGE 4	514
		AGE 3	521
	TERRENEUVIAN	AGE 2	529
		FORTUNIAN	541

PRECAMBRIAN

EON	ERA	PERIOD	BDY. AGES (Ma)
PROTEROZOIC	NEOPROTEROZOIC	EDIACARAN	541 – 635
		CRYOGENIAN	850
		TONIAN	1000
	MESOPROTEROZOIC	STENIAN	1200
		ECTASIAN	1400
		CALYMMIAN	1600
	PALEOPROTEROZOIC	STATHERIAN	1800
		OROSIRIAN	2050
		RHYACIAN	2300
		SIDERIAN	2500
ARCHEAN	NEOARCHEAN		2800
	MESOARCHEAN		3200
	PALEOARCHEAN		3600
	EOARCHEAN		4000
HADEAN			

The Cenozoic, Mesozoic, and Paleozoic are the Eras of the Phanerozoic Eon. Names of units and age boundaries follow the Gradstein et al. (2012) and Cohen et al. (2012) complications. Age estimates and picks of boundaries are rounded to the nearest whole number (1 Ma) for the pre-Cenomanian, and rounded to the one decimal place (100 ka) for the Cenomanian to Pleistocene interval. The numbered epochs and ages of the Cambrian are provisional.

References Cited: Cohen, K.M, Finney, S., and Gibbard, P.L., 2012, International Chronostratigraphic Chart: International Commission on Stratigraphy, www.stratigraphy.org (last accessed May 2012). (Chart reproduced for the 34th International Geological Congress, Brisbane, Australia, 5–10 August 2012.)

Gradstein, F.M., Ogg, J.G., Schmitz, M.D., et al., 2012, The Geologic Time Scale 2012: Boston, USA, Elsevier, DOI: 10.1016/B978-0-444-59425-9.00004-4.

A series representing a history that began when the oldest known fossils appeared 3.5 billion years ago would now be 1,400 volumes long.

Or suppose this: suppose a tree was planted that grew only at the rate of one-tenth of an inch every thousand years. At that rate, planted when the oldest fossil traces of living things appeared in the rock record 3.5 billion years ago, it would now be about 29,167 feet high: higher than Mount Everest. Planted when the first member of our species, *Homo sapiens*, appeared 195,000 years ago, it would now be 19.5 inches high. The same tree, planted at the start of the Christian era, would now be two-tenths of an inch in height.

Or suppose this. Suppose we think of a cosmic calendar in which January 1st, New Year's Day, represents the origin of the Earth, 4.6 billion years ago, and midnight on the last day of the year, December 31st, represents the present moment. Each second of our cosmic calendar would then represent 146 years of Earth history, each hour 525,000 years, each day some 12.6 million years.

The oldest indirect evidence of life would appear on March 4th, and the oldest known forms of living things—bacteria and algae—would appear about March 27th. The oldest abundant hard-bodied animals—mollusks, arthropods, and so on (500 million years ago)—would appear on November 21st, the oldest reptiles (340 million years ago) on December 4th. The oldest mammals (195 million years ago) would appear on December 15th and the oldest members of our own species on December 31st at 11:30 pm. The whole of recorded human history would then take place in the last 40 seconds before midnight on December 31st.

So ancient is the Earth on which we dwell: ancient almost beyond our imagining. So long ago were the resources formed that we now consume at an ever-growing rate. So remote are the events that first made this planet habitable and set in sequence the processes that undergird our existence and well-being today.

3

"FROM SO SIMPLE A BEGINNING"

THE EMERGENCE OF LIFE

Earth teems with life. We are surrounded by it: every nook and cranny of the planet supports its own distinctive assemblage. Above us, around us, below us, within us, life pervades Earth's minutest niches in a dazzling variety of forms: from bacteria to corals, sharks, frogs, eagles, horses, humans, and whales, not to mention roses and oak trees. And, as we trace the origins and emergence of each of these various groups, we confront even more challenging and comprehensive questions. How did life emerge from nonlife? How did the earliest living things develop? How did life itself evolve? One person who wrestled with just those questions was Charles Darwin.

On February 1, 1871, Darwin wrote to his closest friend and confidant, Joseph Hooker. "It is often said that all the conditions for the first production of a living organism are now present, which could ever have been present. But if (and oh what a big if!) we could conceive of some warm little pond with all sorts of ammonia and phosphoric salts,—light, heat, electricity, etc.—were present, that a proteine compound was chemically formed, ready to undergo still more complex changes, at the present day such matter would be instantly devoured, or absorbed, which would not have been the case before living creatures were formed."[1] Sir Joseph Hooker was well qualified to comment on Darwin's musings. Naval officer, ship's surgeon, Antarctic and Himalayan explorer, botanist, geologist, and scientific statesman—whose scientific writings on plants extended from the western United States, Britain, Palestine, India, and the Himalayas to Australia, New Zealand, and the Antarctic—Hooker was Curator of the Royal Botanic Garden at Kew and one of Britain's most eminent and influential Victorian scientists. He was married to the daughter of Darwin's Cambridge tutor, mentor, and friend, the Reverend John Stevens Henslow.

But back to Darwin and the "warm little pond." If living things have evolved, as Darwin suggested, what evidence can we find of their earliest history? What did their earliest representatives look like? What, if any, fossil traces can we find for the origin and emergence of life? Those are the questions to which we now turn.

This fascination with the origin of life and its earliest expression confronted Darwin with a nagging dilemma that was particularly acute for someone who had once served as secretary of the Geological Society: where was the fossil evidence? Darwin readily acknowledged this dilemma.

The sixth and last edition of *The Origin of the Species* contained fifteen chapters, and two of the fifteen were devoted to the geological record. One of these—chapter X, twenty-six pages in length—was titled "On the Imperfection of the Geological Record," and a substantial part of it was devoted to "the sudden appearance of Groups of allied Species in the lowest known Fosiliferous Strata." Darwin reviewed the issue of the nature of the earliest fossils as follows:

> There is another and allied difficulty, which is much more serious. I allude to the manner in which species belonging to the main divisions of the animal kingdom suddenly appear in the lowest known fossiliferous rocks. Most of the arguments which have convinced me that all the existing species of the same group are descended from a single progenitor, apply with equal force to the earliest known species. . . . To the question of why we do not find rich fossiliferous deposits belonging to these assumed earlier periods prior to the Cambrian system, I can give no satisfactory answer. . . . The case at present must remain inexplicable; and may be truly urged as a valid argument against the views here entertained.[2]

More than a century after his death, Darwin's dilemma has not yet been fully resolved. Certainly, there has been progress: more than 500 deposits containing Precambrian microfossils have been found; progress has been made in the reconstruction of late Proterozoic geography, oceans, climate, and atmosphere; the fossil record has been extended back to 3,500 million years ago, some seven times longer than that known to Darwin; experimental studies have shown some success, though we have yet to achieve a comprehensive explanation or model of the origin of life. Our understanding of Phanerozoic life is based on two centuries of study, whereas, until the 1960s, Precambrian life was all but unknown and widely regarded as unknowable. And Precambrian sedimentary rocks make up, perhaps, only 5 percent of the rock record.

This search inevitably involves the need to define "life." Though we ourselves embody and enjoy it, life itself has proved surprisingly difficult to define. The dictionary offers little help; it is defined, for example, as "the quality or property that distinguishes living from dead organisms and inanimate matter," or "the state of being which begins with generation, birth or germination and ends with death," or "the time during which this state continues."

To the cynic, life is one unwelcome thing after another; to the biologist, it's the ability of systems to reproduce, metabolize, move, respond, or grow. Some writers, especially Freeman Dyson,[3] have sketched the possibility that reproduction (in which cells produce approximate copies by exchanging genetic material) may have followed replication (in which molecules, as well as plant grafts, can produce exact copies). Thus genes replicate but organisms reproduce. James Lovelock[4] has suggested that life on Earth functions as a single organism, influencing its environment to support its survival.

There is, in fact, no simple, broadly accepted definition of life, even though the need for one is of increasing importance because of growing interest in "creating" life or profoundly modifying living organisms, and also because of the search for extraterrestrial life and intelligence (SETI). And one reason the problem of definition is so difficult is that we have nothing with which to compare it. How do you differentiate something for which you have a sample size of only one?

THE ORIGIN OF LIFE

Given this difficulty in defining life, it is scarcely surprising that the search for the origin of life has long proved elusive. For centuries the prevailing view, in this as in so many other matters, was influenced by Aristotle (384–322 BC), who concluded that life arose, not only from parent organisms, but also from spontaneous generation within decaying matter—such as wood, soil, meat, dung, and vegetable matter, or even within the bodies of other animals.

This view of spontaneous generation of life was challenged by Francis Redi (c. 1621–1697), an Italian physician, who demonstrated by using covered and uncovered containers of decaying meat, that maggots did not arise spontaneously, but from eggs deposited by flies on the uncovered meat.

But belief in spontaneous generation still persisted for centuries. It was not until Louis Pasteur convincingly demonstrated in 1864 that spontaneous generation did not occur, at least under present terrestrial conditions, that the matter was resolved. But if living things did not indeed arise spontaneously from nonliving materials, there must in the remote past have been at least one, and perhaps many more, occasions when spontaneous generation *did* occur. The puzzle was to understand how, given Pasteur's constraints, this could have taken place, though modern inquiry involves the search, not for complete organisms such as flies and maggots, but for the simplest known forms.

The question of life's origin remains one of the most haunting and perplexing that we confront. It is haunting because it presents us with basic questions of identity, significance, probability, and, to some, even of meaning and of purpose. And it is perplexing because it is not only remote in time, but also occurred under conditions that were unlike those that exist today. But, puzzling as it is, it is a question we can address by following three broad lines of inquiry that can provide clues to life's earliest forms.

- First, what are the modern Earth's simplest living things like, where do we find them, and what clues do they give us about their most distant ancestors?
- Second, based on our best efforts to reconstruct Earth's earliest history, in what environment did life originate, and can we devise both theoretical and experimental models that might shed light on its origin?
- Third, when we search Earth's oldest rocks, what fossil evidence do they yield of either organic remains, as such, or indirect evidence—burrows, tracks, chemical signatures and so on—of the existence of life?

So, first, *what do Earth's simplest living things look like today*; what should we be searching for in the fossil record? Earth's simplest living forms are bacteria, algae, and protists ("protozoans"), including minute microorganisms too small to be seen by a microscope. We should be searching, then, for simple structures such as these, rather than for more complex forms, though even the simplest living things today are probably far more complex than were the earliest forms of life. The earliest forms might have been little more than strands of nucleic acid that had the ability to grow and reproduce.

Consider now the simplest existing living things. Based on their cellular structure, we can recognize three broad groups or domains of living things. The first category of living things, *Archaebacteria*, are represented by only a few living forms today, but they have the capacity to exist in the most extreme environments, such as hot springs, ultrasaline salt flats, and the intestines of higher animals. These "extremeophiles" may, perhaps, suggest what some of the earliest living things looked like.

The second major category of living things is the *Eubacteria*. These simple organisms are the great enablers for other living things, carrying out a huge variety of functions from fermentation and oxygen production to nitrogen fixation. Cyanobacteria (sometimes called "blue-green algae") are eubacteria, found among the earliest fossils in the form of stromatolites. These are layered structures, formed in shallow marine water by the trapping and cementation of particles by their films of microorganisms, especially cyanobacteria (see below). They obtain their energy by photosynthesis, and sunlight, too, must have been a readily available source of energy on the young Earth, even in earlier Precambrian times.

Both archaebacteria and eubacteria are known as *prokaryotes* ("preseeds"). The DNA within their cells is not enclosed by a membrane. They tend to be small in size and simple in structure.

The third great group of living things consists of those having cells with nuclei, which carry their genes enclosed within a membrane within the larger body of the cell. This group—the *eukaryotes* ("EK's")—are the "true seeds," to which we ourselves and most other groups of organisms belong. The viruses, incidentally, do not "fit" into any of these groups and their true nature remains a matter of some debate.

The lowly creatures represented by the archaebacteria or eubacteria are a far cry from what we should regard as characteristic animals or plants, but it is for such simple organisms and structures as these that we should probably direct our search for the oldest fossils.

If the simplest existing living cells are certain to be much more complex than the earliest living things, what are the bare essentials that would define a functioning cell? It seems likely that it would need to be *enclosed* by some sort of semipermeable *membrane* to protect the liquid *cytoplasm* of the interior. Within this liquid there would need to be molecules that provide energy (ATP: adenosine triphosphate), others that provide replication (DNA: deoxyribonucleic acid), and others (RNA: ribonucleic acid) to support the assembly of essential proteins. Virtually all proteins are constructed of the same twenty amino acids. Several suggestions have been made that the earliest cells may have included only RNA, rather than both DNA and RNA.

But even such organisms as these would be far from "simple" and probably far removed, both in form and structure, from the earliest living things. In fact, it is quite possible that the earliest forms of life—if we were able to observe and study them today—might provoke a lively debate as to whether or not we should classify them as 'living' at all. It is also important to remind ourselves of Darwin's view that this emergence of living things in a "warm little pond" or "primordial soup" must have taken place in the absence of other life that might otherwise have consumed them.

This brings us to our *second major line of inquiry*. Any conceptual model for the origin of life must include not only the question of *how* it developed but also that of *where* (in what environment on the early Earth) it developed. A suitable environment would need to provide at least four conditions to support the earliest development of life: *protection* from excessive heat and radiation, *energy, concentration* of the essential chemical ingredients that living cells require, and *enzymes,* to assist in their activity. It is probable that the earliest cells would have contained some structures to support such functions.

Protection would require some sort of coverage; suggestions include that which might have been provided by deep ocean water or liquid water below icebergs. *Energy* could have been provided by volcanic heat, sunlight, lightning, cosmic rays, or extraterrestrial impact. *Concentration* could have been provided by aggregation as "scum," freezing, or evaporation. *Enzymes* may at first have involved assembly assisted by clay mineral templates.

In his 1871 letter to Joseph Hooker, Darwin had conceived of some "warm little pond" as a possible site for life's origin. But later authors have suggested several other possible sites, based in part on the assumptions about the physical and chemical conditions they would have provided. Conditions on the surface of the early Earth, for example, probably involved continuing heavy bombardment by asteroids and comets, conditions that may have been lethal to early organisms.

Some writers, as we have seen, have suggested development in near freezing water below ice sheets, while others have suggested deep ocean hydrothermal vents as a suitable setting. Still others suggest the Earth's earliest organisms may have arisen elsewhere in the solar system or beyond it, being subsequently transported to Earth by the impact of asteroids or comets.

The most popular current hypothesis is that the huge amount of oxygen now locked up in the iron minerals of rocks called banded iron formations (BIFs) came from oxygen released by cyanobacteria, which combined with iron dissolved in the ancient acidic oceans to precipitate iron oxides. With the gradual rise in atmospheric and oceanic oxygen created by photosynthesizing cyanobacteria, the conditions that allowed the worldwide deposition of BIFs came to a close. But this did not happen until about 2,400 million years ago, long after Earth and its life had come into being. Thus, because cyanobacteria evolved much later than when life began—their form of photosynthesis being the source of oxygen on which later life depends—the earliest Earth, when life originated, must have been devoid of free oxygen.

Based on such models of the physical conditions on the early Earth, there is one other kind of model we might consider: the theoretical model of materials and processes that might have been involved in the earliest development of living things. A. G. Cairns-Smith[5] has suggested seven clues to this process. First: evolution requires the transmission of genetic information, so the earliest organisms may have been something like "naked genes," acting simply as template replicators. Second: DNA or RNA nucleotides that are the present basis for inheritance, may well have themselves evolved from other carbon-based materials of a simpler kind. Third: an earlier construction "scaffolding," no longer present, might have played an essential role in the genetic material of the earliest organisms. Fourth: organisms based on one kind of genetic material may have evolved into organisms based on a quite different material. Fifth: the "technology" and materials of the earliest organisms could have been, and probably were, far simpler than those of later organisms to which they gave rise. Sixth: Cairns-Smith suggested that the earliest genes could have been "crystal genes" of inorganic-crystalline form rather than organic-molecular form, based on the sophisticated characteristics of DNA and RNA, depending on the well-known process of inorganic-crystalline growth from a supersaturated solution. It should be noted, however, that there is little or no support among contemporary students in this field for this suggestion, or for Smith's seventh suggestion that clays may have acted as templates and catalysts in early evolutionary development with, perhaps, a kind of natural selection in the growth patterns of clays themselves.

All these are no more than clues, as Cairns-Smith himself describes them. Twenty years from now, other clues may replace them. They are speculative and provocative, as clues and models should be, because scientific models serve a particular purpose: to suggest new experiments, new directions, and new thinking that can lead, in turn, to new knowledge. Meanwhile, debate and discovery continue.

Andrew Knoll,[6] for example, has provided a somewhat different hypothesis, together with a wide-ranging review of the earliest fossils. Freeman Dyson[7] has provided a theoretical model based on the view that the chief characteristic of living things is homeostasis (an organism's ability to create a constant internal state of equilibrium in response to environmental change) rather than replication. All such theoretical models are based on assumptions about Earth's early atmosphere.

In the 1920s, two scientists, Aleksandr I. Oparin, a Russian biochemist, and J. B. S. Haldane, a British geneticist, independently suggested that Earth's original atmosphere might have consisted, not of oxygen—which would have been toxic to early life (breathing pure oxygen can induce nausea, convulsions, and even death in human beings, for example)—but of a reducing mixture of nitrogen, carbon dioxide, ammonia, and methane. If this suggestion was correct, it required radical rethinking of the process by which the first living things could have arisen. After all, most life forms—ourselves included—require oxygen, so it had been generally assumed that early life would have been life of the kind we know. This of course is not true of bacteria that thrive in our stomachs, where there is hardly any oxygen at all, but in the early 1900s such microbes were poorly known. So, the question remained: Without oxygen, how could life have existed? Oparin and Haldane provided the answer. Their key insight was that on the very early Earth, before life emerged, CHON-based organic compounds—generated by purely nonbiological processes—could have formed a life-producing and life-nurturing primordial soup.

The earliest attempt to devise an *experimental model* of such a process was a now classic experiment, as remarkable for its simplicity as it was dramatic in its results. In 1952, Stanley Miller, then a twenty-four-year-old graduate student of Nobel laureate Harold Urey at the University of Chicago, designed an experiment to test the Oparin-Haldane hypothesis. Miller constructed a sealed apparatus containing a mixture of hydrogen, methane, ammonia, and water, designed to simulate the composition of the early atmosphere. Using sparks to represent lightning and heat, Miller showed that after a few days the resulting product contained five amino acids, the building blocks of protein.[8] Later study of Miller's stored material, using more refined analytical techniques, identified more than twenty amino acids. In separate studies, amino acids have also been detected in meteorites.

Subsequent studies by others using the Miller-Urey type apparatus with different mixtures of carbon dioxide, nitrogen, hydrogen sulfide (H_2S), and sulfur dioxide (SO_2) to represent the early atmosphere that might have arisen from volcanism, have produced other molecules. Other experiments have produced a variety of carbohydrates, different amino acids, and nucleotides, which build DNA or RNA, and some of these have also been detected in meteorites. Any consideration of the origin of life must include its ability both to metabolize—that is to grow and sustain itself—and also to reproduce itself. All known living things achieve that by the interaction of nucleic acids and proteins.

But we are far from replicating the production of living things. Miller's experimental results depended on the exact proportions of the component gases, but other geochemists suggest the early atmosphere was less reducing than that of the Miller experiment, and perhaps had much greater concentrations of carbon dioxide and nitrogen than those that Miller selected. And "Miller-type" experiments by other specialists have not yet fulfilled the dreams of the early 1950s. We are still a long way from showing how these amino acids and other products can be combined to create even the simplest cells.

In the years immediately following the Miller-Urey experiment, there was a great sense of optimism that we were on the threshold of "solving" the riddle of the origin of life, or perhaps, even, of re-creating that origin in the laboratory. Half a century later, there is rather less triumphalism, less optimism that we can "explain" or re-create that process. Certainly, many more early fossils have been found, more understanding of the biochemical process has been gained, but even as the quest continues, perplexing questions still arise and obscurities remain. And, as we reason by analogy from the simplest living bacteria and protozoans, we have to acknowledge that even these are probably far removed from what must have been Earth's earliest forms of life. Nor do we know whether replication, reproduction, and metabolism developed together or, if they did not, which preceded the others.

THE EARLIEST RECORDS OF LIFE

The third line of inquiry in the search for potential evidence of life's earliest forms comes from the fossil record itself: what direct or indirect evidence is there of early remains of these early forms of life?

The search for the earliest traces of cellular structure, as opposed to the bacterial mats of stromatolites, has proved to be both elusive and controversial. Various claims have been made based on the size, structure, and composition of supposed microfossils. One of the most thoroughly reviewed has been that of Donald R. Lowe, and other authors, who have described biogenic microstructures from cherts of the 3.4-billion-year-old Strelley Pool Formation of Western Australia, whose size, shape, wall structure, and association strongly suggest an organic origin. They include, not only stromatolites and biolaminates, but also encrusting carbonaceous mats, carbonaceous granules, and oncolites, deposited under conditions ranging from subaqueous to nearly subaerial and locally evaporitic: a remarkably ecologically and probably biologically diverse microbial community. Later authors have reported other biogenic microstructures from the same formation. J. William Schopf, who with Stanley Tyler and Elso Barghoorn is one of the "fathers of Precambrian paleobiology," has recently provided (along with his colleague, Anatoliy B. Kudryavstev) a comprehensive review of these fossils and concluded that they are indeed evidence of a flourishing biocommunity 3.4 billion years ago (2012).[9] There are also several descriptions of filamentous cyanobacteria cells from the Warrawoona Group, rocks from Western Australia that are 3.5 to 3.4 billion years old. Younger rocks, such as the 3.4-billion-year-old Fig Tree Formation of South Africa and the 1.873-billion-year-old Gunflint Chert of the region around Lake Superior, also contain well-preserved bacteria, as well as stromatolites.

The oldest evidence for the existence of living things, however, is indirect and comes from highly suggestive chemical signatures found in 3.85-million-year-old shallow-water marine rocks in Greenland containing graphite, which has a carbon isotope (^{12}C:^{13}C) ratio, which is strongly suggestive of an organic origin. Rocks of slightly younger age yield similar evidence. Since the age of the Earth is generally estimated to be some 4.55 billion years, this marks a surprisingly early suggestion of life on the young planet. Smaller, cellular-like structures in the same rocks have been claimed as microfossils, but these claims are not generally accepted.

Stromatolites continued to form reeflike structures in other areas from 3.55 billion to about 550 million years ago but were then greatly reduced in the fossil record, probably becoming victims to grazing mollusks, echinoderms, and other invertebrates. They were long assumed to be extinct until the discovery of living forms in shallow, super-saline marine waters of Shark Bay, Western Australia. They have since been found elsewhere. These living forms consist of thinly bedded mats formed by cyanobacteria, which trap sediment and so form layered structures of calcium carbonate.

Another possible indirect indication of the early existence of living things is the widespread presence of a rock type known as banded iron formations (BIF) in rocks as old as 3 billion years, but especially in rocks ranging from 2.8 to 1.8 billion years in age. These distinctive rock types, which are found very rarely in younger deposits, consist of a great thickness of paper-thin, alternating bands of iron ore and chert (SiO_2) that extend laterally over hundreds of miles. These formations are known from North America, Australia, Brazil, and other areas and now provide the bulk of the world's iron resources. How were these thick deposits of iron ore and chert formed? How can we account for this great and almost unique episode of iron formation? It now seems clear that they were deposited at a time when the water of the world's oceans was rich in iron and silica, but deficient in oxygen, allowing repeated seasonal precipitation of iron and chert. It seems clear that the gradual buildup of oxygen released by photosynthesis of early organisms allowed its combination with iron dissolved in seawater to create these great deposits. There is confirmation of the almost total absence of atmospheric oxygen in 2.5-billion-year-old rocks from the presence of uraninite, a mineral that can exist only in an oxygen-free environment.

But why are such deposits so restricted in time? Three factors seem to be involved. First, the Earth's early atmosphere lacked significant free oxygen. Earth's present oxygen-rich atmosphere is the product of photosynthesis. Second, the abundant oxygen and silica represented in the rocks of the BIF were available in the ancient oceans because of the earlier absence of silica utilization by sponges and other invertebrates, which utilize it today. Third, oxygen produced by cyanobacteria oxidized iron dissolved in the seawater, combining to form the BIF. As time went on, the bacteria presumably developed a tolerance for oxygen and began to use it in the process of respiration.

These hypothetical mechanisms receive additional support from the first appearance of sedimentary red beds, which are rocks rich in iron oxides, about 2.3 billion years ago. Their presence supports the suggestion of a sharp rise in oxygen levels as a product of photosynthesis (the "oxygen revolution," Great Oxygenation Event, or GOE) soon after this period.

It is against such a series of background changes that we can best understand the little we have as yet discovered about Earth's earliest inhabitants.

The search for life's origins remains elusive, perplexing, and frustrating. We understand, of course, the constraints. We appreciate some reasons for the absence of physical evidence—the rarity of ancient rocks, the hazards of preservation, the Earth's changing oceans, atmosphere, and so on. Though one can only be pleased by the advances of the last half-century, the answers to some of our questions are not yet known. But, in spite of these unanswered questions, it is remarkable that we can at least trace a pattern for some fragmentary parts of life's early development. And we need always to consider the immensity of the time factor involved.

We are not yet close to understanding life's origins. Though we can draw broad conclusions about later patterns of life, our understanding of its origins remains speculative. It is clear that Earth and life have interacted together in "coevolution"; it is clear also that patterns of life and ways of living are not only competitive but also cumulative, that new structures and new environments have created new possibilities as they have emerged together over time.

None of these factors implies that the search for life's origins is futile or that we should discourage the continuing search. It is a challenge as great as any in science and of profound significance to our deepest humanity.

A century and a half after Darwin, we are only a little more enlightened than he was as to exactly how "from so simple a beginning, endless forms, most beautiful and most wonderful have been and are being evolved." The search for that "simple beginning" continues.

4

CLASSIFICATION: THE DIVERSITY OF LIFE

THE VARIETY OF LIFE

We have already discussed some aspects of the nature and origin of life (chapter 3) and have reviewed some of the tantalizing questions they pose. Those questions remain a challenge, not only to our scientific curiosity—important as that is—but also to our hunger to understand more about ourselves and our place in nature. And there are also other aspects of life we need to consider—its exuberant abundance, its dazzling diversity, its patterns of form, and its distribution, for example—if we are to comprehend its origins and its past.

THE ABUNDANCE OF LIFE

Earth teems with life. Around us, below us, above us, within us, a host of living things subsist in a vast variety of ways. The abundance of some of these creatures is remarkable. The Eastern or Atlantic Oyster (*Crassostrea virginica),* for example, can produce from 15 million to 114 million eggs in a single reproductive cycle. The Pacific Oyster (*Crassostrea gigas)* produces 50 million to 100 million eggs per cycle. Both live in great abundance in coastal and estuarine areas, and adults can live up to twenty years. The mortality rate, however, is high. It has been estimated that if all the eggs of a single generation of oysters were to be fertilized and developed, and the offspring developed under the same conditions, the great, great grandchildren of one generation would number 66 decillion, or 66×10^{33}, and their shells would make a mountain eight times the size of the Earth. Nor is the reproductive capacity of some vertebrates any less significant: a single female salmon may lay up to 10,000 eggs at a single spawning, some 30 percent of which hatch into alevins (immature salmon). The female Atlantic Bluefin Tuna can lay more than 10 million eggs in a season.

The land is equally prolific. The upper 1-inch layer of soil near Washington, D.C., for example, has been shown to contain more than 1 million macroscopic animals and more than 2 million macroscopic seeds per acre. A meadow soil at the same latitude proved to contain more than 13 million animals and 34 million seeds per acre. Estuarine and other environments show similar abundances. Add in the microscopic flora and fauna, and the numbers are staggering.

The impact of the abundance of individual species, not only on the broad ecology of life, but also on the surface of the land, was demonstrated by Charles Darwin in his last book—*The Formation of Vegetable Mould through the Action of Worms*—published in 1881, shortly before his death. It was a topic that had long fascinated him and about which he had read a paper to the Geological Society of London, more than fifty years before. Darwin showed that the soil near his home in Kent contained 53,767 earthworms per acre and that they had a profound influence on the formation and reworking of the soil. "It may be doubted," he concluded, "whether there are many other animals which have played so important a part in the history of the world as these lowly, organized creatures."

Subsequent detailed studies of earthworms in other environments and climates have demonstrated even more abundance and variety.

Uncultivated prairie land in Illinois, for example, has been shown to support an average earthworm population of 589,000 worms per acre, and almost three times as many in one case.[1] Other environments are equally prolific. The rainforests of Southeast Alaska support some 500 tons of living things per acre.[2] Bacteria are the most abundant of all organisms, numbering more than 100 million per gram of soil.

This combination of abundance and variety comprises, as we will see (chapter 17), two of the driving forces that Darwin identified in the action of natural selection.

THE DIVERSITY OF LIFE

The prolific development of life is reflected, not only in its abundance, but also in its diversity. Darwin's lowly earthworm, for example, is generally thought to have been *Lumbricus terrestris,* although this was not directly named by him and is but one of at least ten species of earthworms that exist in southern England, quite apart from about 7,000 species that exist worldwide.[3] The extravagant variety of both animals and plants is reflected in the number of species that have been recognized and named. After 250 years of taxonomic classification, at least 1.5–1.8 million species of animals and plants have been identified and named, although the exact number is uncertain, because of the vast literature, variety of languages, and adequacy of published descriptions. About 15,000 new ones are being added every year. Recent studies estimate that the actual number—as opposed to the number already described—may be about 10 million. Other estimates yield strikingly larger numbers.

Of the animals, there are some 5,500 described species of mammals, some 10,000 species of birds, about 8,000 reptiles, 5,500 amphibians, 8 species of lobe-finned fish, about 970 species of cartilaginous fish, and about 27,000 bony fish. Invertebrates make up the rest, with insects making up some 75 percent of all known animal species. Mollusks (snails, clams, and kin), other arthropods (crabs and spiders), protozoans, sponges, corals, wormlike animals, bryozoans ("moss animals"), echinoderms (starfish and sea urchins), and other smaller groups make up the rest. Together, the invertebrates make up 95 percent of all known animal species.

The reason we can give only ranges, as opposed to precise numbers of species, is that there are so many of them. In addition, the status of a substantial number of those already described is questionable. Out of the more than 1 million flowering plants already described and named, some 600,000 have recently been deleted, because of duplication in names. The tomato, for example, has been given 790 different names, and the oak and its varieties 600. On average, a typical plant might have been given two or three names.[4] So ambiguity abounds, and it affects the tentative numbers included here.

The reason these numbers are not precise reflects not only the fact that new organisms are steadily being recorded, while others are being duplicated or deleted, but also the fact that there is still significant debate as to whether some are "true species"—defined and separated by reproductive isolation—or subspecies, or something else.

And these large numbers concern only vertebrate and invertebrate animals, and plants. If we include the four other kingdoms (Fungi, Protista, Archaea, and Bacteria), the number is far greater.

But those staggering numbers include only living organisms. Add the dimension of time, and the 3.6 billion or so years of life's existence, and the numbers become overwhelming.

Such is the extravagant richness and diversity of life on our planet. Whether anything comparable to this diversity exists elsewhere in the vastness of the universe, we do not know, but one thing is clear: given this extraordinary variety of life—both living and extinct—of which we do know something, we need an orderly system to describe, classify, and name its various components. It is to this need for classification that we now turn.

THE PATTERNS OF LIFE

Whenever we wish to refer to objects, or ideas, or places, or people we use names. It does not help, for example, to talk about that curly-haired boy or that fair-haired young woman. We need to be precise. We need to talk about Bill or Jane. But even that may not be adequate. We may know two Janes, both of whom have blonde hair, so we may need two names—family names—to identify them: Jane Carlson and Jane Fuchs. That's why we need names to identify and talk about animals and plants, no less than about people. Adam, I suppose, was the first taxonomist when he coined names for all the creatures presented before him. It is striking that the various "kinds" of animals tend to be recognized not only by trained biologists, but also by untrained casual observers. Ernst Mayr, the great evolutionary biologist, recorded that the "primitive Papuan of the mountains of New Guinea recognizes as species exactly the same natural units that are called species by the museum ornithologist."

Now, it's one thing, of course, to identify "kinds" of animals and plants: it's another thing to name them. Early peoples probably divided animals into "dangerous or harmless" and plants into "edible or poisonous." Aristotle showed how animals could be classified by their habitat, structure, and habits. Later students categorized animals by their feeding habits (omnivorous vs. carnivorous vs. herbivorous) or physiological characteristics (warm-blooded vs. cold-blooded), for example.

All these various methods of classification are valid and each is useful to the extent that it serves some particular need or purpose. But if we are to describe the exuberant diversity of living things, both past and present, we need a system of naming or nomenclature that meets three essential requirements. First, it must provide a unique name for each "kind" or species of organism, and the name must identify both its distinctiveness *and* its close affinities. Thus, if I say that I saw a "hawk," that would be of interest, but an ornithologist would ask, "Which hawk?" For there are several "kinds" of hawk, each of them given a distinctive name. The red-shouldered hawk, for example, is called *Buteo lineatus,* the broad-winged hawk, *Buteo platypterus,* the short-tailed hawk, *Buteo brachyurus,* and the red-tailed hawk, *Buteo jamaicensis.* All these are "hawks," all have a broad resemblance and thus relationship to one another, but each is distinctive, and they remain distinctive because of their reproductive isolation from one another.

The second requirement for any classification is that it must be universally applicable—so that local and dialect names are avoided—and it must be well founded, based on a representative range of type specimens that are well studied and well described. This standardized system of nomenclature must then be developed and enforced by an agreed code of practice, so that, for example, multiple names for the same "kind" of species (synonyms) are avoided.

Third, such a classification should be able to reflect degrees of genetic relationship, not only as in the case of closely related groups, such as the hawks described above, with all four species belonging to the same genus, *Buteo,* but also in still broader similarities and relationships.

BIOLOGICAL CLASSIFICATION

For almost 2,000 years after the death of Aristotle (384–322 BC) a system of biological classification based largely on his concepts was used. The first attempt to develop a more appropriate and useful system began with the English naturalist John Ray (1627–1705), a clergyman and philosopher who published catalogs of plants, birds, mammals, fish, and insects, seeking to identify a system of classification reflective of the Divine Order of Creation. It was, however, the work of a later Swedish naturalist, the great botanist Carl von Linne (Carolus Linnaeus 1707–1778) whose *Systema Naturae* published in 1758 laid the foundation of our present system of classification. Linnaeus graduated from and taught at Uppsala University, of which he ultimately became Rector.

It is the Linnaean system of classification that is universally used today. Linnaeus divided living things, both animals and plants, into species, to each of which he gave a distinctive name. To this system he added a hierarchy of higher and progressively more inclusive categories: genus, order, class, and kingdom.

The "baseline" of this Linnaean classification is the species, defined as a group of actually or potentially interbreeding individuals, reproductively isolated from other such groups

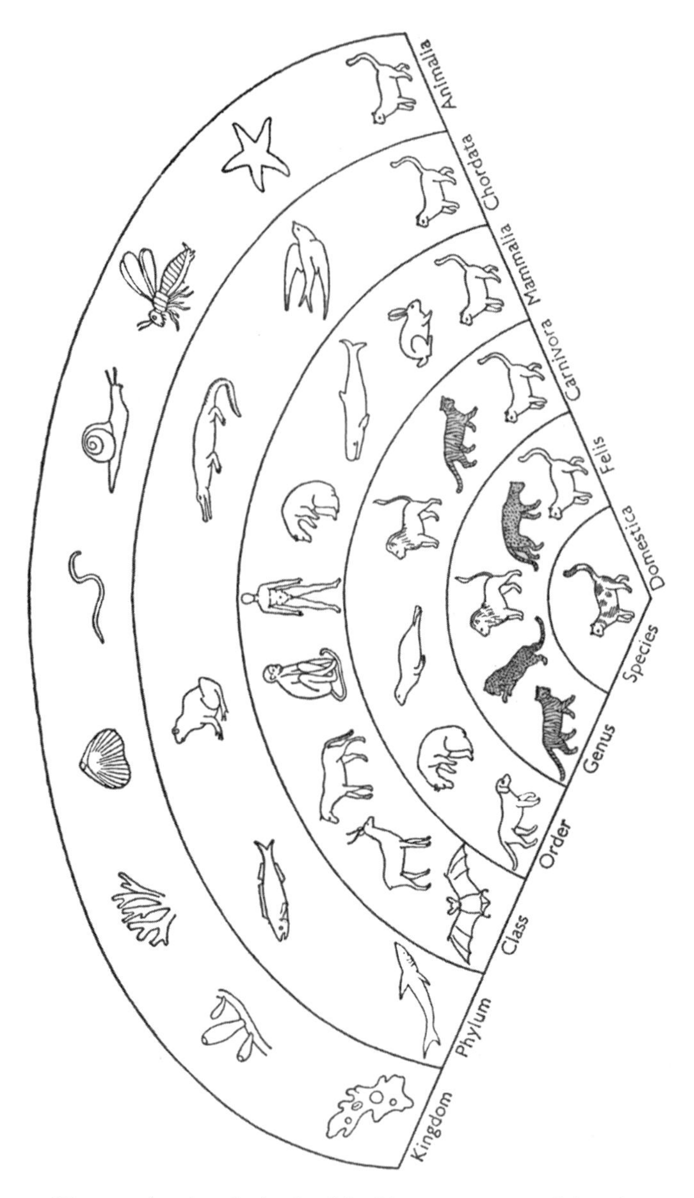

Figure 4.1. Diagram showing the basis of the Linnaean system of classification applied to kingdom Animalia, illustrating the increasing similarity of members of the various taxonomic categories ranked from kingdom down to species.

(From Frank H. T. Rhodes, *Evolution of Life* [New York: Penguin, 1976], fig. 1, p. 22. Reproduced with permission.)

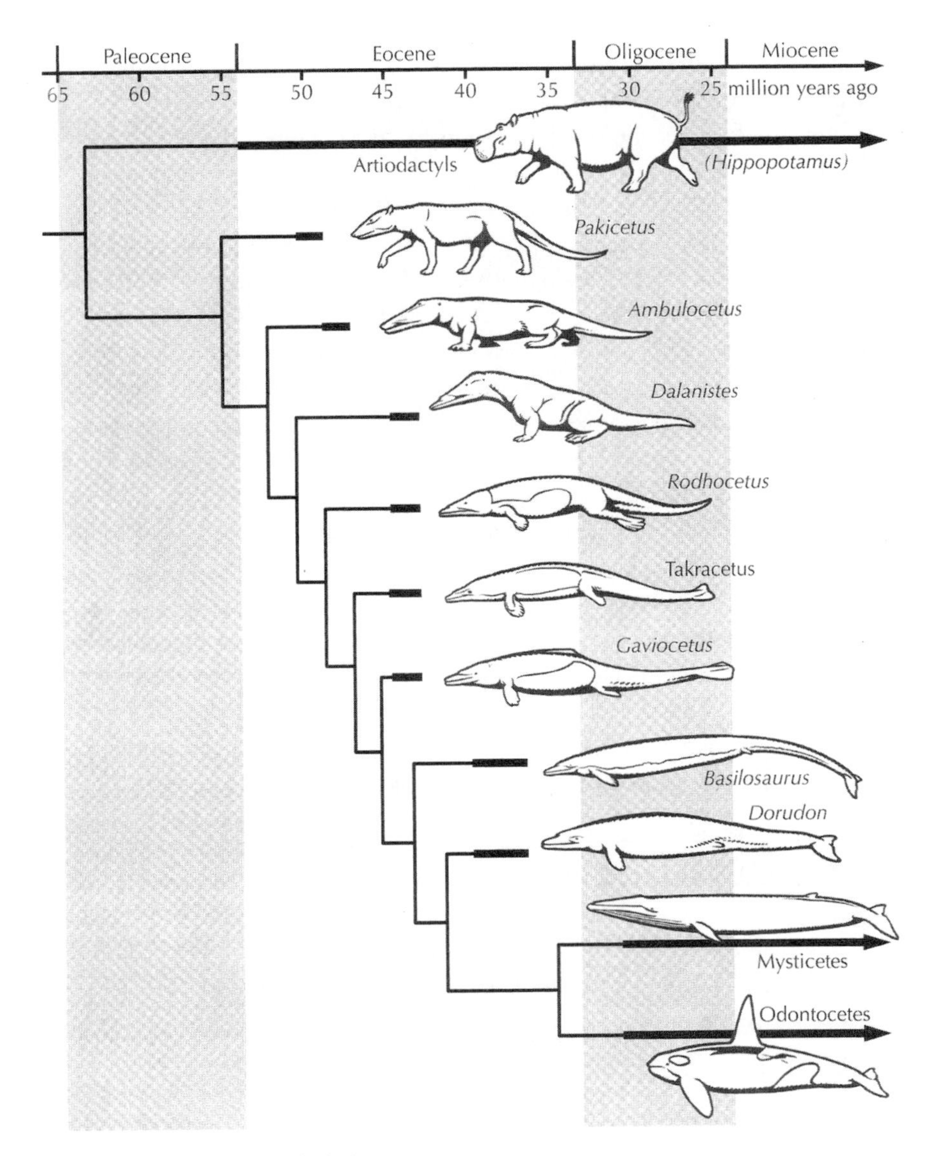

Figure 4.2. Evolution of whales.

(Drawing by Carl Buell. From Donald R. Prothero, *Evolution: What the Fossils Say and Why It Matters* [New York: Columbia University Press, 2007]. Copyright Columbia University Press. Reprinted with permission of the publisher.)

The most inclusive group—the kingdom—is today most often represented by five different categories: Animalia, Plantae, Fungi, Protista and Archaea. These divisions are shown in figure 4.1.

We should note one other question of classification. We have so far tacitly assumed that the characteristics on which we classify animals are self-evident. "Why," we might ask, "are whales classified as mammals and not as fishes?" After all, they are fully aquatic, beautifully equipped for life in the oceans. They have fins and fishlike bodies and they lack the body hair of mammals. Admittedly they are warm-blooded, are air-breathing, and have mammary glands and lungs. But what makes them mammals, rather than fish? From fossil studies of their geological history, it is clear that whales descended from terrestrial mammalian ancestors. In fact, beautifully preserved fossils show this remarkable transition from 52-million-year-old wolflike fossils from Pakistan, through amphibious forms, through fully marine forms that still retained small weak hind limbs. Their fishlike characteristics are thus adaptations imposed on their fundamentally mammalian bodies. Taxonomists—scientists who classify and name organisms—have concluded (following Darwin) that the best and most meaningful basis for classification is one based on ancestry, so the accepted basis for classification has become the phylogeny or the evolutionary history of the group. This is not, of course, the only possible basis for classification, but it has been adopted as the most useful.

THE MAJOR GROUPS OF ORGANISMS

Though we will meet them again in subsequent chapters, it will be useful to give a brief outline of those groups most commonly represented as fossils. They represent only a fraction of the thirty-five recognized phyla of living animals and the eleven phyla of living plants.

Kingdom Protista

Mostly unicellular (one celled), aquatic organisms having a nucleus, including radiolaria, foraminifera, diatoms, and dinoflagellates (p. 68).

Kingdom Animalia

PHYLUM PORIFERA	Sponges, with cavities surrounded by porous body walls, commonly made of silica, calcite, or spongin (p. 66)

PHYLUM COELENTERATA	Corals, jellyfish, and sea anemones, in which a single body cavity performs all vital functions (p. 68)
PHYLUM BRYOZOA	Small colonial animals, with calcareous skeletons, "moss animals" (p. 79)
PHYLUM BRACHIOPODA	Bivalved, marine shellfish, "lamp shells" (p. 79)
PHYLUM MOLLUSCA	Highly developed creatures, both terrestrial and aquatic, including snails, clams, oysters, squid, nautilus, and slugs (p. 75)
PHYLUM ANNELIDA	Highly developed, segmented worms, including a number of terrestrial and aquatic groups (p. 77)
PHYLUM ARTHROPODA	Segmented animals, with jointed appendages on each segment, including lobsters, crabs, and insects (p. 79)
PHYLUM ECHINODERMATA	Spiny-skinned animals, including starfish, sea urchins, and sea lilies (p. 79)
PHYLUM CHORDATA	Animals with a notochord, including all vertebrates (fish, amphibians, reptiles, and mammals) (p. 91)

Kingdom Plantae

PHYLUM "THALLOPHYTA"	A heterogeneous group of nonflowering, seedless plants, lacking roots, stems, and leaves, including algae, bacteria, lichens, and fungi (p. 108)
PHYLUM BRYOPHYTA	Plants with leaves and stems, including mosses and liverworts (p. 109)
PHYLUM TRACHEOPHYTA	Vascular plants, fully adapted to terrestrial life, including ferns, flowering plants, trees, and grasses. The largest group of plants (p. 111)

THE DISTRIBUTION OF LIFE

One of the most striking features of living things is the way in which their various structures adapt them to different ways of life: the gills of a fish, the wings of a bird, and so on. But frequently the more perfectly an animal is adapted to a particular way of life, the less well-suited it is to other ways of life: jellyfish cannot walk and pigs don't fly, for example. Every living organism thus has its own distinctive lifestyle and distribution, controlled largely by its structure and physiology. The relationship between an organism and its environment is known as its ecology (from the Green *oikos*: dwelling place). One of the most striking things about all organisms—of both the present and the past—is how widely they have become adapted to different and often distinctive ways of life. Wherever we go—from the snows of the highest mountains to the depths of the oceans, from the deserts to the jungles, Earth teems with life.

But this dazzling richness and variety of life in every nook and cranny of the Earth is a feature of the present. Earlier organisms were more limited in variety and more restricted in distribution. In fact, this slow extension of the biosphere, as one environment after another was colonized, is one of the most striking aspects of the long history of life.

FACTORS CONTROLLING THE DISTRIBUTION OF LIFE

Although it is often easy to observe the general way of life and broad pattern of distribution of a given species, it is much more difficult to evaluate the various factors that influence its habits and control its range. Nor is this difficulty diminished by the fact that the factors controlling importance in one environment (say a meadow) may be quite different from those of importance in another (such as the waters of a lagoon). What are these factors? We may think of them as being of three broad types: physical, chemical, and biological. The physical conditions may include such things as temperature, pressure, light and atmospheric conditions, depth of water, viscosity and diffusion of air, water currents, tides, topography, geographical position, and composition of land surfaces, physical barriers to migration, conditions of sedimentation, and so on. Chemical factors include salinity, hydrogen ion concentration, gas and organic content of water, the oxygen, nitrogen, and carbon dioxide content of air, and the presence or concentration of a large number of compounds and trace elements. The external biological factors (as opposed to the "internal" factors represented by the organism's physiology and structure) are represented by the numbers and kinds of associated organisms; their interrelationships as food, prey, competitors, parasites, etc.; mobility; population size and rate of change; and birth and mortality rates of species, as well as various other factors.

Few of these factors are constant, and most of them exhibit seasonal, diurnal, or random variation, which is often considerable. Clearly, therefore, the equilibrium that exists between organisms and their environment is both dynamic and intricate, and a slight variation in any one factor may have profound effects on the general stability. Such environmental stimulus and change may affect individual organisms in various ways. They may, for example, be driven out of their original niche, they may be destroyed, or they may continue with or without some modification. These modifications themselves may also vary. The greater development of branches and leaves on the sunny side of a tree near a forest margin and the adjustment of the pupil of the eye to light of varying intensity are familiar examples of nongenetic adjustment known as acclimatization. Other changes favored by environmental pressure may be inherited in successive generations, however, and these are known as adaptations. The process of adaptation is a common one, and is of the highest importance in evolutionary development.

So far we have tacitly assumed that organisms respond as individuals to such changes. This is strictly true, but almost all individuals exist in some kind of association with others, and the response of a group of organisms may be quite different from that of any given individual member. The relationships between individuals naturally vary. A population is any association of one or more species that constitute a closely knit, interacting system. These interactions involve such things as food competition, parasitic relationships, and so on. Communities are more complex natural groupings, usually involving several populations that inhabit a given area, such as a lake or a forest. All the various factors influencing individual organisms also influence communities: a community undergoes developmental, diurnal, and seasonal changes, for example.

Within such a community, individuals may share various relationships. They may, as we have seen, be competitors for food, they may be prey and predator, they may be host and parasite, and each of these broad relationships includes endless degrees of difference. An example will make this clearer. Many organisms live together in intimate mutual association, referred to as symbiosis. This may involve the joint association of animals, or the joint association of plants, or the association of animals with plants, and the relationship itself may be one of three kinds. First it may be a voluntary association that is mutually beneficial and without damage to either partner. This mutualism is found, for example, in the frequent attachment of sea anemones to the shells of crabs, which provides concealment and the defense of the anemone's "stinging cells" for the crab and locomotion for the otherwise sessile anemone. Commensalism is a voluntary union from which only one of the partners benefits, although the other is unharmed. It may be permanent or temporary, intimate or free living. Small fish and crustaceans often exist in such a way within the tentacles or body cavity of coelenterates, thereby gaining protection and some assistance in food gathering. The third degree of relationship is represented by the involuntary association of parasitism, in which the benefit of the parasite results in progressive damage and sometimes death to the host. Here again, parasitism may be of various kinds, temporary or permanent, or nonsexual or sexual (between the male and female of a single species, such as the angler fish, in which the smaller male is permanently attached to the female).

THE DISTRIBUTION OF LIFE

The net effect of these multitudes of various factors is to limit the range of species and communities. We have already seen that this limited distribution is reflected in the distinctive populations of different physical environments, such as freshwater lakes, or swamps, or estuaries, and this is one important method of classifying environments.

There are other ways too, however, in which the distribution of living things is reflected. One of the most obvious of these is the presence of broad climatic zones, which vary with both latitude and altitude, each supporting a distinctive fauna or flora. The general sequence of zones is traversed both from the equator to the poles and also from the base to the summit of mountains.

The Earth may also be divided into six zoogeographic realms, historically based (at least for terrestrial organisms) primarily on mammals and birds, although they also support other animals and plants that are more or less characteristic. The Palearctic realm embraces Europe, northern Asia, and North Africa and is characterized by the reindeer, hedgehog, European bison, Marco Polo sheep, and wild ass. North America comprises the Nearctic region, characterized by the caribou, porcupine, pronghorn antelope, American bison, and mountain goat, while South America, the Neotropical realm, supports the giant anteater, sloth, tapir, and New World monkeys. The Ethiopian realm includes Africa and the Eastern Mediterranean lands and is inhabited by the zebra, gorilla, giraffe, gnu, aardvark, and African elephant. The Oriental realm of the Far East supports the gibbon, the Malay tapir, the tiger, water buffalo, and Indian elephant, and the Australian realm the marsupial kangaroo, koala bear, wombat, and Tasmanian wolf. These faunas are not everywhere sharply separated, and there is some intergradation, although the realms are generally, though not entirely, separated at present by either oceanic or land barriers (the Sahara desert separates the Ethiopian and Palearctic realms, for example).

Now these differences cannot be explained in terms of different environments. Broadly similar ecological conditions are found in different realms—South Africa and Australia, for example—but the faunas are quite distinct, although in each case the animals tend to assume similar modes of life and even sometimes have a broadly similar appearance. These differences and some similarities are, in fact, explicable only in terms of the geological past. Why, for example, is the fauna of North America more like that of Northern Asia than that of South America, to which land it is now joined by the Isthmus of Panama? The study of fossil faunas from these areas reveals that until relatively recently, North America was joined to Eurasia by a northerly connection but was separated from South America. These, and other problems, we will discuss later.

5

SPINELESS WONDERS

THE INVERTEBRATE REALM

We live in societies dominated by vertebrates. We share our teeming cities not only with members of our own species but also with countless numbers of other creatures—from pet cats and dogs to gerbils, parakeets, and goldfish, as well as a rich assortment of wild birds—that are also vertebrates. In the countryside, we encounter the same situation: cattle, sheep, horses, chickens, deer, raccoons, badgers, bats, snakes, mice, foxes, bears, and birds. And in ponds, lakes, streams, and rivers, we find frogs, toads, turtles, fish, and more, all of them vertebrates. Large, abundant, and conspicuous: these are the animals that catch our attention, that dominate the food chain.

But before our teeming human species, before the cat and the tiger, before the dog and the wolf, before the cattle and the horse, before the eagle and the finch, before the bat, before the snake and the turtle, before the salmon and the cod, before the rose and the thistle, before the willow and the oak, before the grasses and the herbs, long, long before all these, Earth's teeming creatures were of a different clan. The world's oceans were the home of a vast array of other creatures—spineless wonders—every bit as varied, beautiful, successful, and significant as their vertebrate descendants.

It is easy to overlook this other great group of multicellular animals—the invertebrates—on whom the whole bio-economy of our planet depends. These creatures exist in countless numbers and represent a dazzling variety of different forms: flies, mosquitoes, butterflies, bees, snails, bed bugs, scorpions, earthworms, termites, ants, centipedes above, on, and in the land, and in the oceans, clams, oysters, more snails, crabs, shrimps, lobsters, jellyfish, sea pens, corals, sea lilies, sea urchins, starfish, and more. In lakes, ponds, streams, and rivers, we find still more. These invertebrate creatures form a critical link, not only in the food chain but also in the long history of life.

Furthermore, they far outnumber vertebrate animals, both in the variety of kinds and in the overwhelming numbers of individuals by which their countless species are represented.

In chapter 4 we saw the uncertainties and ambiguities concerning the precise number of existing species. One recent count concluded that, of the estimated 1,367,550 described species of living animals, 1,305,250 are invertebrates, and, of these, more than a million are insects, many represented by countless numbers of individuals.

Nor should we ignore the teeming numbers of unicellular microscopic organisms—the bacteria and related forms—that are present in every environment, around, upon, and within us, existing in such abundance that their combined weight has been estimated to be at least ten times that of all macroscopic creatures combined.[1] The history we are about to consider—the history of Earth's creatures—is their story, and, conspicuous though their vertebrate relatives later became and still are, the invertebrates, as well as the bacteria and all their unicellular allies, continue to play a huge role in the larger economy of the living world.

Traditional classification of all these various groups has undergone profound change in the last thirty years or so and is still in a state of flux, being far from settled. In such an account as this, intended for the general reader, I have not followed the formidable nomenclature of the latest taxonomy but have retained some of the traditional and more familiar names of major groups as informal groupings, rather than formal taxonomic categories.

With that as background, let us turn to the geologic history of these spineless wonders.

INDIRECT INDICATORS OF EARLY ORGANISMS

When we turn to the fossil record for clues to the early development of life, we face an immediate handicap, for many of the oldest sedimentary rocks have been metamorphosed, and outcrops are rare. The oldest minerals known, as we have seen, are zircon crystals more than 4 billion years old, and these themselves are the detrital remains of still older sedimentary rocks.

The earliest indications we have of the presence of living things on our planet are represented by three rather different traces. There are, first, body fossils, which are the remains of earlier forms of life. As we have seen, these generally represent only the hard parts (shells and skeletons) of once-living beings, which, even then, have often undergone some chemical modification or mineral replacement. Other fossils are represented by microscopic cellular structures, such as those described in chapter 3. Second, we sometimes find in sedimentary rocks traces and indications of the former presence and activities of living creatures: things like trails, tracks, and burrows of ancient organisms, and these are frequently preserved, even in the absence of the creatures who produced them. Such creatures may, in fact, have been soft-bodied and have left no other record of their existence. The third type of fossil evidence available is provided by chemical signatures of rocks, which are indicative of the former existence of living things. Carbon isotope ratios (^{12}C:^{13}C) are such an example.

Having described these types of fossils, we should note that fossilization is a rarity. The vast majority of all organisms that have lived and died have left no trace, and even the patchy record we have is heavily biased toward creatures having hard parts and living in environments favorable to preservation. There are huge gaps, for example, in the fossil record of birds, because of both the light structure of their bones and their way of life.

The earliest traces of indirect evidence for the existence of living things, comes from highly suggestive chemical signatures found in 3.85-million-year-old rocks of the Isua Complex in western Greenland. These represent the battered remains of ancient shallow-water marine rocks and they contain graphite, which has a carbon isotope ^{12}C:^{13}C ratio strongly suggestive of an organic origin. Rocks of slightly younger age yield similar evidence. Since the age of the Earth is generally estimated to be some 4.55 billion years, this marks a surprisingly early indication of life on the young planet.[2] Though microscopic cellular-like structures in the same rocks have been claimed as microfossils, these structures are not universally accepted as of organic origin.

THE "EDIACARAN" FAUNA

One of the many other major glimpses we have of life before the Cambrian comes from rocks ranging in age from 635 to 542 million years, just before the dawn of Cambrian times. Fossil remains of this age were first discovered in 1868, but their significance was not realized until the discovery in 1957 of a distinctive fossil—*Charnia*—in rocks of undoubted Precambrian age in England. This fossil was also known from the Ediacara Hills of the Flinders Range of South Australia and has since been described, together with other characteristic fossils, from some thirty other localities, with especially rich collections from White Russia, Canada, Namibia, and Europe.[3] These various localities expose sedimentary rocks of similar age, but of different marine depositional environments, each with slightly different faunas, though all of them were benthic in habit.

The Ediacaran *Iagerstätten* type of preservation of exceptionally fine details resembles that of the younger Middle Cambrian Burgess Shale, as well as the Solnhofen Limestone of Bavaria. Most members of the Ediacaran fauna are found immediately above ancient glacial deposits (tillites), emphasizing a profound change in global climate, as the "Snowball Earth" experienced a series of ice ages between 725 million and 541 million years ago, with each glacial period separated by a warm interglacial period.

Figure 5.1. Ediacaran fossils. The Ediacaran fauna consists of the impressions of soft-bodied fossils without skeletons, whose biological affinities are still controversial. This image is a reconstruction of the Ediacaran community, with most of the fossils assumed to be related to jellyfish, sea pens, and worms.

(Photo from the National Museum of Natural History, Department of Paleobiology. Courtesy of the Smithsonian Institution. Reprinted with permission.)

In most parts of the world the fossil record of these distant times is very fragmentary, because of gaps in the succession of fossiliferous sedimentary rocks. In contrast, China contains several localities that seem to have unbroken rock sequences, and teams of Chinese geologists are now analyzing these, not only studying the fossil sequences but also attempting to use carbon isotopes and other chemical studies to reconstruct the changing patterns of ancient ocean chemistry.

Collections from these various localities include more than thirty recognized genera, represented by some forms resembling quilted bags, others resembling sea pens, including *Charnodiscus*, which was about a foot and a half in length, while still others *(Dickinsonia)* resembled circular, flattened, segmented, wormlike or coelenterate-like creatures, and still others *(Spriggina)* were wormlike with well-developed head shields. Others resemble arthropods, and some, such as *Tribrachidium,* are unlike any other known living or fossil groups.

Not surprisingly, this collection has generated speculation about the affinities of these fossils. One respected paleontologist—Adolf Seilacher, of the University of Tubingen—regards the whole fauna as unrelated to any subsequent forms, having been wiped out by a massive wave of extinction and leaving no descendants. Most paleontologists, however, regard the soft-bodied Ediacaran creatures as representative of an unusual period of earth history, before the widespread existence of burrowing animals and other predators. Some organisms were, perhaps, protected by a covering of microbial mats. In this interpretation, the Ediacaran fossils were preserved during a rare "taphonomic window" and are best understood as early, soft-bodied forerunners of later marine invertebrate groups, including cnidarians, worms, and arthropods. Ediacaran forms persisted into the Early Cambrian, but most disappeared around the dawn of Cambrian time, about 541 million years ago. Only a few of the various Ediacaran genera bear a close resemblance either to later Cambrian forms succeeding them or to any known invertebrate phylum.

Nor is their lack of similarity to later forms the only puzzling feature of these fossils. By the time we get our first glimpse of these multicellular Ediacaran animals, more than 80 percent of the history of life was already over. Yet there are, as yet, no traces of obvious ancestors in older rocks.

Paleontologists continue to differ in their interpretation of the relationships of these creatures. The only hard-bodied remains in this rich assortment of Ediacaran fossils are nested calcareous tubes, about an inch in length, which seem to have housed some kind of polyp. On many surfaces of the Ediacaran rocks, trails and burrows are preserved, marking the existence of a more active group of soft-bodied creatures that are otherwise unknown to us.

Though the Ediacaran creatures fill a major gap in the fossil record, they raise other tantalizing issues, not only over their affinities, about which lively debate continues, but also over their lack of resemblance to the slightly younger creatures of the earliest Cambrian time that succeeded them. To the fossil remains of these early Cambrian organisms—the Ediacaran creatures show little overall resemblance.

Collectively, these Ediacaran fossils, so perplexing in their affinities, have been found at thirty different localities throughout the world. They raise at least five basic questions:

- Affinities. What are these creatures, which appear without ancestors or descendants? The several forms have been referred by different authors to a remarkable range of known groups, including algae, microbial mats, protists, sponges, jellyfish, mollusks, and worms. Some impressions are so extraordinary that, as we have seen, Adolf Seilacher has separated the whole assemblage as a distinct kingdom (which he named the Vendoza). Others have suggested that they may be an abortive experiment in multicellular life that failed in competition with Cambrian competitors.
- Preservation. How does this collection of soft-bodied animals come to be so well preserved and represented in more than a score of widely scattered areas? Perhaps the absence of mobile, browsing scavengers was a major factor, with the seafloor essentially covered in microbial mats. Perhaps the later rise of mobile, epi-, and infauna meant that soft-bodied Ediacaran forms were no longer easily preserved. Were physical conditions in the oceans such that the typical decay and disintegration of soft parts was somehow prevented?
- Distribution. How did these creatures, some of them clearly fixed and sessile, come to gain such rapid, worldwide distribution? Did their distribution, and even, perhaps, their origin, depend on changes in ocean chemistry, or in atmospheric oxygen levels? As mentioned, they existed during a prolonged period of alternating glacial and interglacial episodes that were part of a worldwide glaciation (the Gaskiers glaciation). Could this have been a factor in their rapid rise and distribution?

• Ancestry. How can we account for the relatively sudden appearance of so many, so varied forms in different parts of the world with no apparent ancestors? Some earlier egglike structures have been recorded, and these may represent the embryos of multicellular groups, but the nature of these is still not fully clear. How can we account for the absence of any less "complex" forms, or of any obvious evolutionary change over a period of some 43 million years?
• Extinction. How can we explain the sudden worldwide disappearance of the Ediacaran fauna at the dawn of Cambrian times? Though a few forms have been reported from rocks of Cambrian age, most of the Ediacaran fauna did not survive into the earliest Cambrian, some 541 million years ago. The search for extinction mechanisms is seldom easy and almost never conclusive: competition, predation, environmental changes have all been suggested, but the disappearance of the Ediacaran fauna remains as much an enigma as does its appearance.

EARLY CAMBRIAN FOSSILS

The earliest Cambrian strata are marked by an increase in burrowing organisms and then the worldwide appearance of a distinctive group of microfossils known as Small Shelly Fossils (SSF), most of them only a millimeter or so in size. These represented a major step of the development of "hard" parts, which provided both protection and support for their invertebrate owners and permitted the subsequent development of new modes and patterns of life. Mineralized remains of these hard parts also greatly increased the chances of fossilization, as well as allowing increase in size. Most of the major invertebrate groups that make up the fossil record appear early in the Cambrian. Some of these had hard parts that were internal, some external; some were made of silica, some of calcium carbonate, and still others were phosphatic; some were microscopic, others reached a foot or so in size; all were marine, at least in their earliest appearance.

Remarkable as is the sudden appearance of so many well-differentiated, hard-bodied Cambrian forms, a series of discoveries in rocks of Early and Middle Cambrian age reveal the even more striking variety of forms lacking mineralized skeletons. Extensive collections from the Middle Cambrian Burgess Shale of British Columbia and Utah, as well as from Early Cambrian rocks of China, Greenland, Poland, and Canada, reveal a menagerie of contemporary soft-bodied Cambrian creatures, all exquisitely preserved, some clearly representing existing groups—such as arthropods and worms—but others of forms so unfamiliar as to defy categorization. Thus, one is so bizarre that it is aptly named *Hallucigenia*. Another, *Opabinia*, has five eyes and a long, grasping proboscis, while still another form—an 18-inch-long predator, bears the name *Anomalocaris,* indicating the frustration of attempting to classify its features in relation to other groups. Early Cambrian fossils from China and from the

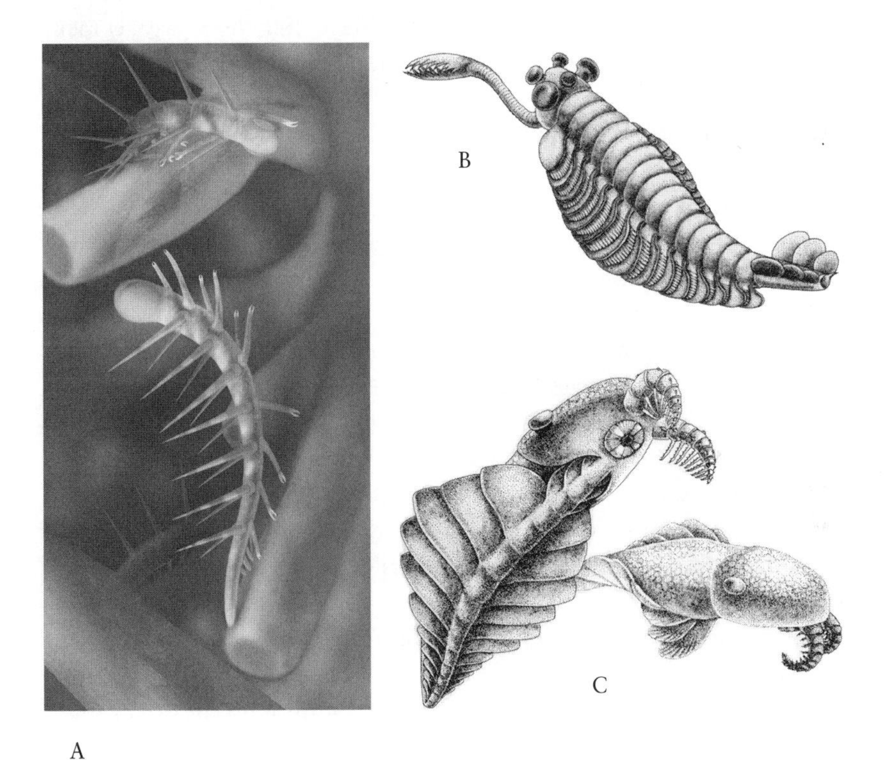

Figure 5.2. *Hallucigenia, Opabinia, Anomalocaris.* (A) *Hallucigenia* (courtesy of Smithsonian Institution; painting by M. Parrish). (B) *Opabinia*, showing the frontal nozzle with terminal claw, five eyes on the head, body section with gills on top, and the tail piece in three segments (drawn by Marianne Collins; used with permission). (C) The two known species of *Anomalocaris*: top, *A. nathorsti* as seen from below, showing the circular mouth, misidentified by Walcott as a jellyfish, and the pair of feeding appendages; bottom, *A. canadensis* as seen from the side, in swimming position (drawn by Marianne Collins; used with permission).

Burgess Shale also contain elongated forms resembling the living *Amphioxus,* a relative of chordates.

Something of the difficulty of interpreting these early forms of life can be gathered by considering the history of the description and interpretation of these fossils from the Burgess Shale, a remarkable 540-million-year-old deposit exposed in British Columbia. The Burgess Shale contains exquisitely preserved remains of ancient creatures that were buried by an underwater avalanche of fine mud that preserved details of their soft parts and structure. It was first discovered by Charles Doolittle Walcott in 1909, who collected from it regularly for fifteen years. Assisted by his family, Walcott collected more than 65,000 specimens, which he described in a series of papers, generally interpreting them as early members of existing phyla. In 1894 he became the director of the U.S. Geological Survey and, later, the Secretary of the Smithsonian Institution—a truly remarkable career, incidentally, for someone who never finished high school.

Just how extraordinary some of these Burgess Shale creatures were was beautifully demonstrated by H. B. (Harry) Whittington, Woodwardian Professor of Geology at Cambridge and two of his students—Simon Conway-Morris and Derek Briggs—in their redescription of *Opabinia regalis,* whose flattened strongly segmented body, about 2 inches long, had a tail fin and five eyes, together with an elongated, trunklike snout, ending in a pincerlike jaw.

The Burgess Shale fossils include a remarkable variety of different forms, some quite unlike any known group. Other fossils, however, do clearly represent members of living groups. For example, *Sidneya* is an arthropod, *Canadia* an annelid, and *Choia* a sponge. But a few other forms, in addition to *Hallucigenia,* bear no resemblance to other known groups; these include *Opabinia, Amiskwia, Anomalocaris,* and *Wiwaxia.* Whittington and his coauthors concluded that they represented a greater degree of biodiversity in Cambrian times than exists today.

Stephen Jay Gould (1941–2002), writing in 1989, provided a popular account of these fossils. He was critical of Walcott's interpretation of these remarkable fossils, accusing Walcott of having "shoehorned" them into existing phyla, rather than recognizing their identity as examples of novel and otherwise unknown groups.[4]

Study, reinterpretation, and controversy about these fossils continue. Meanwhile, similarly well-preserved fossils (*lagerstätten*) have been found in other parts of the world, such as Chengjiang in China, as well as the Burgess Shale.

Quite apart from the continuing debate about the affinities of these fossils, it is noteworthy that they represent the product of such an extraordinary burst of diversity so early in the recorded history of life. All the major phyla are represented in what is often described as the "Cambrian explosion," which occurred some 25–30 million years before the deposition of the Burgess Shale.

All these discoveries make the enigma of the "Cambrian explosion" even more enigmatic, for they reveal how little of the true richness and diversity of Cambrian life (or the life of any other period) is represented by the typical fossil record, made up, as it is, chiefly by organisms having hard parts. But the sudden first appearance of so many creatures—hard-bodied and soft-bodied—remains an enigma. How can we account for it? Half a dozen explanations have been offered and we might summarize them briefly.

Some address the problem by asserting that there is no problem, arguing that we have clear evidence of late Precambrian life and that Cambrian fossils emerge gradually over some 65 million years. But, while the first part of the statement is correct—we do have Precambrian fossils—the second is misleading. Although the Cambrian period spanned 65 million years, and all the "fossilizable" phyla do appear in the Cambrian, all but one—the Bryozoa—appear within a much briefer interval in the early part of the Cambrian Period, though all are represented by classes that are now extinct.

Others assert that, though Precambrian multicellular organisms did exist, because they lacked hard parts, most have left no record. Clearly the Burgess Shale and similar deposits lend support to this suggestion.

Others have argued that environmental changes, such as low oxygen levels in Precambrian times, might have acted as a barrier to the development of Precambrian multicellular life. Thus also, an increase in oxygen levels, in both the atmosphere and the hydrosphere, has been proposed as an explanation, as has a possible change in the chemistry of the oceans, such as a sudden increase in calcium, so allowing the development of hard parts. While none of these views is wholly implausible in principle, in practice none seems satisfactory to fit all the facts of the case.

Perhaps one other comment, while not an explanation, needs to be made in considering this issue. The Cambrian explosion took place in marine environments that were relatively unoccupied by the host of competitors that marked the later history of life. That lack of intense pressure from established "occupying groups" must have been one factor, among others, in allowing the Cambrian explosion.

The earliest Cambrian fossils are found in northern Siberia in cliffs along the Kokuikan River,[5] where thick sequences of Cambrian rocks contain well-preserved small fossils. Similar sequences have since also been recorded from China, India, England, Canada, and Australia. Some of these fossils are phosphatic in composition, others calcareous or siliceous, and they include microfossils of spicules and spines, as well as tubelike forms that housed organisms of unknown form or affinity. Other fossils include small, *phosphatic* shells of brachiopods ("lamp shells") and monoplacophoran mollusks, resembling those still living today. Later, conical calcareous spongelike archaeocyathids appeared, forming reeflike structures in places, as well as early echinoderms, bivalve mollusks, and *calcareous* lamp shells (brachiopods) and later a profusion of trilobites, which are arthropods, with distinctive segmented exoskeletons, divided into head, thorax, and tail.

A handful of major groups of organisms came to dominate life in the Paleozoic seas. If we are to understand the development of life, we need to take a look at them, in all their invertebrate richness.

THE INVERTEBRATE CLANS

As noted in chapter 4, taxonomists—those who classify organisms—have created thirty-five major groups (phyla) of animals, but we will content ourselves by describing only half a dozen or so invertebrate groups that are most common in both living and fossil forms: sponges, coelenterates, mollusks, worms, echinoderms, brachiopods, and arthropods.

Sponges (phylum Porifera) are among the simplest of all multicellular animals. Unlike most "higher" groups, they have an extraordinary capacity for regeneration of their parts. Sponges, when crushed through the meshes of a sieve, can reassemble their cells to form a new sponge. In fact, until the nineteenth century, they were often regarded as plants. They function as living sieves, their body walls perforated by pores or canals

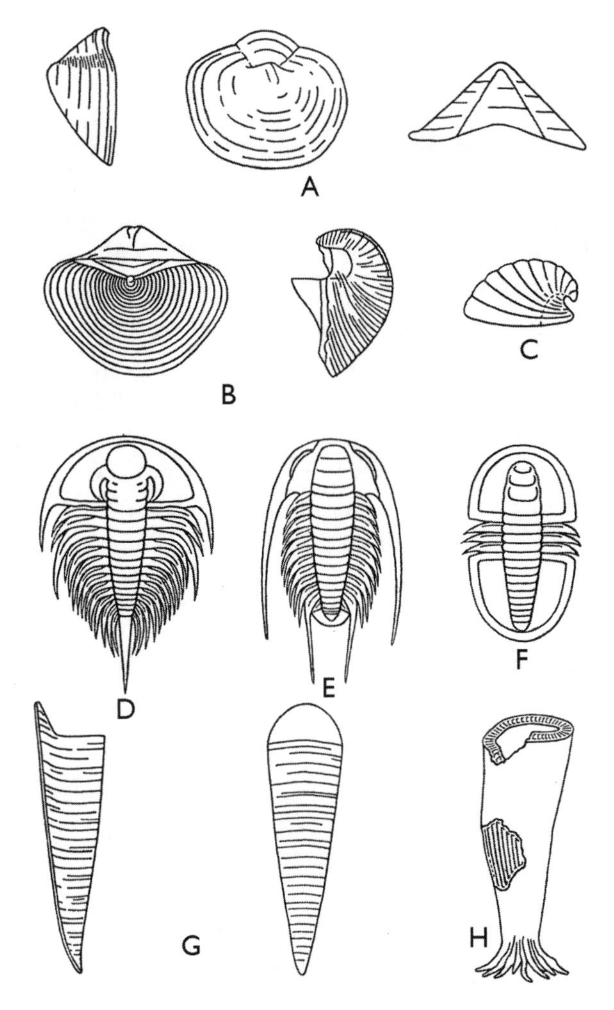

Figure 5.3. Typical early Cambrian fossils. (A–B) Brachiopods (C) Gastropod (D–F) Trilobites (G) Pteropod (H) Archaeocyathid.

(From Frank Rhodes, *Evolution of Life* [New York: Penguin, 1976], fig. 6, p. 81. Reproduced with permission.)

through which water is drawn into the inner body cavity, filtered for food, then expelled from an opening at the top of the sponge. The movement of water through the body of the sponge is assisted by whiplike, flagellate collar cells, while the body is supported by delicately shaped microscopic spicules, which are either discrete or fused into a rigid living framework. Sponges are sessile and mostly, though not exclusively, marine, and they vary greatly in form and color. Though rarely conspicuous as fossils, they are widely distributed.

One distinctive spongelike group found only in Early Cambrian rocks from many parts of the world was the *archaeocyathids* (fig. 5.4), which built rigid, vaselike skeletons of calcium carbonate, with double perforated walls, up to a foot or more in height. Though most were conical in structure, others were candelabra-like, basketlike, or dishlike in form. They were often tethered to the ocean floor by rootlike hold-fasts. Their remains form reeflike structures in some places and, in fact, their resemblance to both sponges and corals has led to the suggestion that they might be ancestral to both, though some regard them as a separate phylum.

Archaeocyathids first appeared near the base of the Cambrian in Siberia and later flourished for a "brief" time in many parts of the world. Yet by early Middle Cambrian times they were found only in Australia and the Siberian fold belt and then soon became extinct. Why they became extinct is particularly puzzling, for they had achieved worldwide distribution and faced no obvious competitors.

Two groups of microscopic fossils are present in Paleozoic rocks. Both are classified as *protists,* a kingdom of organisms that includes both animal-like forms (such as *Amoeba),* and plantlike forms (such as diatoms). Fungi-like forms are also known. But the protists, though "spineless," are not invertebrates in a strict sense. In protists, all the functions of life are performed by a single unspecialized cell.

Protists represent an inconspicuous part of the world of living things, but a part on which all the rest of the economy of life depends. It is a world, not only of extraordinary abundance, but also of astonishing variety, represented by several hundred thousand species. Its inhabitants are found everywhere, from every nook and cranny of the land, as well as in the air and from the depths of the oceans, to the bodies and dark interiors of all Earth's other creatures.

Any account of the history of life is bound to be based largely on fossils, and fossilization generally requires the possession of hard parts—skeletons and shells. Only a few of the myriad of unicellular creatures have hard parts, but four do, and their microscopic remains are common in some sedimentary rocks. In fact, they are so abundant in places that their tiny skeletal tests form significant rock deposits. Not only that, but some form the base of the food chain for other ocean dwellers. So

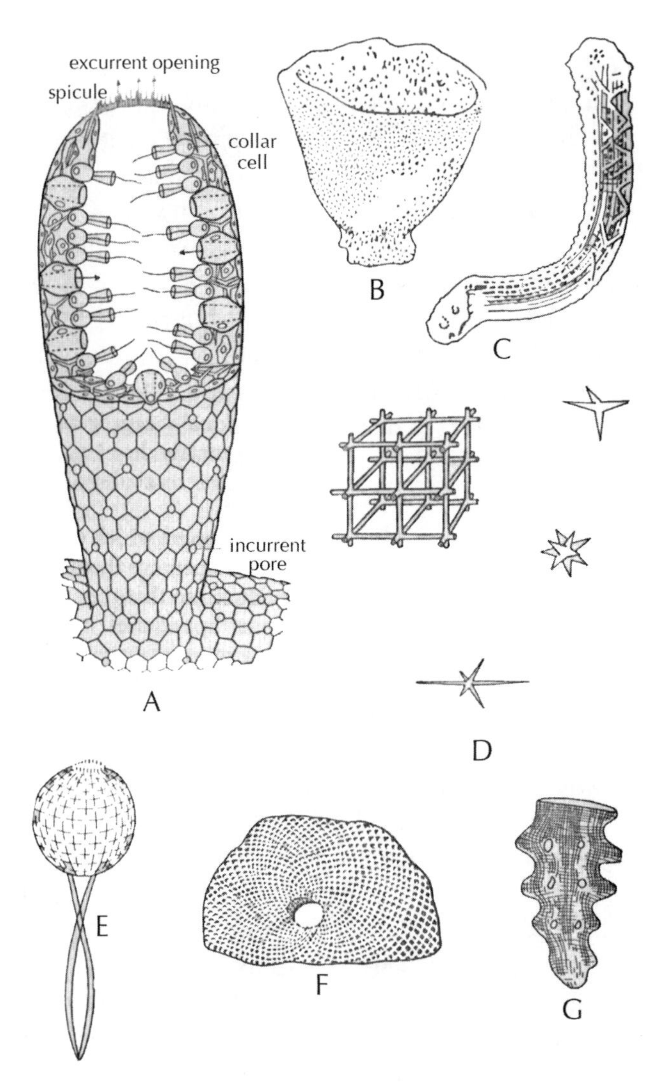

Figure 5.4. Sponges, living and fossil. (A) A simple sponge, showing general structure (after Buchsbaum); (B–C) Living sponges: (B) A horny sponge from the Mediterranean (C) A deep-sea glass sponge, *Euplectella*, "Venus's flower basket"; (D) Sponge spicules, showing variation in shape (highly magnified); (E–G) Paleozoic sponges: (E) *Protospongia*, from the Cambrian, height 4 inches; (F) *Receptaculites*, a spongelike fossil from the Ordovician, diameter 3–4 inches; (G) *Hydnoceras*, from the Devonian, height 8 inches.

(From Frank Rhodes, *Evolution of Life* [New York: Penguin, 1976], fig. 7, p. 101. Reproduced with permission.)

we should take a moment to recognize them, and, in recognizing them, reflect not only on the crucial role such microscopic creatures play in life's economy, but also on their antiquity. Long before the vertebrates, long before the invertebrates, appeared on the scene, bacteria and protists—"simple" unicellular creatures such as these—inhabited our planet. They were not only the dominant life—they were the only life. It was unicellular organisms which contributed to creating the oxygen atmosphere that sustains us. It was some of their members who were responsible for the formation of some of Earth's greatest resources (including, perhaps, iron ores of the banded iron formation and petroleum, for example) on which our society now depends. Apart from the bacteria, they constitute Earth's most ancient forms of life, so that it is from some of their members, in a broad sense, that we ourselves have developed. It is some of their members that today are pathogens contributing to diseases that plague us, even as we depend on the beneficial effects of other protists for our very existence.

In the more visible panorama of life's history, it is easy to overlook these microscopic creatures. To recognize their role, we review four groups. All are eukaryotic, housing their genetic material within the cell nucleus, and almost all are planktonic, floating in the surface waters of the oceans.

The *Foraminifera* ("forams") are amoeba-like protozoans that use their irregular and temporary body extensions (pseudopodia) in both locomotion and food gathering. Some are naked, but others build tiny "shells" perforated by holes ("foramen," hence their name); these calcareous shells are often of elaborate structure and great beauty. Though most are microscopic, a few are visible to the naked eye. One fossil group—fusulinids, from Pennsylvanian and Permian rocks—are the size and shape of large wheat grains and widely used for rock correlation. The great pyramids of Giza in Egypt are built of limestone containing another kind of abundant, button-sized foraminifera called *Nummulites,* once erroneously interpreted as either coins or lentils left from the food of those who built the pyramids.

Fossil foraminifera are widely used in petroleum exploration in the dating and correlation of Late Mesozoic and Cenozoic sedimentary rocks. They are also useful in reconstructing the climate and bathymetry of ancient depositional environments. Some are so abundant that their remains form deposits, such as *Globigerina* ooze, that cover vast areas of the ocean floor.

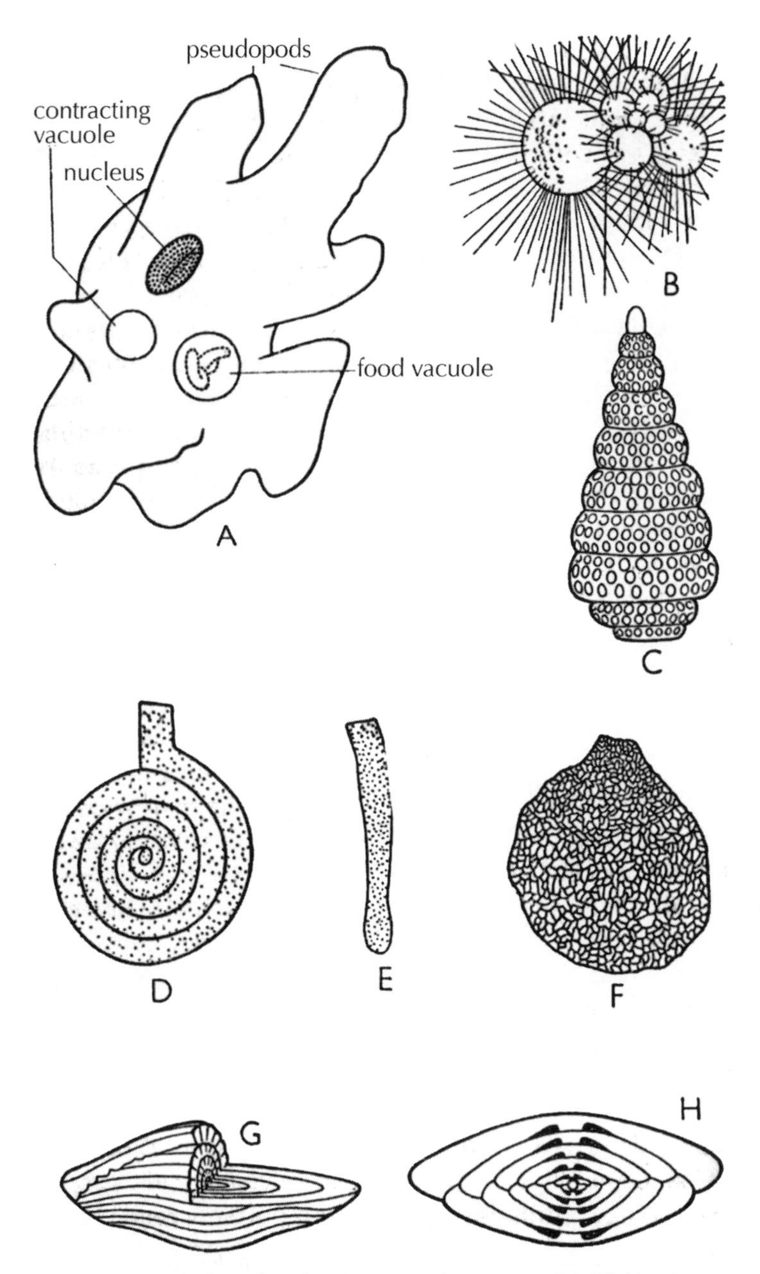

Figure 5.5. Protists. (A) Amoeba, showing general structure; (B) *Globigerina*, a living foraminifera; (C) A radiolarian, *Lithocampe*, from the Devonian; (D–H) Paleozoic foraminifera: (D) *Ammodiscus*, Silurian–Recent; (E) *Hyperammina*, Silurian–Recent; (F) *Saccammina*, Silurian–Recent; (G–H) A fusulinid from the Pennsylvanian-Permian, showing general structure; (H) A view in thin section. (All highly magnified).

(From Frank Rhodes, *Evolution of Life* [New York: Penguin, 1976], fig. 15, p. 126. Reproduced with permission.)

They are found in marine sedimentary rocks ranging from Cambrian to Recent in age; most are benthic in habit. Planktonic forms appeared in the Jurassic. Early forms had agglutinated tests, but most later forms have "shells" of calcium carbonate, though a few later forms continue to have tests of agglutinated materials. Estimates of the number of living and fossil species that have been described vary widely, from about 4,000 marine species to 275,000 living and fossil species.

Radiolaria are microscopic, marine protists that build "shells," many of exquisite beauty and delicacy, composed of opaline silica. They are known in rocks ranging in age from Cambrian to Recent. Many have tests that resemble Christmas tree ornaments, while others have tests that are bell-like, basketlike or helmetlike in form. Some 500 living species are known, ranging from tropical to polar waters. Their remains are often relatively more abundant below the calcium compensation depth (CCD: typically 4,000–5,000 feet in most areas) at which tests of calcium carbonate dissolve in seawater. So abundant are they in places that they form radiolarian ooze, which can later come to be lithified as radiolarian chert. Because they are sensitive to temperature, radiolarians have proved to be useful paleoclimate indicators, while their relative abundance with depth can provide useful information on paleobathymetry.

Like radiolarians, *diatoms* have siliceous tests, but unlike them, diatoms are found in both marine and freshwater environments and are restricted to the shallower zones where they can utilize sunlight to support the photosynthesis on which they depend. They are useful index fossils in the correlation of some Cenozoic rocks; their remains are so abundant that in places they form diatomaceous earth, which becomes lithified to form diatomite and is used extensively in filters.

One more group of protists deserves recognition, not because of their size—they are so minute that they are known, not as "microfossils," but as "nanofossils"—but because of the conspicuous thickness and widespread distribution of their remains. These are the coccoliths, minute calcareous disklike, star-shaped or spiny plates covering the spherical surface of brown algae called coccolithophores. In shallow marine environments, coccoliths are known to be useful paleoclimatic indicators. But—for all their minute size—they have also made a conspicuous contribution, not only to the deposits of the depths of the ocean, but also to the form of the land: great areas of the continents are formed of rock of

Cretaceous age—chalk—which consists largely of their remains. From the white cliffs of Dover, to the cliffs at Etarat that Claude Monet loved to paint, to the Niobraran Hills of Kansas, the lowly coccoliths have left their dazzling white signature on the land.

All the other invertebrate groups are multicellular. Coelenterates or cnidarians (Phylum Cnidaria*)* include corals, sea anemones, and jellyfish.

The group also includes sea fans, sea pens, and the freshwater *Hydra.* These are simple organisms that lack the specialized respiratory, circulatory, and excretory systems of other invertebrate groups. Coelenterates are saclike creatures whose bodies consist of two cell layers separated by jellylike material. The body opening serves for both intake of food, which is wafted in by the tentacles that surround it, and expulsion of waste. This structure is shared by both sedentary corals and free-floating jellyfish. In fact, most coelenterates alternate generations between a medusa (jellyfish) form and a polyp form. The fundamental body plan of the two is similar, despite their strikingly different appearance.

Above the sponges and corals of the Cambrian seafloor, jellyfish pulsated through the surface waters. These delicate creatures have no hard parts, so they are rare as fossils. In spite of the flowerlike appearance of corals and the delicate tracery of jellyfish, both share stinging cells in their tentacles by which they can both defend themselves and capture their prey. The name of the phylum in which jellyfish are placed, Cnidaria, is based on the Greek word for "nettle." All coelenterates are radially symmetrical. Though their soft bodies make them unlikely candidates for fossilization, jellyfish-like fossils are known from rocks as old as those of Ediacara from the late Proterozoic. They have been a remarkably persistent and successful group, continuing unchanged for at least 530 million years. Some living jellyfish reach a diameter of more than 7 feet and have tentacles that are 120 feet long.

Corals and sea anemones, whose living members represent some 2,300 genera and more than 6,000 species, are all marine. Corals are typically colonial, sessile (fixed) creatures, so abundant in many warm parts of the ocean that their calcareous remains form huge coral reefs. Typically their polyp structures do not go through a medusa stage.

Most corals and sea anemones live in clean shallow marine waters. They depend on water movements to bring in food, and the reef-building corals are confined to tropical waters. This restricted distribution makes fossil representatives useful indicators of paleogeography and climate.

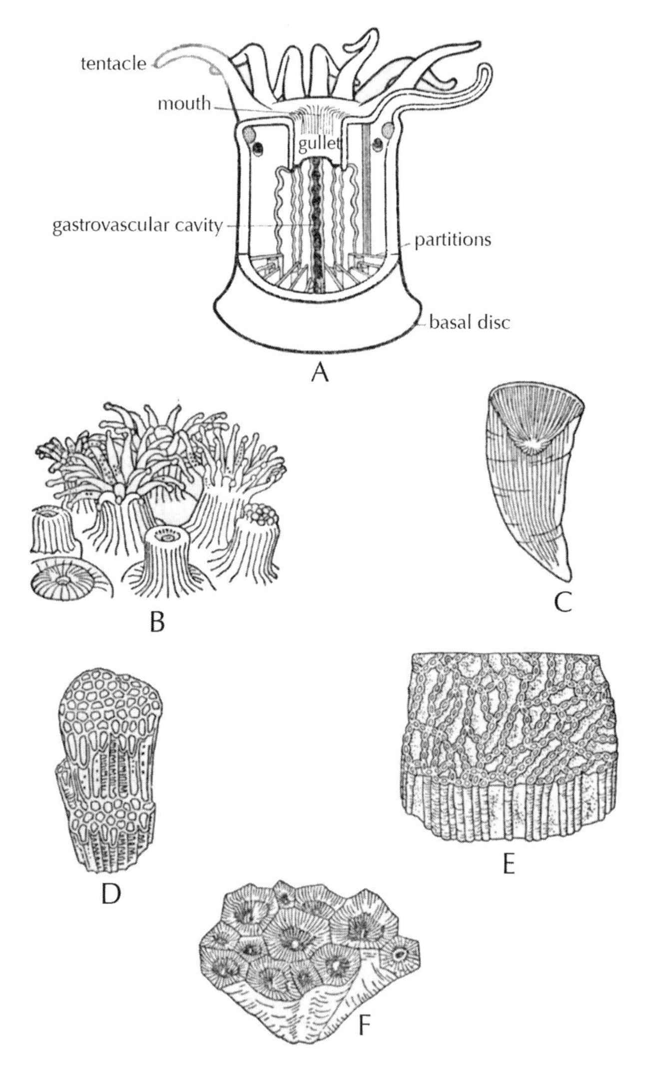

Figure 5.6. Coelenterates. (A) Structure of sea anemone (after Buchsbaum); (B) Living corals; (C–F) Paleozoic corals (all about half natural size): (C) *Streptelasma*, Ordovician; (D) *Favosites*, Ordovician–Permian, the honeycomb coral; (E) *Halysites*, Ordovician–Silurian, the chain coral; (F) *Lonsdaleia*, Carboniferous.

(From Frank Rhodes, *Evolution of Life* [New York: Penguin, 1976], fig. 8, p. 102. Reproduced with permission.)

Other corals are more widely distributed. Two major subgroups of corals are both confined to Paleozoic (Ordovician to Permian) rocks: *tabulate corals*, colonial forms with individual polyps housed in hexagonal, cylindrical, or chainlike tubes, each divided by horizontal tabulae ("little tables"), and *rugose corals,* including both solitary and colonial forms. Like the tabulate corals, rugose corals were widely distributed, and they were especially prolific reef builders in Silurian and Devonian times.

One other coelenterate group, the sea fans, though common in living reefs, have "skeletons" made of a flexible protein (gorgonian) and so are rare as fossils.

The mollusks (phylum Mollusca) are one of the most successful, conspicuous and abundant of all living groups, with countless representatives along the world's shorelines and oceans, as well as in lakes and rivers and on the land. Mollusks show great variation in form and size: some giant squid are reported to reach 70 feet in length, and some living clams (*Tridacna*) weigh up to 500 pounds. Living mollusks include snails, slugs, clams, oysters, mussels, chitons, squid, octopuses, and the pearly nautilus. But in spite of this variety, they share a common and relatively simple body plan. They have a soft, generally bilaterally symmetrical, unsegmented body, typically having a head and a large visceral mass resting on a fleshy foot. The body is surrounded by a fleshy layer (the mantle), which secretes a shell in some members of the group. In more specialized members—clams and squid, for example—some of these typical molluscan features are modified or lost. The foot, for example, has been modified for digging in bivalves and into tentacles in squid.

Clams, mussels, abalone, escargot, calamari (squid) and other mollusks provide food for humans and other animals, and prehistoric shell middens show that our forebears shared our tastes. Mollusks exist in countless numbers, representing some 130,000 species, distributed over all parts of the world.

Of the five major classes of mollusks, two are relatively rare: the chitons (polyplacophorans) are small, algae-grazing creatures of tidal pools, which are covered by an "armor" of eight small overlapping plates, while the tusk shells (scaphopods) have cone-shaped shells which protect the animal as it burrows in the sand. Both are known as fossils, but they are rare.

All of the three other major classes of mollusks are widespread as fossils. Snails (gastropods, "stomach-foot") include marine, freshwater, and land-dwelling representatives. Although their earliest members were exclusively marine, snails are abundant in living communities ranging from tropical to subpolar oceans and are found on land in almost all environments up to an elevation of about 18,000 feet. Most gastropods

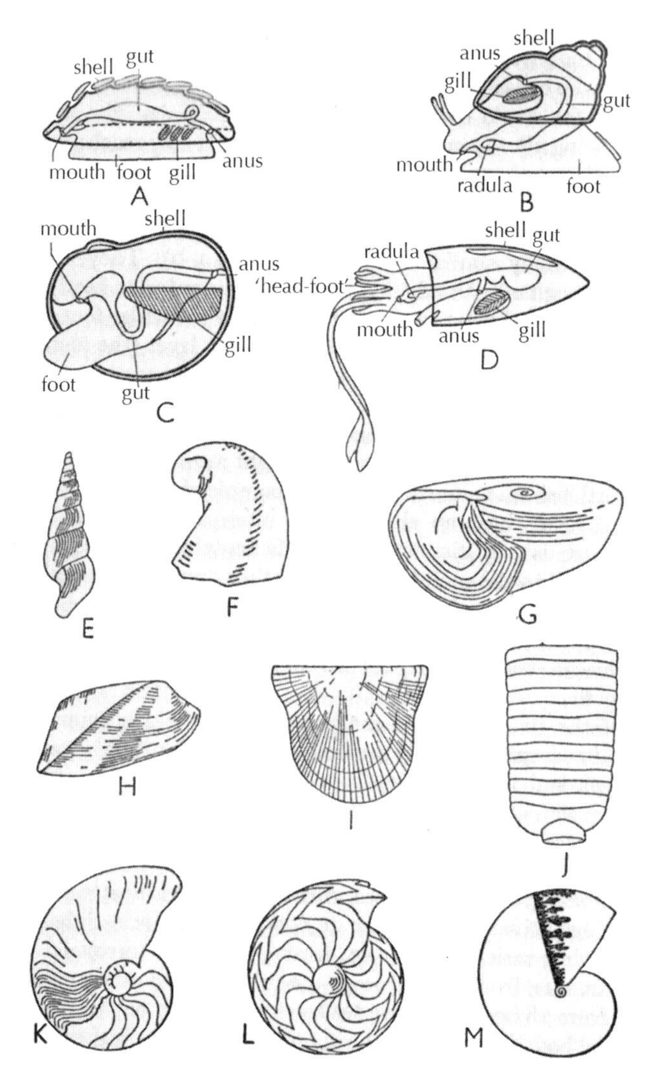

Figure 5.7. Variety in structure and form of mollusks. (A–D) Structural similarities between: (A) Chiton; (B) Gastropod; (C) Pelecypod; (D) Cephalopod; (E–G) Paleozoic gastropods (all one-half natural size): (E) *Loxonema*, Ordovician–Carboniferous; (F) *Platyceras*, Silurian–Permian; (G) *Maclurites*, Ordovician, basal surface uppermost; (H–I) Paleozoic pelecypods (about natural size): (H) *Goniophora*, Silurian–Devonian; (I) *Dunbarella*, Carboniferous; (J–M) Paleozoic cephalopods showing increased complexity of suture line: (J) *Endoceras*, Ordovician (×⅓) (showing siphuncle); (K) *Agoniatites*, Devonian (×¼); (L) *Goniatites*, Carboniferous (×2); (M) *Medlicotta*, Permian (×1).

(From Frank Rhodes, *Evolution of Life* [New York: Penguin, 1976], fig. 9, p. 106. Reproduced with permission.)

have a distinct head, often bearing eyes and other sensory organs and a well-developed foot. Most secrete a coiled, unchambered shell of calcite. Some graze on algae, but most are carnivorous predators, capable of drilling into the shells of other creatures. Others are suspension feeders, detritivores, and even parasites.

Other gastropods, of course, have invaded fresh waters, as well as land, where they compete successfully with other groups.

Closely related to gastropods is a small group known as monoplacophorans ("single shell bearers"), which were widespread in Cambrian to Devonian times. Although monoplacophorans were long thought to be extinct, a living representative species (*Neopilina galatheae*) was found in a deep-water ocean trench off Costa Rica in the early 1950s. This living fossil emphasizes how selective the available fossil record is.

Pelecypods ("hatchet-foot") or bivalves are also abundant in both living and fossil faunas. They include clams, scallops, and oysters, all of which have bodies enclosed in two-piece, hinged calcareous shells. In them the head and most sensory organs have been lost, while the foot has been modified to function in burrowing.

Cephalopods ("head foot") are active, pelagic, predatory mollusks, including the living nautilus, octopus, squid, and many fossil forms. A ring of tentacles surrounds the mouth of most forms, while many have a well-developed head with eyes, beaklike jaws, and two or four gills. The body may be naked—as in the octopus—or have an internal skeleton—as in squid—or a coiled external shell, as in the nautilus. Fossil cephalopods had shells of great variety and are important index fossils in rocks of Mesozoic age.

Worms (phylum Annelida) are as common in the oceans as they are on land, but their soft bodies make them unlikely candidates for fossilization. Worm trails and burrows in ancient rocks are, however, among the earliest fossils known to us, and well-preserved body fossils from the Middle Cambrian Burgess Shale show a rich variety of forms. Many microfossils of scolecodonts (chitinous jaw components of polychaete worms) show them to have been abundant in Paleozoic times. Worms have a mouth and anus at opposite ends of their segmented bodies, with well-developed nervous and excretory systems.

The organization of worms, with a bilateral symmetry and a definite head and tail connected by a digestive tract, gives them a mobility unlike that of some other groups we have so far mentioned, such as sponges and

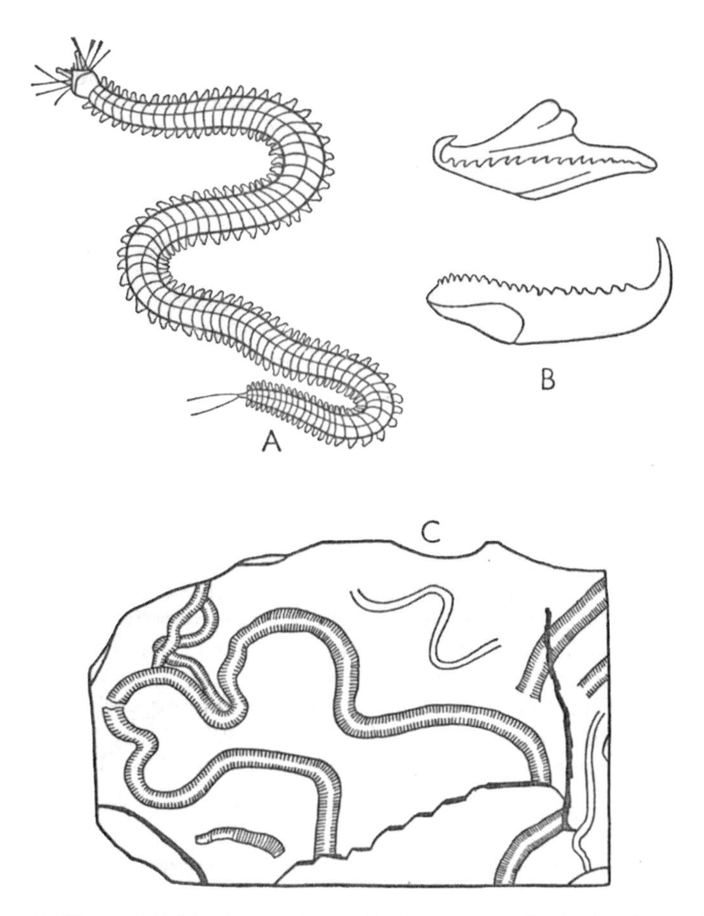

Figure 5.8. Worms. (A) *Nereis*, a modern polychaete worm; (B) Scolecodonts, the minute chitinous jaw components of fossil worms (greatly magnified); (C) Fossil worm-borings in a rock (natural size).

(From Frank Rhodes, *Evolution of Life* [New York: Penguin, 1976], fig. 10, p. 108. Reproduced with permission.)

corals, and thus allows an active mobile existence, rather than a passive sedentary one.

Though inconspicuous as fossils, worms must have played as vital a role in ancient seas as they do today. Studies show that the entire upper 2 feet of coastal sediment passes through worms within a period of two years.

On land, earthworms play an important role in soil formation by both their movement and their digestion. Charles Darwin wrote that "It may be doubted whether there are many other animals which have played so important a part in the history of the world as have these lowly organized creatures."

The phylum Echinodermata ("spiny-skin") includes animals that are exclusively marine: sea urchins, sand dollars, starfish, brittle stars, sea lilies, and sea cucumbers. Fossil representatives include a variety of both sessile, attached forms (blastoids, most crinoids and cystoids) and free-living forms (edrioasteroids and sea urchins). All echinoderms tend to be more or less spiny, having a rigid test ("shell") made of calcium carbonate, or are studded with small calcareous plates. All have a distinctive water vascular system, a network of internal canals that terminates in tube feet, which are used in locomotion and food gathering. Sea cucumbers (holothuroids) are bottom-dwelling creatures, with mouths surrounded by a ring of tentacles, their leathery skin containing minute spicules. Echinoderms often have a fivefold radial symmetry. The sexes are usually separate.

Brachiopods ("lamp shells," from their resemblance to Roman lamps) and bryozoans ("moss animals") are often grouped together as lophophorates, because both have a long, coiled feeding structure called the lophophore. Brachiopods were, perhaps, the most characteristic and abundant of the animals of Paleozoic seas. Brachiopods have a two-part shell, which in the earlier forms is phosphatic in composition, and in later forms calcareous. These shells consist of two hinged valves that are bilaterally symmetrical, although they typically differ in shape and size. Most brachiopods are attached to the seafloor by a short, fleshy stalk or pedicle, and all have circulatory, digestive, and nervous systems.

Bryozoans (phylum Bryozoa), in spite of their superficial resemblance to corals, are quite distinctive in structure. They share a basic body plan very like that of brachiopods, with a lophophore surrounding the mouth, and a U-shaped digestive tract. Unlike the brachiopods, they are colonial animals, whose calcareous skeletons are found in vast numbers in some Paleozoic rocks.

The other great invertebrate group is the arthropods (phylum Arthropoda). Arthropods include such familiar living marine forms as shrimps, crabs, lobsters, and barnacles, and such terrestrial forms as spiders, scorpions, ticks, centipedes, millipedes, and insects. They have become adapted to an extraordinarily wide range of habitats and exist not only in a great variety of forms, but also in great abundance as individuals. Extinct forms include trilobites, eurypterids, and some problematic Cambrian forms. Three-quarters of all known living species in the world are

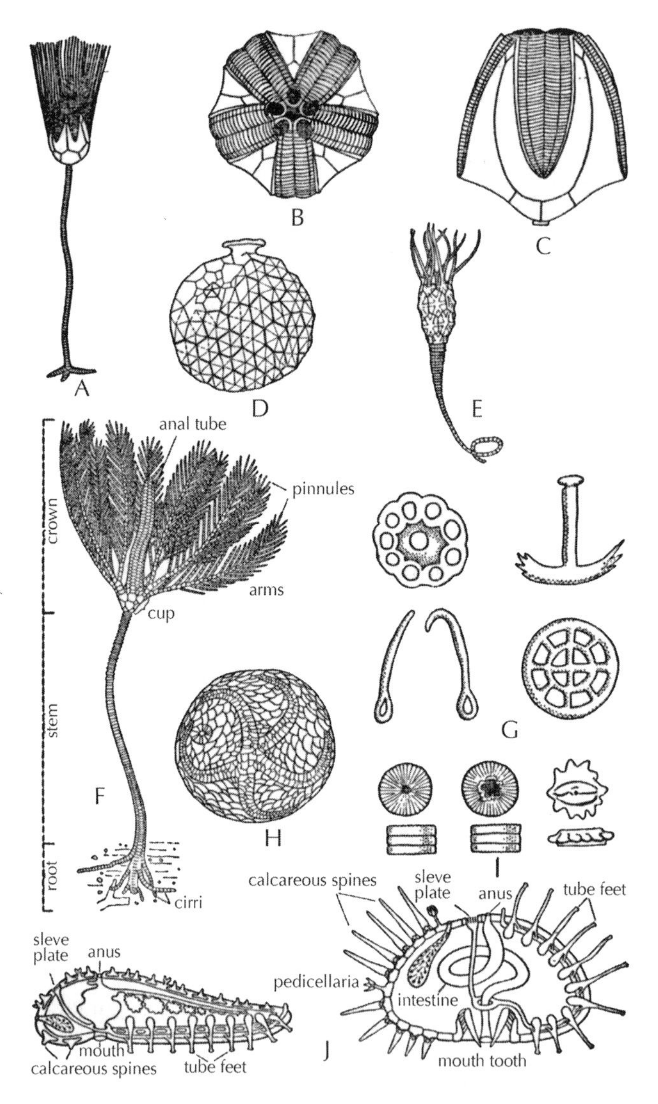

Figure 5.9. Echinoderms. (A–C) Blastoids, *Pentremites*, Carboniferous: (A) Restoration of animal, showing stem supporting budlike calyx, which bears armlike pinnules (about half natural size); (B–C) Upper and lateral views of calyx, showing petaloid ambulacra areas (×4) (after Bather); (D) *Echinosphaerites*, an Ordovician cystoid (natural size); (E) *Macrocystella*, a Cambrian eocrinoid (natural size); (F) *Botryocrinus*, a Silurian crinoid, showing general structure (natural size) (after Bather); (G) Holothuroidean spines (greatly magnified); (H) *Edrioaster*, an Ordovician edrioasteroid (natural size); (I) Crinoid columnals from stems (natural size); (J) Starfish and sea urchin, cross sections to show similarity of general structure (after Buchsbaum).

(From Frank Rhodes, *Evolution of Life* [New York: Penguin, 1976], fig. 11, p. 110. Reproduced with permission.)

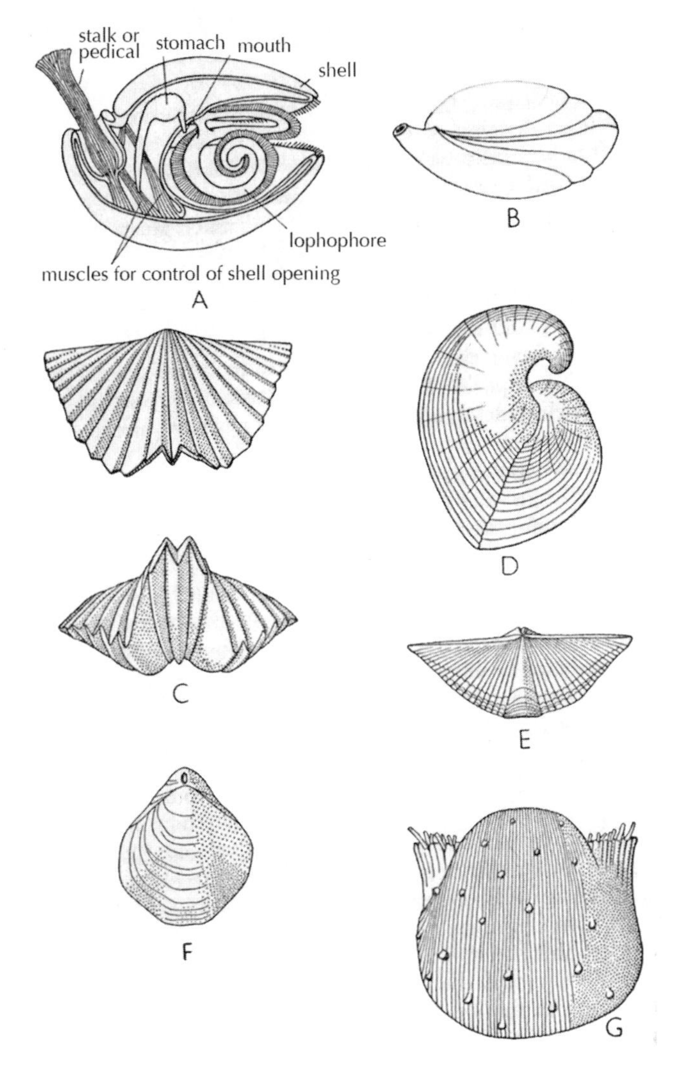

Figure 5.10. Brachiopods. (A) Structure of an articulate brachiopod; (B) Brachiopod shell in same relative position as A; (C) *Platystrophia*, Ordovician–Silurian; (D) *Conchidium*, Silurian; (E) *Mucrospirifer*, Devonian; (F) *Composita*, Carboniferous–Permian; (G) *Linoproductus*, Carboniferous–Permian. (All approximately natural size.)

(From Frank Rhodes, *Evolution of Life* [New York: Penguin, 1976], fig. 12, p. 114. Reproduced with permission.)

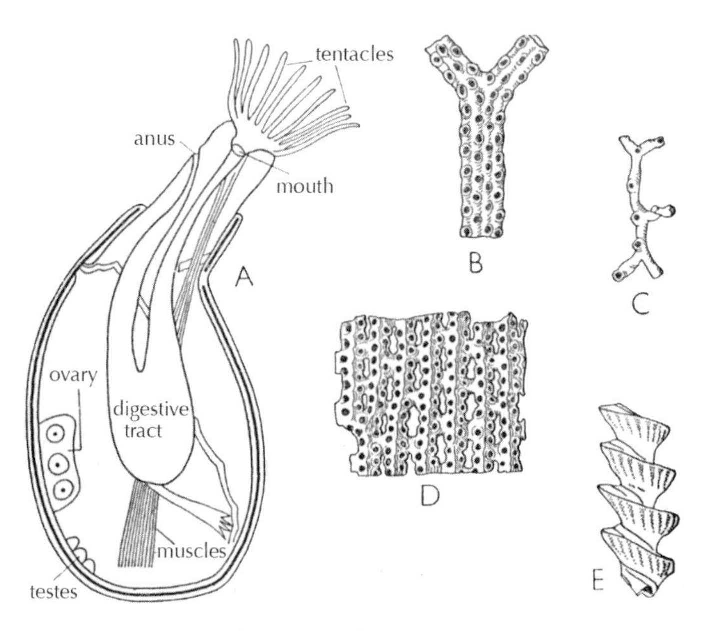

Figure 5.11. Bryozoans. (A) A living individual, showing structure (highly magnified); (B–E) Fossil bryozoa. The minute depressions each housed individual animals: (B) *Thamniscus*, Silurian–Permian (×10); (C) *Stomatopora*, Ordovician–Recent (×20) (D); *Fenestrellina*, Silurian–Permian (×3); (E) *Archimedes*, Carboniferous (natural size).

(From Frank Rhodes, *Evolution of Life* [New York: Penguin, 1976], fig. 16, p. 130. Reproduced with permission.)

arthropods. What unites this seemingly dissimilar group of creatures is the basic body plan they share, consisting of a segmented body with jointed appendages, covered by an external skeleton of varying rigidity and well-developed digestive, nervous, circulatory, and reproductive systems.

So much for an overview of the main invertebrate groups. These creatures represent the major groups of Paleozoic marine invertebrates and protists. They represent a dazzling variety of form and function, exquisitely adapted to their varying modes of life, as filter feeders, bottom scavengers, reef builders, and so on. But the Paleozoic Era covered a vast length of time from 541 to 252 million years ago—longer than the Mesozoic and Cenozoic eras combined, in fact—and not all the groups we have described flourished at the same time. The creatures of the Cambrian seas included worms, small inarticulate brachiopods, primitive echinoderms, cone-shaped hyolithid mollusks, dendroid graptolites

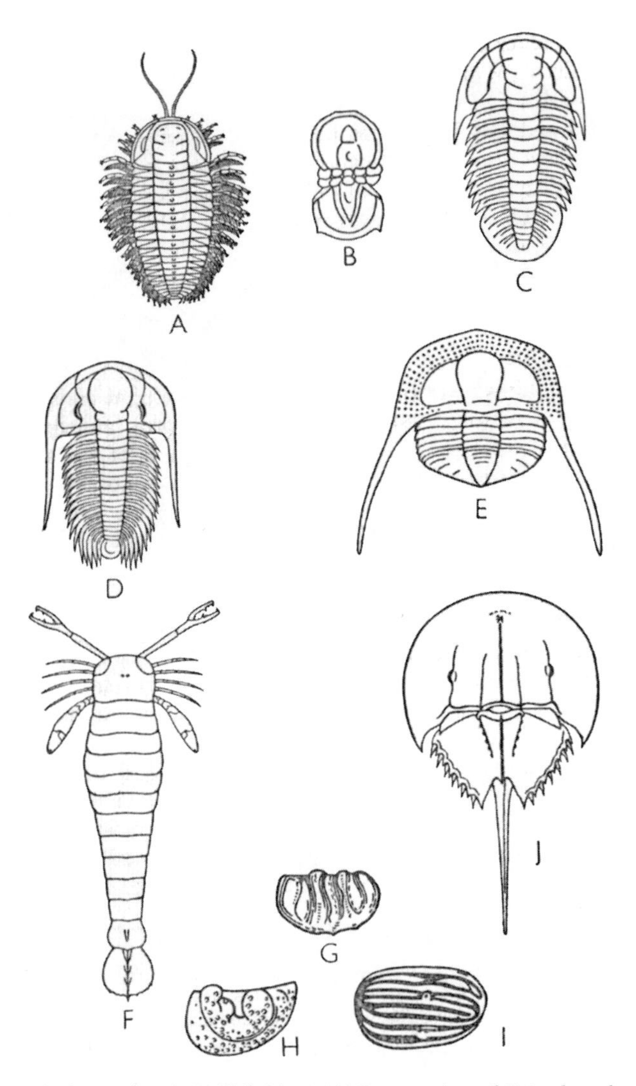

Figure 5.12. Arthropods. (A–E) Trilobites: (A) Restoration of *Triarthrus becki*, a trilobite from the Ordovician of New York (after Beecher and Raymond); (B–D) Cambrian trilobites: (B) *Agnostus*, Cambrian (×4); (C) *Bathyuriscus*, Middle Cambrian; (D) *Paradoxides* Middle Cambrian, length up to 18 inches; (E) *Cryptolithus*, Ordovician–Devonian, a trilobite; (F) *Pterygotus*, Ordovician–Devonian, a eurypterid, up to 7 feet long; (G–I) Paleozoic ostracods (greatly magnified): (G) *Tetradella*, Ordovician–Devonian; (H) *Hollina*, Devonian; (I) *Glyptopleura*, Carboniferous–Permian; (J) The living horseshoe crab, *Limulus*, a "living fossil," length 18 inches.

(From Frank Rhodes, *Evolution of Life* [New York: Penguin, 1976], fig. 13, p. 116. Reproduced with permission.)

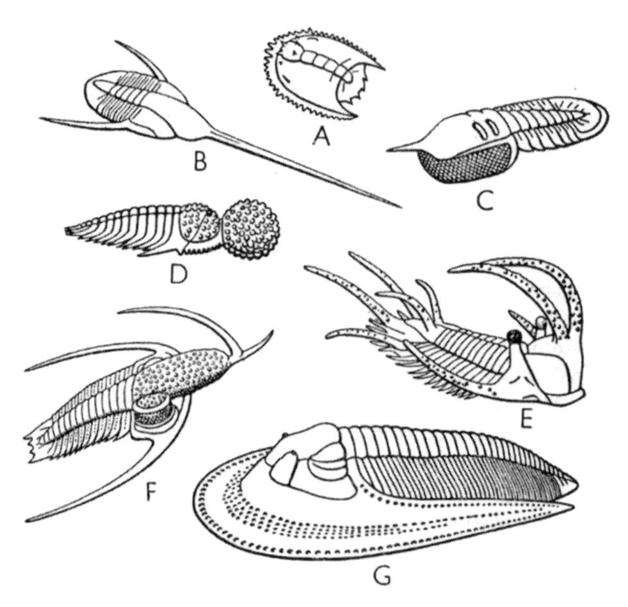

Figure 5.13. Trilobites. (A) Larva of *Acantholoma*, Silurian (×20); (B) *Lonchodomas*, Ordovician (natural size); (C) *Symphysops*, Ordovician (×1.5); (D) *Staurocephalus*, Silurian (×1.5); (E) *Ceratarges*, Devonian (×2); (F) *Teratorhyncus*, Ordovician (×2.5); (G) *Paraharpes*, Ordovician (×3).

(After Hupé. From Frank Rhodes, *Evolution of Life* [New York: Penguin, 1976], fig. 14, p. 123. Reproduced with permission.)

and—above all—trilobites. More than a thousand species of trilobites have been described from rocks of Cambrian age (541–485 million years ago). They range in size from giant 18-inch-long, shaggy-looking *Paradoxides* (what a beautiful name!) to the tiny, blind *Agnostus.* About three-quarters of all described species of Cambrian fossils are trilobites: that is one measure of their dominance. They exhibited adaptations to various modes of life, from crawling and burrowing on the seafloor to being active swimmers.

Impressive as these fossil remains of the known creatures of the Cambrian seas are, the Burgess Shale reminds us that there must also have been a host of other soft-bodied forms, now unknown to us.

One other group was represented in rocks of Cambrian age. These were the graptolites, a group of extinct colonial marine organisms occurring in countless numbers in the dark shales of much of the Paleozoic. The earliest colonies, which appeared in the Late Cambrian, were branching and dendritic in appearance, but later Ordovician and Silurian forms were marked by a decrease in the number of branches. Along the branches were developed a number of cuplike structures (thecae), each housing an individual organism, and these thecae vary widely in form. From a geological point of view the graptolites are of particular interest because of their great value as index fossils—they evolved rapidly and became worldwide in extent, most of them apparently having lived as floating plankton. More than fifty graptolite zones have been recognized. Zoologically, however, the graptolites are of no less interest, for they long represented one of the most tantalizing and perplexing fossil groups. They were variously referred at different times to the bryozoa, coelenterates, and plants, and were even regarded by some as inorganic. Recent studies of perfectly preserved forms extracted from limestones have shown, to everyone's surprise, that their closest affinities are with the invertebrate chordates: the Hemichordata. They are probably most closely related to the tiny colonial hemichordates (pterobranchs) of modern seas. The true graptolites became extinct at the close of the Silurian, but the dendroid types survived until the Carboniferous.

The Cambrian fauna declined at the end of the period, to be replaced in the Ordovician by one in which articulate brachiopods, ostracods, corals, crinoids, and nautiloid cephalopods all increased in variety and numbers and in which, though trilobites were still present, they declined in relative abundance. The filter feeders and seafloor deposit feeders of the Cambrian times gave way to a variety of more clearly defined and finely adapted feeders and predators, especially those that were sessile, suspension feeders. With this expansion in feeding habits, there seems also to have been an expansion in range from shallower to deeper waters.

The end of the Ordovician times saw another turnover in species, with corals, crinoids, and bryozoans becoming more common in Silurian rocks, together with brachiopods and small numbers of new genera of trilobites. In the following Devonian period, they were joined by goniatite cephalopods and a host of new brachiopods, corals, mollusks, and crinoids. By now trilobites had become rare.

The pattern of the decline of trilobites is illustrated by the fact that they had been represented by 219 different taxonomic families in the Cambrian, 151 in the Ordovician, 36 in the Silurian, 44 in the Devonian, 4 in the Carboniferous and 5 in the Permian. Yet, in spite of this steady decline, they were a remarkable group, of high mobility, complex

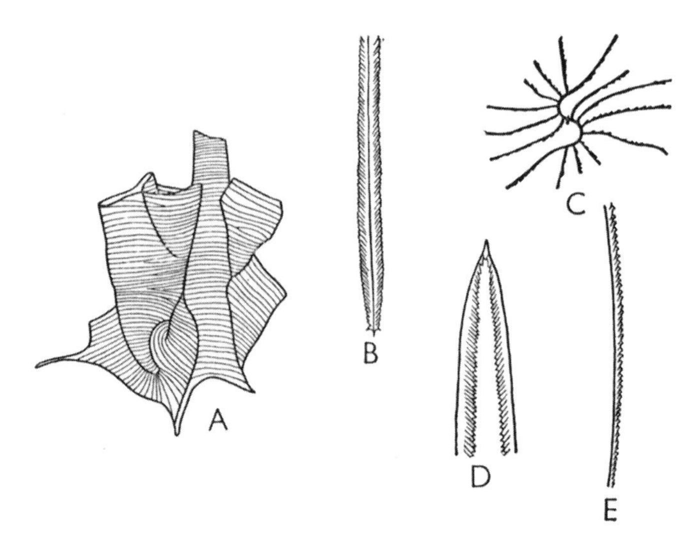

Figure 5.14. Graptolites. (A) *Diplograptus*, Middle Ordovician–Lower Silurian, highly magnified drawing, showing detailed structure of early growth stages (after Wiman); (B) *Diplograptus*, complete colony (natural size); (C) *Nemagraptus gracilis*, Ordovician; (D) *Didymograptus*, Ordovician; (E) *Monograptus*, Silurian (C–E all natural size).

(From Frank Rhodes, *Evolution of Life* [New York: Penguin, 1976], fig. 17, p. 137. Reproduced with permission.)

structure, extraordinary variety, and great longevity. They existed for almost 300 million years.

Early Carboniferous (Mississippian) times were marked by low-lying lands and widespread clear, shallow, limestone-depositing seas. In some areas (parts of Russia and Alaska) coal-forming swamps, lagoons, estuaries, and deltas already existed, and these conditions spread throughout the Northern Hemisphere in later Carboniferous (Pennsylvanian) times. The warm, low-lying lands were bordered by clear, shallow seas, with a rich fauna of invertebrates and fish. In the Southern Hemisphere, conditions were quite different. The southern continents (Africa, India, Australia, South America, and Antarctica) were parts of a single great land mass (Gondwanaland), and only later drifted apart. These continents seem to have stood high above sea level, and extensive late Paleozoic glacial deposits are found in the Southern Hemisphere.

New mountain chains were formed in many areas in Late Carboniferous and Permian times. The warm shallow seas persisted, and immense reefs developed in scattered areas (from England to Texas, for example).

Great thicknesses of salt and red sediments were deposited in land-locked basins.

The marine life of this period of Late Paleozoic times was different in detail, rather than in general character, from that of the Middle Paleozoic. The protozoans, for example, were supplemented by one other group of great importance, the wheat-grainlike fusulinids, which persisted to the close of the Permian. These are among the most widespread of all foraminifera, and their intricately sculptured shells are remarkably complex structures, especially as they were secreted by a single cell. These fusulinids, unlike most foraminifera, are sufficiently large to be clearly visible to the naked eye, and their detailed structure can be studied in thin sections of the limestones in which they occur (fig. 5.5, A and H).

Sponges, corals, and bryozoans continued with only relatively minor changes, but the trilobites—as we have seen—declined to a handful of genera and many new groups of brachiopods, echinoderms, cephalopods, and other mollusks appeared. The overall faunal change was not, however, spectacular, for by this time all the major marine invertebrate groups were well established in their various ways of life, and the changes they underwent were essentially minor in character, involving changes in diversity of different subgroups and a closer adaptation to the environments in which they lived.

The end of the Permian was marked by the greatest wave of extinction the Earth has ever experienced. It is sometimes referred to as the Time of Great Dying; it is estimated that up to 95 percent of all marine species, and up to 70 percent of vertebrate terrestrial species became extinct. During this time, 83 percent of all genera and 57 percent of all families are estimated to have become extinct. Though not the only episode of mass extinction, it was the greatest. We will review the question of extinction in chapter 16.

So far we have been concerned with the great host of marine invertebrates that dominated the faunas of the Paleozoic, and, as we have seen, life was confined to the seas for the first 150 million or so years of the era.

The first undoubted record of the advent of life on land is provided by spores and trace fossils from the Ordovician and a primitive land plant from the Upper Silurian, about which we will say more in the next chapter.

It was also in Silurian times that the first air-breathing animals appeared. Evidence of their presence is provided by a millipede-like creature in Silurian rocks. The complete colonization of the land by animals could have taken place only after the spread of freshwater and land plants, although the earliest terrestrial vertebrates seem to have been carnivores. It is no accident that both land plants and land animals appear at roughly the same period of geological time. Some of the earliest terrestrial arthropods probably depended on the plants for food, just as some plants were later to depend on animals for pollination and still other animals were to depend on both for food. This interdependence of plants and animals is a reminder of the similar interdependence that exists throughout the world of living things. It is one of the major keys to understanding the evolutionary process.

The arthropods seem to have been the first animals to colonize the freshwater environments. One ancient aquatic group we have already discussed is the eurypterids, at least some of which seem to have inhabited freshwaters. Closely related to them, but differing chiefly in the presence of pincers on the second pair of limbs and "combs" on the first trunk segment, are the oldest scorpions, found in the Silurian of both Europe and North America. These are thought by some to be the oldest air breathers, but it seems more likely that they lived in inland waters. By Carboniferous times, however, descendants of these forms were clearly air breathers, and the Carboniferous also saw the appearance of the first true spiders, although both spiderlike and ticklike creatures have been recorded from the Early Devonian Rhynie Cherts of Scotland. The oldest insects are Lower Devonian and they were already well established by Carboniferous times. We will consider them in chapter 10.

It is worth pausing to note the significance of these arthropod successes in the conquest of the land: by Devonian times ticks and spiderlike creatures, by Carboniferous times true spiders, as well as scorpions, cockroaches, and primitive dragonflies, had all established themselves firmly ashore. The extent of the success can best be measured by comparing them with other groups, for the ability to live on land is an uncommon one among other animals. Apart from the vertebrates and arthropods, only worms and gastropods are common air-breathing animals. In the arthropods, however, the habit was taken over with spectacular success by every large order. Some indeed are so well adapted that they are equally at home in air or water—land crabs, for example. Now, how is it that the arthropods should have been so successful in a competition in which other groups have achieved so much less? The answer seems to lie in a number of distinctive characteristics, each of which gave them a flying start. Two of these are of particular importance. First, the arthropods were well protected by their chitinous covering against the hazards of desiccation that land life involves. The body functions depend on a constant internal supply of fluid. In water there is little danger of any moisture loss except through osmosis, but on land, evaporation is continuous. Some existing land animals have only partly solved this problem, for they are still confined to moist and damp environments (frogs are an example) but the arthropods' chitinous envelope supplied a ready-made answer to the problem.

Second, any rapid movement on land requires strong limbs, for there is no help in bearing the weight of the body from the surrounding water. Here, too, the arthropods were preadapted to their new way of life, for their numerous, varied, and chitinous-covered appendages provided all that was required. As we turn to other more familiar conquests of the land, it is easy to overlook the early and widespread success of the arthropods, but, by any standards, theirs was a remarkable achievement.

Far less conspicuous than the insects were the two groups of Paleozoic mollusks that established themselves ashore (though one group still clung to the water). The earliest known freshwater mussels are found in the Upper Devonian rocks of both Ireland and New York. These were the ancestors of the teeming beds of mussels that thrived along the edges of the later coal swamps. Their crowded shells occur in countless "mussel bands" throughout the Coal Measures and have proved to be of the greatest value in the correlation of coal seams over wide areas.

The oldest land snails are known from rocks of Upper Carboniferous age in Nova Scotia, where they are found inside the fossilized stumps of ancient trees. In some later periods, freshwater snails became so common that their remains make up most of the rock in which they occur. The beautiful Jurassic Purbeck Marble, used extensively in interior church decoration, is crowded with the remains of one genus, *Viviparus*.

It would be good now, perhaps, to glance quickly back over the Paleozoic, for we have covered around 300 million years of invertebrate history in the course of these few pages. As the long Paleozoic Era drew to a close, virtually all the familiar invertebrate groups were already well established. Many others lived with them. But the end of the Paleozoic times was, in one sense, the end of the dominance of the invertebrates, for during those long ages they were dominant, yet not alone. Across the once barren Earth that saw their advent, a mantle of green had slowly spread. In the swiftly flowing streams and in the restless seas other forms of life had come into being and had, in some respects, taken their place. Wood and bone—these were the commodities that were to alter the whole history of life on the land. With their advent, a new age dawned.

6

BONE, SCALES, AND FINS: THE EARLY VERTEBRATES

CHORDATE CHARACTERISTICS

The detailed structures of many of the major groups of invertebrate animals we have reviewed are unfamiliar to most of us, but when we come to the chordates, a phylum that includes the vertebrates, things change. We are on our home ground, the structures as familiar to us as are our own bodies: we share these distinctive bodily characteristics with about 64,000 other named species of living vertebrate animals.

Although chordates constitute only some 3 percent of all living species, they are widely distributed, marvelously adapted, numerically abundant, and strikingly successful. All share a few distinguishing characteristics for at least part of their development: a single hollow nerve cord, which is differentiated in vertebrates into the brain and spinal cord; a notochord, a flexible rodlike axis, running underneath and supporting the nerve cord; pharyngeal gill slits; a tail extending beyond the anus; and segmentation, reflected in the muscles and the vertebral column.

The major group to which vertebrates are assigned—phylum Chordata—includes two small groups (the tunicates and the cephalochordates) and one large group: the vertebrates (or Craniata). Most chordates are bilaterally symmetrical, with complete, relatively complex digestive tracts. There are nine major groups of vertebrates, five of them "fishes" and four of them tetrapods.

Before we describe these major groups, let us say a word about their origin. Origins are one of the most important, but also one of the most puzzling aspects of paleontology. The Ediacaran fauna, you may recall (fig. 5.1), appears abruptly, fully formed, with no obvious ancestors. So do the faunas of the Early Cambrian. The same is true for many later groups, even though numbers of persuasive transitional forms are also known. In the case of groups without obvious fossil ancestors, the tendency, in looking for ancestral forms, is to cast around for other organisms from which distinctive characters might have been developed.

We need to bear two points in mind as we make this search. First, we can expect to find more clues and closer similarities when we consider the simplest, rather than the most complex, members of the new descendant or derived group. Second, the embryonic or larval early stage of any creature is often quite unlike the adult. The tunicates, or sea squirts, for example (fig. 6.1E), are sedentary creatures enclosed in a sac-shaped body and live in shallow seas. Anything less like a vertebrate is hard to imagine, for, although they have gill slits, they have no nerve chord, no vertebrae, nor even a notochord. But the larval form is quite different: it is free-swimming with a tadpole-like body that has both a notochord and a nerve cord above it.

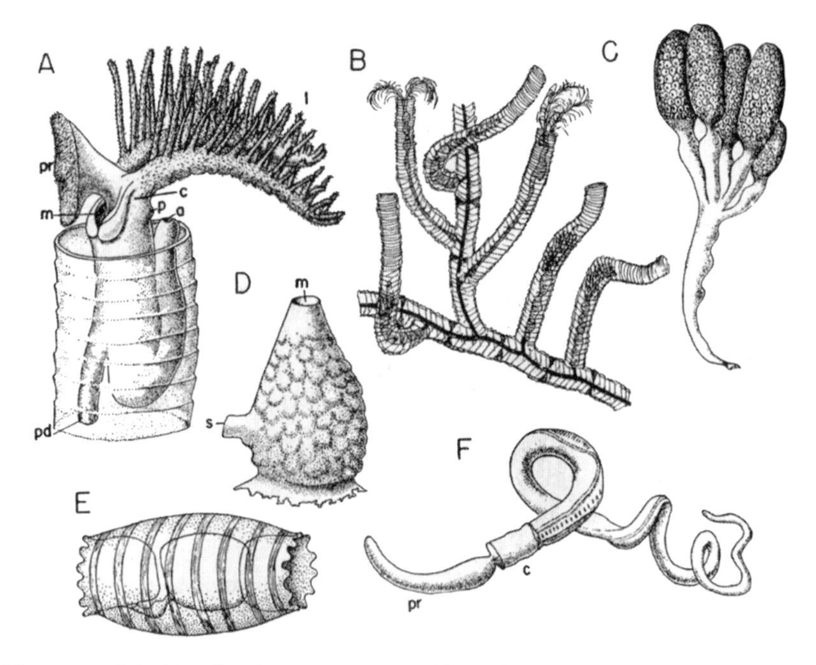

Figure 6.1. Primitive chordates: tunicates and hemichordates. (A) An individual from a colony of *Rhabdopleura* projecting from its enclosing tube; (B) A colony of *Rhabdopleura;* (C) A colonial sessile tunicate; (D) A solitary tunicate; (E) A free-floating tunicate, or salp; (F) An acorn worm, *Balanoglossus*. Abbreviations: a, anus; c, collar region; l, lopophore; m, mouth; p, pore or opening from coelom; pd, peduncle (stalk) by which individual is attached to remainder of colony; pr, proboscis; s, siphon to carry off water and body products.

(After Romer 1970. From Donald Prothero, *Bringing Fossils to Life* [New York: Columbia University Press, 2003]. Reprinted with permission of the author.)

Within the chordates, as we shall see, there are two such "primitive" groups related to vertebrates: the urochordates, which include the sea squirts or tunicates, which we have just described, and the cephalochordates, which include *Branchiostoma* (*Amphioxus*), the lancelets, small, eel-like animals a couple of inches long that live in shallow temperate and tropical marine waters. These creatures exhibit all the classical chordate structures but lack the backbone of vertebrates.

This brings us to a second major line of evidence in our search for origins: the fossil record. Surprisingly, in spite of their soft-bodied structure, fossils of forms that are remarkably similar to the lancelet have been found in the Early Cambrian of South China (*Yunnanozoon*) and in the Middle Cambrian Burgess Shale of British Columbia (*Pikaia*, see fig. 6.2A). *Pikaia*, when it was first discovered in 1911, was regarded as a worm, but more recent studies suggest it to be a lancelet-like chordate. These are the earliest creatures known to have had a notochord, and this anatomical evidence suggests that they are related to vertebrates, perhaps broadly ancestral. So too, perhaps, were the conodont animals (see below).

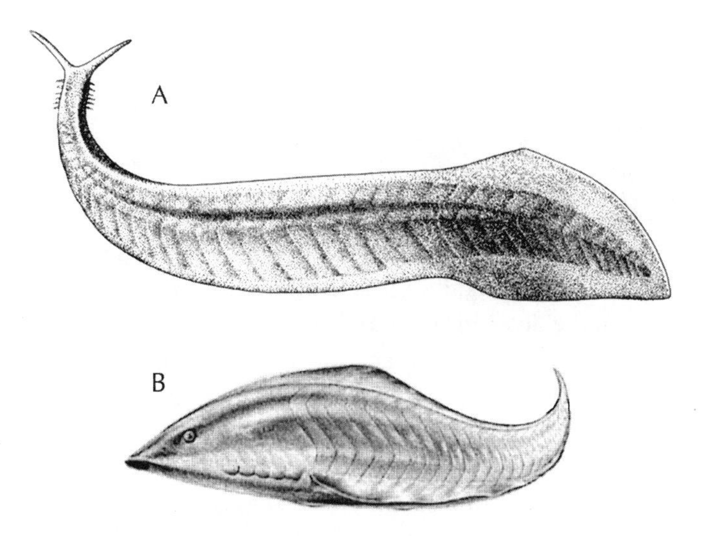

Figure 6.2. Primitive chordates: *Pikaia* and *Myllokunmingia*. (A) *Pikaia*, the world's first known chordate, from the Burgess Shale. Note the features of our phylum: the notochord or stiffened rod along the back that evolved into our spinal column, and the zigzag muscle bands. (B) *Myllokunmingia*, Lower Cambrian chordate.

([A] drawn by Marianne Collins; used with permission. [B] illustration by John Sibbick; used with permission.)

This convergence of deduction from the form and structure of living forms and their earliest fossil representatives—even if not necessarily their direct ancestors—provides some encouragement and support for this suggested ancestry of chordates.

There is one other group that we should mention in passing.

The conodonts are extinct (Cambrian to Triassic) toothlike fossils that are valuable index fossils and have been shown to be related to chordates. Made of calcium phosphate, microscopic in size, and abundant in many sedimentary rocks, conodonts are now thought to be the "jaw" parts of elongated, eel-like animals, about 1.5 inches long, with large eyes, segmented bodies, and dorsal and ventral tail fins (fig. 1.1). They had a broad resemblance to living hagfish, or lampreys.

Whether *Pikaia* or conodont animals were ancestral to chordates remains a puzzling question, as does the related broader question of chordate origins. The nearest chordate relatives are the echinoderms, which share their deuterostomic characteristics, though they lack other chordate features. Some have suggested that vertebrates arose from the larval form of tunicates or sea squirts, or from cephalochordates (lancelets and their relatives); others have suggested they arose from the mitrates, an ancestral fossil group of echinoderms, or that they shared a common ancestor with this group.

BONE

Almost all vertebrates share one other feature: bone. Bone not only supports and protects the body's soft parts, but also produces red and white blood cells and plays a significant role in mineral storage. It has been a major factor in the diversification and success of the vertebrates.

Bone has provided a marvelously strong but supple supporting spine and a protective braincase, thus allowing a variety of modes of life and freedom of movement, as well as allowing growth in brain size. These features have profoundly influenced the evolution of the vertebrates. But bone, composed chiefly of calcium phosphate, is the vertebrate answer, not only to the challenges of bodily support and function, producing blood cells and storing minerals, for example, but also sometimes to active defense (antlers and tusks, for example). Bone can also provide protection. Many of the earliest fish, for example, had an external covering of hard bony plates and scales. And it is these that constitute the oldest known fossil bone. In rocks of Early Ordovician age in Australia and the Middle Ordovician Harding Sandstone of Colorado, microscopic dermal bony plates and scales are present, but there are no teeth. These fossils represent agnathan (jawless) fish such as *Astraspis* and *Thelodus*, which lived in what is generally regarded as shallow-water, marine environments, though a recent study has suggested a fresh water origin.[1]

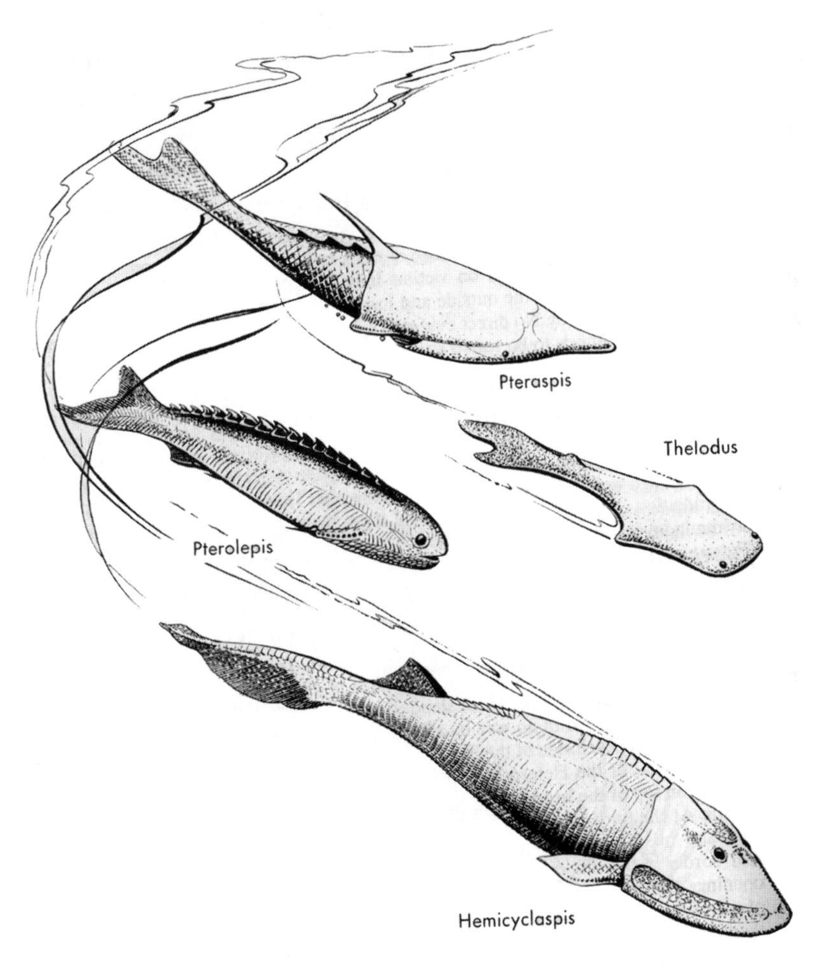

Figure 6.3. Four ostracoderms, jawless vertebrates of Silurian and Devonian age, drawn to the same scale. *Hemicyclaspis* is a cephalaspid, *Pterolepis* an anaspid, *Pteraspis* a heterostracan, and *Thelodus* a thelodont. *Hemicyclaspis* was about 8 inches long.

(Prepared by Lois M. Darling. From Edwin H. Colbert, *Colbert's Evolution of the Vertebrates* [New York: Wiley-Liss, 2001], 32. Reproduced with permission of the publisher. Permission conveyed through Copyright Clearance Center, Inc.)

JAWLESS FISH *("AGNATHA")*

The appearance of these earliest fish represents a major event in the long history of life. Without it, we should not be here. Nearly 33,000 species of living fishes are known and they play a significant role in our daily lives, both as a major source of protein and as members of the larger food chain, with 100–300 new species added every year.[2]

All the earliest fishes—the "agnathans"—lacked true jaws, though many were protected by bony armor. Living jawless fishes (the lamprey and the hag fish) are members of the same group, though they lack the bony protection their ancestors enjoyed. These early jawless fishes included the "ostracoderms," which are armored. Most of these, such as *Pteraspis* (see fig. 6.3) had flattened bony head shields, lateral eyes, and flexible, scale-covered bodies, though some—like *Jaymoytius* —were laterally compressed. They were typically 6–10 inches long and, though their Ordovician representatives are found in marine rocks, later groups were common in both marine and fresh or brackish waters during Silurian and Devonian times (some 444–359 million years ago).

Ostracoderms seem to have been bottom dwellers, and the finely developed internal sensory network within the head shield of some of these creatures, such as *Cephalaspis* and *Kieraspis*, has been exquisitely preserved and studied in some specimens.

The Anaspida (anaspids) are a related order of ostracoderms. They were small, 4 to 6 inches long, and *Birkenia* was a typical member. These fishes were more laterally compressed, with their heads covered, not with a heavy shield, but with small, separate scales that also covered the rest of the body. *Birkenia* was probably an active swimmer.

From Ordovician until Early Devonian times, the ostracoderms prospered, but by the end of the Devonian all the major groups of jawless fishes had disappeared. The decline of the agnathans possibly reflects the unequal competition between them and the more highly developed fish.

All the other traditional vertebrate classes—Acanthodii, Placodermi, Chondrichthyes, Amphibia, Reptilia, Aves, Osteichthyes, and Mammalia—are gnathostomes, or jawed vertebrates. We should note, however, that alternative classifications involve more inclusive groupings; for example, not dividing the placoderms from the acanthodians.

The evolution of jaws, flexible bodies, and paired pelvic fins provided gnathostome fishes such a high degree of feeding efficiency and mobility that by the end of Devonian times (359 million years ago) they underwent rapid

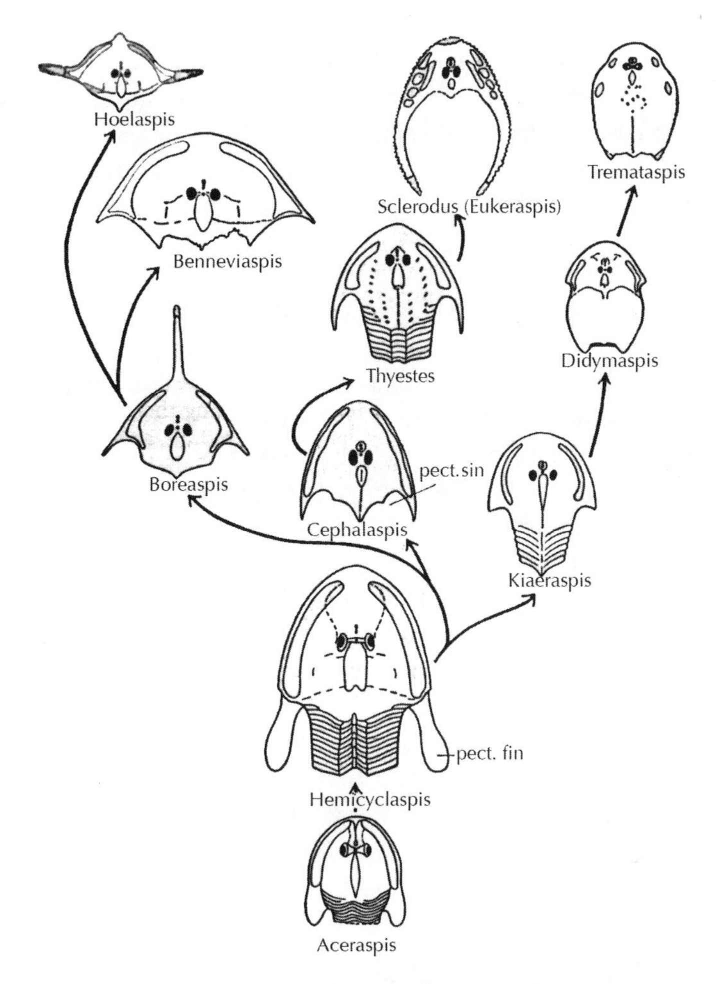

Figure 6.4. Variation in head shields of cephalaspid ostracoderms.
(After W. K. Gregory, *Evolution Emerging* (New York: Macmillan, 1951). From Frank Rhodes, *Evolution of Life* [New York: Penguin, 1976], fig. 23, p. 166. Reproduced with permission.)

evolution and rich diversification of form, essentially replacing the once-ubiquitous Agnatha, The only Agnatha known from the great span of post-Devonian time are the living lampreys and hag fish. They may well have survived because of their specialized mode of life. They feed by attaching themselves to other fish by means of a funnel-like mouth armed with a rasping tongue, by which they feed on their victims, as well as by being scavengers.

The jawless, finless ostracoderms were limited to a bottom-dwelling, mud-grubbing existence, but the development of jaws and paired fins allowed the jawed fish—and us, their descendants—to exploit a host of new sources of food and new ways of life. One indication of the extent of this subsequent post-Devonian explosion and diversification of jawed fish is given by the fact that there are some 32,000 living species. Without jaws and paired fins, the vertebrates could never have left the water for the land.

ACANTHODIANS (CLASS ACANTHODII)

The acanthodians were small, streamlined fishes, typically up to 8 inches long, with one or two dorsal fins, up to six pairs of pointed ventral fins, and an elongated, heterocercal tail. They are often called "spiny sharks," though they are not related to sharks. The substantial head was covered with bony plates and had large eyes; the body was covered with small, distinctive square scales. Some acanthodians lacked teeth. *Climatius* (fig. 6.5D) was a typical member of the group. Acanthodians include the oldest known gnathostomes (vertebrates with jaws). The development of jaws was one of the great events in vertebrate history, allowing the subsequent exploitation of new sources of food and new environments.

Acanthodians first appeared in the late Ordovician, became abundant in the Devonian, and lingered on until the Early Permian. The true affinities of acanthodians have been a matter of some debate. They were once classified with sharks, then with placoderms, and later with bony fishes. We here treat them as a separate class.

PLACODERMS (CLASS PLACODERMI)

Before describing the placoderms, which were among the first jawed vertebrates, we should say a word about the origin of jaws. In most vertebrates, the jaws are made of cartilage or bone and are derived from the two anterior gill arches supporting the gills. These gill arches are believed to have become the jaw itself in fish and the hyoid arch, which suspends the jaw. It has been suggested that the original benefit of such jaws may have been to improve respiration, rather than feeding. In placoderms the jaw structure was simpler than that of most later fish, in that the upper portion of the first gill area (the hyomandibular) was not incorporated into the jaw. Placoderms were the dominant fish of Devonian times, appearing in the early Silurian (some 440 million years ago), flourishing in both marine and nonmarine environments in Devonian times, and becoming extinct at the end of that period: the only class of vertebrates to suffer that fate, unless one regards the acanthodians as a class. They were the first vertebrates to develop paired pelvic fins.

Placoderms were heavily armored, jawed fishes that came to dominate life in the seas, rivers, and lakes for 60 million years. The name "placoderms" ("plated-skin") describes one of their most striking features: a bony shield that covered the head and "trunk" with a joint between the two, which helped to increase the gape of the jaws. Some 200 genera have been described, and they include some forms that reached up to 30 feet in length. From the great variety of placoderms—of which there are five or six different orders—we here describe two representative groups: the antiarchs, and the arthodires.

The antiarchs (order Antiarchi), such as *Pterichthyodes* (fig. 6.5C) were characterized by a high, boxlike head and neck armor, a ventral mouth, small eyes, strong bony pectoral fins, and a scaly body covering. Their general structure suggests that these creatures were bottom feeders. They are especially common in rocks of Middle to Late Devonian age and a few survived into the Early Permian. A more streamlined member of the group, *Bothriolepis*, had strong pectoral fins and an elongated slender heterocercal tail (one in which the vertebral column turns upward into the larger of the two lobes, in contrast to most modern bony fishes which have a homocercal tail, with lobes of equal size). Most members of the genus, of which there are twenty species, were a foot or so in length, though one grew to some 3 feet long. These were dominantly freshwater bottom feeders but may also have been capable of living in shallow marine environments.

The arthrodires (order Arthrodiri) were active, carnivorous placoderms, of which 200 genera are known. They had a heavy, bony head shield consisting of fused plates, with large eyes and widely articulated jaws that lacked true teeth. However, the jaws themselves were sharp cutting blades. The articulation suggests that both the upper and lower jaws could be moved to widen their bite. Behind this heavy armor, the rest of the body seems to have been unprotected. One genus of this group, *Coccosteus* (fig. 6.5A), was some 16 inches long, but another, the giant *Dunkleosteus* (formerly *Dinichthys*) (fig. 6.5B), grew to a length of more than 30 feet and must have been the most formidable marine carnivore of its day.

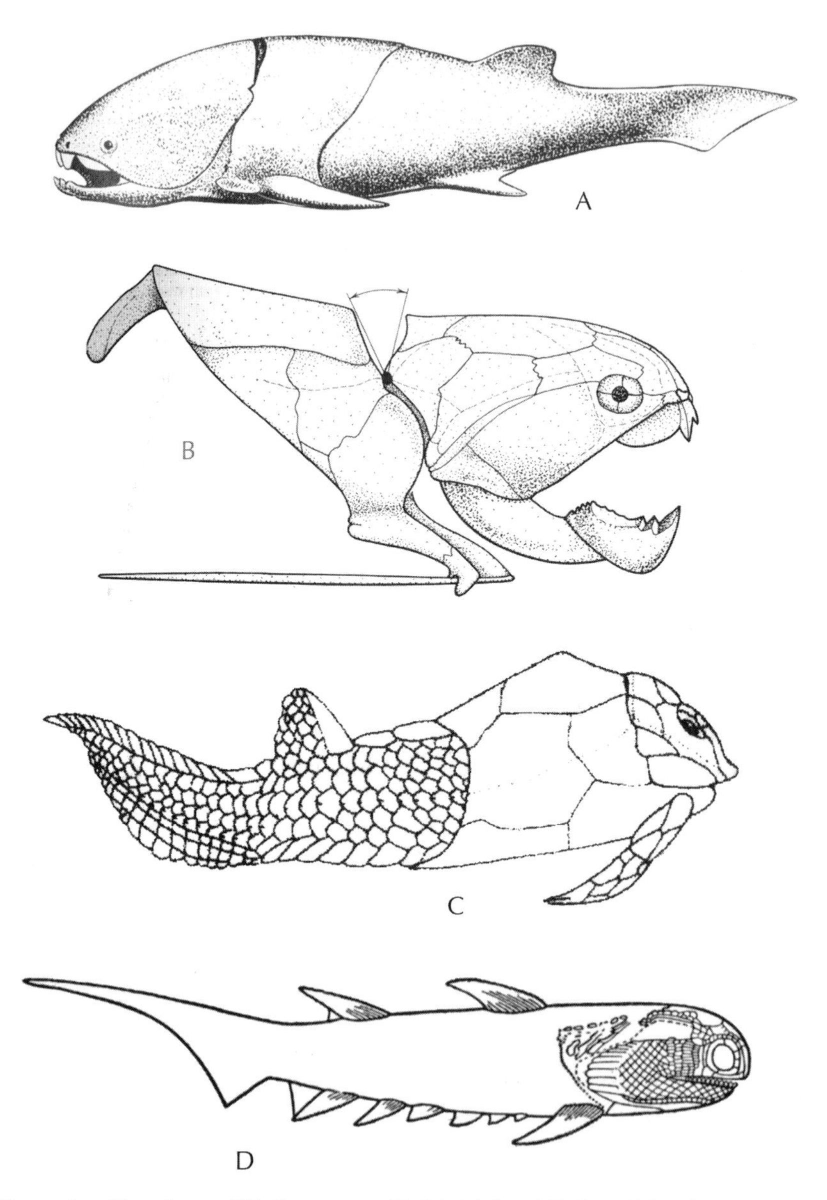

Figure 6.5. Placoderms. (A) *Coccosteus*; (B) *Dinichthys;* (C) *Pterichthyodes*; (D) *Climatius.*

([A] and [B] from Edwin H. Colbert, *Colbert's Evolution of the Vertebrates* [New York: Wiley-Liss, 2001], 41, 43. Reproduced with permission of the publisher. Permission conveyed through Copyright Clearance Center, Inc. [C] from Michael Benton, *Vertebrate Paleontology*, 3rd ed. [Oxford, U.K.: Wiley-Blackwell, 2004], 472. Reproduced with permission. [D] from Frank Rhodes, *Evolution of Life* [New York: Penguin, 1976], fig. 24D, p. 168. Reproduced with permission.)

Placoderms were the dominant fish of Devonian times, but they declined greatly by the end of the period. Perhaps their heavy armor and largely bottom-dwelling existence made them unequal competitors with their swifter cartilaginous and bony contemporaries.

SHARKS AND RAYS (CLASS CHONDRICHTHYES)

The fourth class of fish is the Chondrichthyes, all of whose members lack true bone and have a skeleton of cartilage, rather than bone. The group includes sharks, rays, and chimaeras, or ratfish, the latter tending to be deep-water dwellers. Almost all of them are marine, and most are predatory fish, though some, such as the whale sharks and basking sharks, are filter feeders. Unlike the placoderms, most sharks are beautifully streamlined, powerful swimmers, having relatively small eyes but otherwise highly sophisticated sensory systems that can detect small electric charges generated by muscle contractions of their prey, small mechanical disturbances of the water, and minute concentrations of chemicals. Some living species reach more than 50 feet in length.

Because they lack bony skeletons, sharks and their allies are poorly represented as fossils, though the teeth, dermal denticles, and sharp spines of some forms are common fossils in places. The teeth of sharks are continuously shed and replaced, growing in whorls. The teeth reflect the diets of the various groups, ranging from flat or rounded grinding forms to sharp biting and serrated cutting forms. A single shark may have several hundred teeth. Unlike those of other vertebrates, the teeth of the sharks are set into the gums, rather than into the jaws. Sharks appeared in Silurian times and fossils became abundant in the Devonian. Sharks differ from the placoderms, not only in having more advanced jaws and true teeth, as opposed to bony extensions of the jaws, but also in having two pairs of paired fins and in the absence of heavy body armor. Male sharks bear clasping devices (as did many placoderms) behind the pelvic fins, allowing internal fertilization of the female, which then either lays eggs in distinctive egg cases ("mermaid's purses" are the egg cases of sharks, skates, and chimaeras), or gives birth to live young. In this latter case they differ markedly from bony fish, many, but not all of which lay hundreds of eggs, few of which survive to maturity. Chondrichthyans also lack a swim bladder. Sharks have five to seven open gill slits located on the side of the body, through which water passes, from which they extract oxygen.

Living sharks and rays are grouped together as elasmobranchs. Early sharks include *Cladoselache* (fig. 6.6A), which was about 3 feet in length, whose remains are well preserved in the Late Devonian Cleveland Shale of northern Ohio and whose external form broadly resembles that of living sharks. The freshwater *Xenacanthus* was an elongated, spined relative, nearly 2 feet long, the remains of whose tapered, spear-shaped body are widespread in the freshwater deposits of the Late Paleozoic coal swamps. *Pleuracanthus* (fig. 6.6B), a late Paleozoic shark, was about 2.5 feet in length.

Sharks developed a striking variety of forms in the Late Paleozoic, being known from both marine and nonmarine rocks. They are superbly successful marine predators, some of which (such as the extinct *Carcharodon megalodon*) reached a length of about 40 feet. There are more than 400 species of living sharks.

Unlike the streamlined bodies of sharks, skates and rays have flattened bodies, with pectoral fins developed into large, flaplike projections

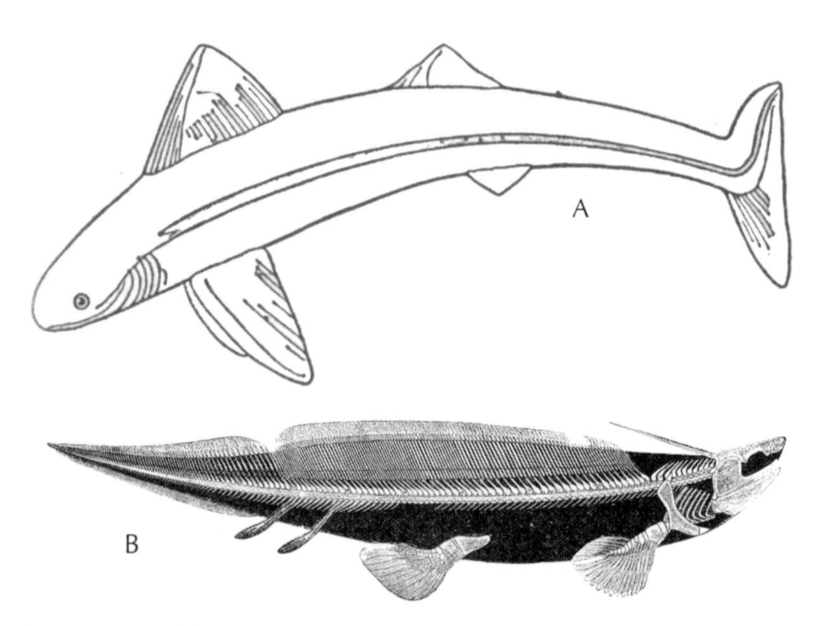

Figure 6.6. Fossil sharks. (A) *Cladoselache*, a late Devonian shark; (B) Reconstruction of *Pleuracanthus*, a late Paleozoic shark.

([A] from Frank Rhodes, *Evolution of Life* [New York: Penguin, 1976], fig. 25A, p. 170. Reproduced with permission. [B] from Ray Lankester, ed., *A Treatise on Zoology* [London: Adam and Charles Black, 1909].)

that allow them to "fly" through the water, with eyes on the top of their heads, and mouths and gill slits on their undersides. Their teeth are blunt, mollusk-crushing structures, but they also feed on crustaceans and small fish. Rays, skates, and their allies are represented by more than 500 living species. Some mantas reach almost 20 feet in breadth, while the smallest is only 4 inches. Unlike the swift, predatory hunting style of the sharks, rays and skates spend much of their time waiting for food on the floor of the ocean, though a few are freshwater dwellers. As with their sisters, the sharks', skates', and rays' lack of structural change over long geologic time reflects the success of their distinctive pattern of life.

BONY FISHES (CLASS OSTEICHTHYES)

Bony fishes (class Osteichthyes) include two major divisions: the Actinopterygians (ray-finned fishes) and the Sarcopterygians (the lobe-finned fishes, including tetrapods). These two groups have a very different structure and history, but both differ from the Chondrichthyes in having a bony skeleton, as opposed to one made of cartilage. Ray-finned fishes outnumber all other vertebrates combined in the variety of species (estimated at some 30,000). The lobe-fins, excluding tetrapods, in contrast, are represented by only seven living species, the coelacanth (*Latimeria chalumnae*) and six species of lungfish. Almost all bony fishes are cold-blooded and are characterized by an operculum covering the gills, scales or plates, paired fins, air-filled organs, which control buoyancy (or lungs, in the case of lungfishes and bichir), paired nostrils, and a relatively complex brain. They show enormous variation in form, ranging from sturgeon, paddlefish, gar, elephant nose fish, tarpon, herrings, anchovies, trout and catfish to sea horses, pirhannas, angler fish, swordfish, sunfish, plaice, flying fish, and eels. They inhabit an extraordinarily wide range of environments, from ocean depths of more than 6 miles to the high Andes, from the equator to polar waters, from hot springs to freezing ponds and even caves. Most are restricted to either fresh or marine water, but some (such as the stickleback) can survive both, while some saltwater forms (such as the salmon) migrate to freshwater for spawning, and others, such as the American and European eels, leave their freshwater habitat and migrate to the ocean for spawning. Some, such as the tuna and billfish, are migratory in habit, traveling more than 5,000 miles. Bony fishes can be herbivorous, carnivorous, or omnivorous.

Fishes are not only the oldest vertebrates, preeminently successful in so many aquatic environments, but they are also the ancestors of the terrestrial vertebrates. There are, as we have seen, two main groups of bony fish. The fins of the first group, the Actinopterygians (ray-finned fishes), are based on a web of stretched skin, supported by rays of keratin with bony supports at the base. The second group, the Sarcopterygians or lobe-fins, have altogether stronger and more powerful fins, with a strong, bony axis, articulated with the pectoral girdle. It was from this second group of bony fish, the Sarcopterygii, that the Amphibia and all other vertebrates evolved. Living lobe-finned fish include the lungfish (or Dipnoi) and the coelacanths.

The lungfishes, though abundant in Devonian times, are represented today by only three genera (one each in South America, Africa, and Australia). They are restricted to freshwater; the Australian lungfish can use its sturdy pectoral and pelvic fins to crawl across the bottom of the lakes and ponds in which it lives. The African lungfish lives in areas subject to seasonal drought; when the ponds dry up, it burrows downward into the mud and breathes through its lungs. The Australian lungfish, thought to be more primitive, breathes by means of a single lung. The other two have paired lungs.

Now, how do these creatures fit into the evolutionary history of the vertebrates? It is tempting to hail them at once as the link between the aquatic and the land-living vertebrates—but that would be too hasty a verdict. Both living lungfishes and their Devonian forebears are altogether too specialized to fill such a transitional role: they all show reduction in the amount of skeletal bone, the bones of the skull are quite unlike those of other vertebrates, the teeth are highly modified, and so on. But, all in all, the lungfishes, though not the direct ancestors of the terrestrial vertebrates, have anatomical characteristics showing they were related.

The second group of lobe-finned fishes is the coelacanths. The coelacanths are represented by two living species: one, *Latimeria chalumnae* (see fig. 6.7), from off the coast of Africa, and a second species recently discovered from Indonesia. Coelacanths, though well known from 120 fossil species, were long thought to have become extinct in Cretaceous times—65 million years ago—but in 1938, a fisherman caught a living coelacanth from deep water off the coast of East London, South Africa. Since then, additional specimens have been caught off the coasts of the Cormora Islands, Kenya, Tanzania, Madagascar, Mozambique, Indonesia, and South Africa.

Coelacanths are 5 or 6 feet long, weighing up to 150 pounds, heavily built, wide-mouthed, deep blue fish that are nocturnal hunters. They are variously known as "Old Fourlegs," "Living Fossils," and a "Lazarus Taxon."

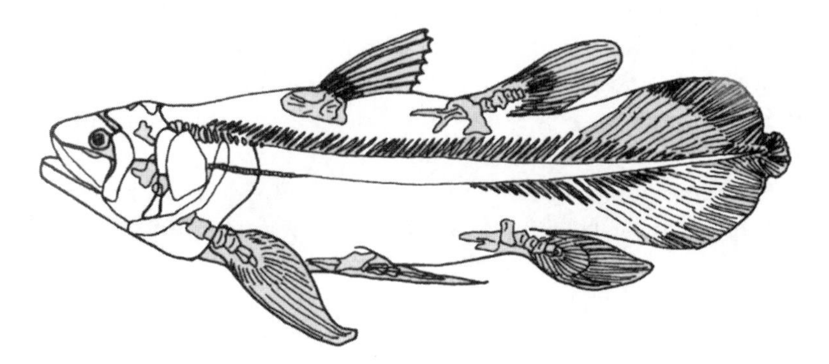

Figure 6.7. *Latimeria chalumnae.*
(From Jennifer Clack, *Gaining Ground* [Bloomington: Indiana University Press, 2002], 51. Reproduced with permission of the publisher.)

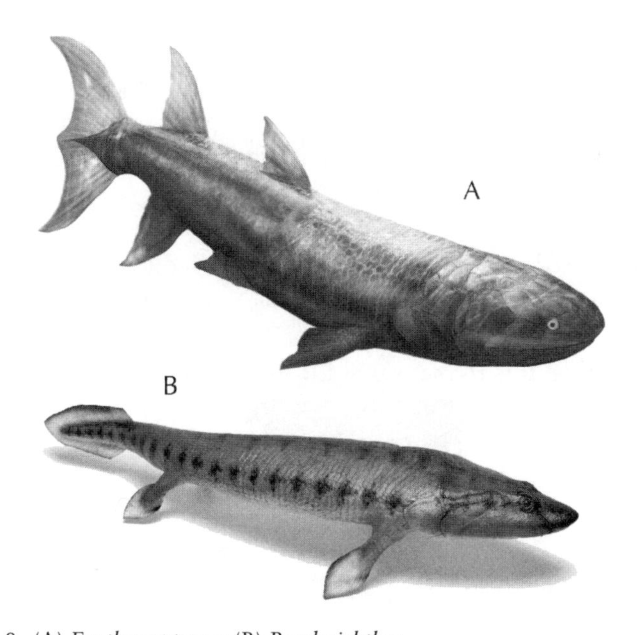

Figure 6.8. (A) *Eusthenopteron*; (B) *Panderichthys.*
([A] illustration by Karen Carr; used with permission. [B] illustration by Charlie McGrady, from devonianlife.com; used with permission.)

The coelacanth has several remarkable features, including a striking spear-shaped tail, strong fins, a distinctive, fluid-filled, sharklike cartilaginous backbone, and live-bearing reproduction. So, we have to ask again: have we met in this strange and beautiful fish, our remote ancestors? And, again, the answer seems to be "No." Molecular studies of living forms suggest that neither they, nor the lungfish, qualify for that role, but that both are sisters, rather than ancestors, of the early tetrapods, which evolved some 400 million years ago. But if we go back to the fossils of that period, a third group of extinct lobe-finned fishes—the crossopterygii—make more likely candidates for that ancestral role.

Eusthenopteron (fig. 6.8A) is a lobe-finned fish from the Devonian with fins that included bones similar to those found in tetrapod arms, wrists, legs, and ankles, though they could not have supported the body in walking. A roughly contemporary Devonian fish, *Panderichthys,* from Latvia, though known from fragments since the nineteenth century, has only recently been fully described. It was 3–4 feet long and had many features that are transitional between fish and amphibians. Its flat, triangular head was amphibian-like, as were the component bones of the skull, and it had strong pectoral fins that may have allowed it to crawl. It lacked the typical dorsal and anal fins of other fishes.

The osteolepids, another group of lobe-finned fish, were the only group to have had a bifurcated set of bones within the paired fins, thus resembling the tetrapods. And some osteolepids had internal nostrils, though whether this feature was the result of parallel evolution, rather than close genetic affinity, is debated.

A discovery in 2010 of well-preserved tetrapod tracks from 397-year-old tidal flat sedimentary rocks from Poland indicates, however, that, though *Panderichthys* probably resembles the ancestral group from which amphibians emerged, it itself occurred too late in time to have played that role. It is to the search for the origin of the amphibians that we turn in chapter 8.

7

THE GREENING OF THE LAND

THE PLACE OF PLANTS

It is all too easy to regard plants as secondary players to the main actors in the overall evolution of life—the animals—but their impact on both the evolution of animals and the development of the planet has been profound. Satellite studies show, for example, that plants synthesize 105 billion tons of biomass a year from carbon dioxide taken from the air.[1] The microscopic phytoplankton produce about half of this, though they make up only 1 percent of total plant biomass, while terrestrial plants represent more than 90 percent of the Earth's total biomass. And, just as plants have influenced the history of life, they have also exercised a major influence in shaping Earth's climate by removing carbon dioxide from the atmosphere. Nor is this all, for the growth of plants exercises a profound influence on rock weathering and soil formation, while plant material provides the essential food for the vast majority of invertebrate and vertebrate animals.

One of the major hazards of our present burgeoning population and our continued clearing of forests is the 7 billion or so tons of carbon dioxide we are now adding to the atmosphere every year, contributing to a steady buildup that may threaten the future ability of forests to absorb this greenhouse gas.

The development of leaves was a factor of huge significance in the evolution of plants and in their climatic impact. The first vascular plants were, as we have seen, leafless, with naked, branching stems, not more than a foot or two in height. They existed for some 30 million years. Apart from a small, enigmatic, 390-million-year-old plant, *Eophyllophyton,* from China, the oldest known plant with leaves—*Archaeopteris* (fig. 7.1C)—grew to about 30 feet in height. Leaves also appeared independently in other groups of plants, including sphenopsids, pteridosperms, and lycopsids, even though the leaves of the lycopsids were small. With the development of leaves, "flat solar panel[s] for capturing sunlight and powering photosynthesis"—as David Beerling calls them[2]—plants emerged as major components of the biosphere. They now cover more than 70 percent of Earth's land area.

This surge in development of land plants, which has been called the "Devonian Explosion," took place when levels of carbon dioxide were appreciably higher, but these levels plunged over the period between 400 and 350 million years ago, falling by as much as 90 percent. This fall weakened the beneficial sheltering of the greenhouse effect and ultimately produced an ice age. It was against such dramatic change that land plants exploded in abundance and Earth's leafy canopy developed.

Earth's Early Paleozoic lands were lifeless and barren: a desolate wilderness of harsh, rocky landscapes, devoid of soils, devoid of vegetation, devoid of life. It was the emergence of plants, as they slowly spread and colonized the land, that made possible the subsequent evolution of terrestrial animals and softened the harsh contours of the planet that, much later, became our home. In the long history of the Earth, few events were of such enduring significance as the development of land plants. With their emergence, a new world came into being. Without it, we should not be here.

Land plants developed from green, filamentous algae, with which they share several common features, and which continue to thrive today in aquatic environments. The transition of such plants from water to land involved four major adaptations:

- The need for some supporting structure, to replace the buoyancy provided by water
- Protection against desiccation
- Water for survival
- The need for new reproductive strategies for spore dispersal, to replace the earlier means of dispersal through the water

These modifications were substantial, and we can trace their slow emergence and development in the fossil record of early plants. The spores, for example, show the development of a protective covering. The plants themselves developed an outer waxy, waterproof covering (the cuticle) as well as an internal structure of xylem and phloem—the vascular system—to support the stem and transport food and water.

What are the earliest indications we have of plants? Three indirect indications of the early existence of photosynthesis can be identified. The first is based on the isotopic signature of two carbon isotopes (^{12}C and ^{13}C), the lighter of which is selectively taken up during photosynthesis. Plants are characterized, therefore, by a higher ^{12}C ratio, and evidence of this is found in ancient rocks, as old as 3.8 billion years. These ratios probably reflect the activity of cyanobacteria, also known as blue-green algae, which obtain their energy by photosynthesis.

A second indirect indication of the existence of photosynthesis is the widespread presence of banded iron formations, whose deposition from 3.5 to 2.0 billion years ago reflects the oxidizing effects of molecular oxygen produced by cyanobacteria. This effect reflects, in turn, the major role played by cyanobacteria in converting Earth's atmosphere from a reducing to an oxidizing environment, thus allowing the development of ancient life.

Still a third indirect indication of the presence of cyanobacteria comes from the widespread development of stromatolites in rocks of Precambrian, largely Proterozoic, age, spanning the period from 545 to 320 million years ago. These structures are also the product of mats of cyanobacteria.

There is also direct—as opposed to indirect—evidence of early filamentous cyanobacteria found in such ancient rocks as those of the Bitter Springs Formation of central Australia. But none of these concern land plants, and it is to these plants that we now turn.

LAND PLANTS

The earliest direct indications that we have of the existence of *land* plants come from the Ordovician period: 472-million-year-old rocks from the Argentine Andes and 462-million-year-old rocks from Oman and the Czech Republic contain fossils of microscopic spores, representing the reproductive structures of liverworts—simple plants, lacking both stem and roots—which were probably ancestral to all later plants, and which still survive today. These fossil spores, from widely separated parts of the world, represent at least six different species of liverwort, emphasizing both the early widespread geographic distribution and the emergent diversity of these simple plants. We have, as yet, no remains of plants themselves from such ancient rocks, which is not, perhaps, surprising, since spores, with their thick protective covering, were presumably far more resistant to erosion and destruction than were the more fragile structures of early plants. One beautifully preserved Early Devonian microfossil from Shropshire, England, represents the spore sac of one of these early plants. How the earliest developed is not clear, but they may well have emerged from soil-forming algae or other microbes.

The oldest glimpse we have of the remains of vascular plants themselves comes from rocks of Silurian age (about 425 million years old) found in Wales and elsewhere. One early and typical representative, *Cooksonia* (fig. 7.1A), was a low-growing, leafless land plant, 2 or 3 inches in height, that had thin, bifurcating stems carrying reproductive sporangia at their tips. These stems included xylem, which both provided support and allowed fluids to circulate within the plant. The stems had no true root system. There is, however, some controversy as to whether *Cooksonia* was a true vascular plant.

Baragwanathia, known from Late Silurian rocks (about 420 million years old) of Australia, was a robust vascular plant, some 10–12 inches in height, with needlelike leaves that were spirally arranged around branching, stubby, vertical stems. The closely related genus *Protolepidodendron* (fig. 7.1D) from the Middle Devonian had leaves that were forked at the tips. Some related genera grew to one and a half feet or more in height. Related forms were widely distributed in North America, Europe, China, and Australia.

Early Devonian times saw the rapid diversification of land plants. Among these, the best known are *Rhynia* (fig. 7.1B) and several related forms, whose remains have been preserved in exquisite detail in Early Devonian (410-million-year-old) chert (silica) deposits formed by ancient volcanic springs at Rhynie, Aberdeenshire, Scotland. Even the cellular structure of these early land plants has been preserved. Also preserved are the remains of contemporary Devonian animals, all of them arthropods.

Rhynia (fig. 7.1B), which was broadly similar to *Cooksonia*, had erect, Y-shaped, branching, leafless stems, some 6 inches or so in length, bearing sporangia at their tips. It had no true roots but did have tiny rootlike structures growing from horizontal, subsurface extensions of the stem. These plants and their relatives were widely distributed in Devonian times. The living liverworts are probably descended from these very early plants.

Asteroxylon, another plant found in the Rhynie Chert, is related to living club mosses. Still another plant from the Rhynie Chert, *Aglaophyton,* was a nonvascular plant. These simple plants grew to 18 inches in height, with features intermediate between those of the bryophytes (such as liverworts and mosses) and vascular plants.

By Late Devonian times (374 million years ago) several groups of stronger, taller, and more structurally advanced plants appeared. For example, *Psilophyton* (fig. 7.1E), had a trunk with far more conductive tissue (xylem) than its predecessors, allowing it to grow significantly taller than some of its contemporaries. Other contemporary plants grew to giant size, supported by their woody trunks, strengthened by the addition of xylem, anchored by roots, and multiplied by the development of female megaspores and male microspores.

By Late Devonian times, land plants showed substantial diversity. Lycopods (lycophytes: the ancestors of the living, lowly club mosses) with scalelike bark, grew into mighty trees, some 150 feet high, later contributing to the great coal deposits of Pennsylvanian times. Sphenopsids (sphenophytes: "jointed" plants, which include the living horsetails) were represented by trees some 100 feet high. We shall meet many of these groups again in our discussion of later geologic periods, but their advent must have had a substantial impact on both the landscape and the atmosphere, utilizing CO_2 in photosynthesis, and increasing oxygen concentration. Although today's club mosses and sphenopsids are of modest size and make up a minute percentage of living plants, in Carboniferous times they were among the dominant plant groups of the coal-forming forests.

The first true leaf-bearing plants were fernlike forms that also appeared in Late Devonian times. Leaves provided a far more efficient basis for photosynthesis than the bristle and hairlike appendages of *Rhynia* and its allies. Plants such as *Rhacophyton* (a spore-bearing plant) from the Late Devonian show the emergence of leaflike structures, while *Archaeopteris* (fig. 7.1C), of similar age, was a spore-bearing plant with fernlike leaves supported by a trunk and coniferous in structure. These conifer-like trees grew to a height in excess of 20 feet or so and, in many of their features, are intermediate and therefore perhaps ancestral, between ferns and conifers, though this is far from certain.

Our knowledge of these ancient plants owes much to Sir (John) William Dawson (1820–1899), one of the first to study them. Born in Nova Scotia of Scottish parents, he graduated from the University of Edinburgh in 1842, as one of the students of Robert Jameson. It was Jameson, it might be noted, who was also one of Charles Darwin's professors at Edinburgh and of whose lectures Darwin wrote that he found them "incredibly dull. The sole effect they produced upon me was the determination never as long as I lived to read a book on Geology or in any way to study the science."[3] Returning to Canada, Dawson worked first in exploration for

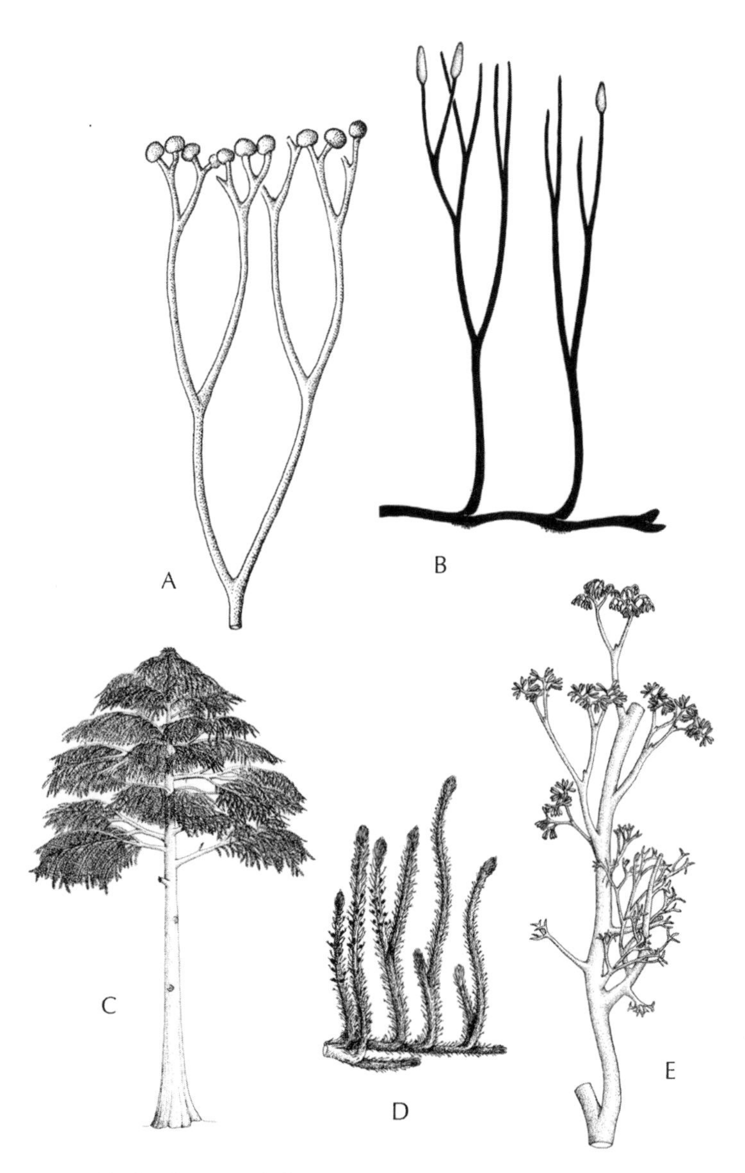

Figure 7.1. (A) *Cooksonia caledonica*. Fertile specimen. Lower Devonian. (B) *Rhynia*, primitive vascular plant of the Rhynia type. (C) *Archaeopteris*. Reconstruction of *Archaeopteris* sp. A tree about 13 feet high. Upper Devonian. (D) *Protolepidodendron scharyanum*, Lower Devonian. (E) *Psilophyton dawsonii*. Restoration of aerial axis with sterile and fertile branches.

(All images from Wilson N. Stewart and Gar W. Rothwell, *Paleobotany and the Evolution of Plants*, 2nd ed. [Cambridge: Cambridge University Press, 2010]. [A] redrawn from Edwards, 1970. [C] based on reconstruction by Beck, 1962. [D] redrawn from Kräusel and Weyland, 1932. Reprinted with permission of Cambridge University Press.)

coal and then became the Province of Nova Scotia's first superintendent of education, playing a major role in educational reform. His travels within the province allowed him to continue his passion for geology and in 1855 he was appointed Professor of Geology and Principal of McGill University, a post that he held for thirty-eight years. He twice guided Charles Lyell in the field and named one of the earliest reptiles in his honor.

It was during these years that Dawson began a series of monographs on the fossil plants from the Silurian, Devonian, and Carboniferous rocks of Canada, and he has come to be regarded as one of the founders of paleobotany.

Dawson was a man of extraordinary talent and prodigious energy. His prolific publications range from agriculture to education, from architecture to economic geology. On his appointment as principal at McGill University, he found the institution in a sorry state, with a low enrollment, ambivalent faculty, and decaying buildings on its tiny campus. He immediately addressed these problems, while also agreeing to serve as principal of a newly created normal school and teaching a full load (up to twenty hours per week!) of courses at both institutions. Under his devoted leadership, McGill blossomed and within a decade established a place as the leading university of Canada.

Dawson was also a pioneer in the education of women and partly responsible for the establishment of a women's college associated with McGill and for the subsequent admission of women to McGill.

But his administrative duties never displaced his commitment to fieldwork, and Dawson devoted the summers to continuing his studies of eastern Canada. In all, he published 400 scientific works, including descriptions of Silurian, Devonian, and Carboniferous plants, a treatise on Acadian geology, and many popular books on geology and paleontology. He discovered and described what he regarded as the earliest known organism—*Eozoon canadense*—although it has since been shown to be of inorganic origin.

Dawson was also a vigorous opponent of Darwin's theory of evolution and published extensively against it, as well as writing books on broader theological topics.

Dawson's achievements were widely recognized during his lifetime. He was awarded the Lyell Medal of the Geological Society. He was elected to the Royal Society and was knighted in London by Queen Victoria. He served as president of both the American and the British Associations for the Advancement of Science, the only person ever to have served in both capacities.

It was William Dawson whose pioneering studies provide the foundation of our knowledge of Paleozoic plant life.

By Late Devonian times, the oldest seed-bearing plants appear in rocks from Belgium (*Runcaria*), West Virginia (*Elkinsia*), and elsewhere. In them, these earliest seeds are not fully enclosed in an ovary but are "naked." (It was from gymnosperm plants such as these that the giant seed ferns developed, whose remains also contributed to the world's great coal-forming forests of Carboniferous times.) These and other plants seem to have thrived in low-lying, coastal, swamplike environments, probably in or near equatorial areas. And, much later, it was probably from these or some similar plants that the flowering plants, in all their glorious profusion, subsequently arose.[4]

But all that is still far ahead of us in time, and before we leave these lowly Devonian plants it is worth taking a moment to look back at the transforming influence they had on our planet. With their coming, the barren, rock-strewn surface of the land became clothed in green. With their coming, the way was opened for animals to venture ashore, as a source of food became available. Associated with the plants of the Rhynie Chert, for example, are several species of spiderlike forms, as well as a wingless insect and a form of tick.

It is no mere chance that Devonian times marked, not only the spread of land plants, but also the establishment of terrestrial arthropods and vertebrates. It was only when food sources were available that terrestrial animal life emerged, just as it was the subsequent development of insects that later made possible the pollination of many flowering plants.

Plants also changed the chemical budget of the Earth, not only supplementing the ancient pattern of chemical weathering, but also influencing the carbon cycle. All this emphasizes again the interdependence and intricacy of evolutionary changes.

PLANT LIFE OF THE CARBONIFEROUS PERIOD

The Carboniferous Period (359 to 299 million years ago) is named for the abundance of plant remains that form the vast coal deposits that characterize the latter half of the period (the Pennsylvanian). These deposits, which are found in midwestern and eastern North America, northern Europe, and Asia, formed the basis of the Industrial Revolution and continue to provide a significant portion of the world's energy supplies. Other substantial coal deposits are also mined in China, South Africa, India, and Australia.

During the Carboniferous period the Earth's continents were aggregated into two great land masses: Eurasia in the north and Gondwana in the south. The southern continent was centered around the South Pole, and advances and retreats of ice caps led to rhythmic changes in sea level in the coastal basins where coal deposits were formed. These probably resembled the low-lying paralic swamps and tide-level basins that today border the Gulf Coast and the estuarine level swamps of India and parts of the Far East.

Many of the fossil trees themselves have features (such as the smoothness and thickness of the bark, their prolific growth and luxuriant foliage, the absence of seasonal growth rings, the large size and thin walls of many of the cells, and so on) which suggest that the peaty soil in which they grew was almost permanently waterlogged. The climate was apparently warm or hot and humid, with abundant and perennial rainfall. It was the rain that maintained the swamp conditions, and these, in turn, that prevented the rapid decay of the plants and so led to the formation of coal. Under such conditions in existing forests, trees grow very rapidly—often 10 feet in a year. In such spreading tangled forests, the great trees of the coal age flourished, trees that are quite unfamiliar to our modern eyes. The flora of the inland (limnic) coal-forming basins was less diversified, although not greatly different in general character.

The classification of these nonflowering plants is complex and still in something of a state of flux, with little consensus on the details. We will here use a simple informal classification that seeks to avoid some of the present complexities and ambiguities.

The *bryophytes* include the living liverworts and mosses, represented together by some 20,000 existing species. They are small nonvascular plants that have been able to establish themselves on land by remaining small and inhabiting damp environments. Fossils resembling bryophytes are known from rocks of Late Silurian age, and they are also found in Devonian and Carboniferous rocks.

All the remaining plants we describe here are vascular plants.

The first, the *lycopods* (Lycophyta) include the modest club mosses and ground pines among their living members. These are small, trailing, herbaceous plants that bear spores in terminal cones. They have true roots, leaves, and stems consisting of woody tissue and pith, but the stems are jointed. Like the horsetails (see below), they are today a much-reduced group, their forerunners in the Carboniferous coal swamps being the giant scale trees. In these, long awl-like leaves were attached to the stem, on which they left a raised polygonal scar when they shed. These mighty trees, often with a diameter of 6 feet and reaching a height of more than 100 feet, are among the most important constituents of coal.

Two genera are particularly common. *Lepidodendron* had a tall slender trunk that branched repeatedly and dichotomously near the top. These upper branches were covered with slender pointed leaves and the spore cases were borne on the tips of the limbs. The diamond-shaped leaf scars were arranged spirally. *Sigillaria* had an altogether different appearance, with a stouter, usually unbranched trunk whose upper portion bore a brushlike fuzz of longer bladelike leaves, scars of which were arranged in vertical rows separated by ribs on the bark. More than a hundred species of each of these two genera are known, though not all were of such gigantic proportions. The typical roots (*Stigmaria)* of the scale trees were massive spreading structures.

The second group of vascular plants are the *pteridophytes* or *pterophytes*—the ferns—which are non-seed-bearing plants that tend to have large, compound leaves, the undersides of which bear sporangia. Ferns were abundant in Carboniferous times and are well represented as fossils. Living ferns tend to be most abundant in temperate and tropical areas. In coal forests, they formed the undergrowth below the canopy.

Some authors include the horsetails with the ferns, but we will treat them as a separate phylum.

The arthrophytes or sphenopsids include the horsetails and their relatives. Living horsetails grow only a few inches in height, but their forebears (*Calamites)* grew to a height of 80 feet or more and were conspicuous members of the Carboniferous forests. They grew in imposing stands and bore long graceful leaves (known as *Annularia*) that were unlike the stubbier leaves of living forms.

Sphenophyllum, a slender shrub or climbing plant, was probably a relative of the arthrophytes.

THE SEED-BEARING PLANTS

All the plants we have so far considered were or are somewhat imperfectly adapted to life on the land, for the critical stages of their reproduction are dependent on the presence of water. The sperms are minute flagellated cells that must swim through a liquid in order to fertilize the eggs. Because of this dependence on water, many ferns and all the early plants were restricted to damp lowlands. The seed ferns (*Pteridospermophyta)*, however, and all those that follow are true seed-bearing plants. In them the male microspore is represented by pollen grains that germinate to produce sperm, and the female megaspore is retained and germinates to produce an ovule fertilized within the protective and nutritive structures that together are known as seeds. This gives the newly fertilized "plants" a much better chance of survival than that afforded by the spore-bearing plants by freeing the plants from the necessity of the moist environment needed for sperm to reach ovules by swimming in free-living gametophytes, thus enabling the plants to survive even in arid conditions. Seed plants are thus not only freed from the necessity of having a film of water available to allow successful fertilization, but they also "escape" from temporarily adverse environmental conditions because germination of the new plant, which occurs immediately in spore-bearing plants, is deferred until the levels of moisture needed for successful growth are available in the environment. The seed also allows greater distribution of plants because of the protection and delay in germination afforded the embryo. The seed, then, provided a quantum leap for land plants, separating them from their dependence on a restrictive uniformly moist environment, allowing them to invade new environments and to evolve within these environments as they encountered new and different selective forces. The development of seeds was clearly a major step in the adaptation of plants to life on the land, although in the earliest plants of this kind the seeds were not completely enclosed and protected, as they are in modern flowering plants.

Pteridosperms or seed-ferns were a heterogeneous group of gymnosperms that reproduced by seeds, rather than the spores of typical ferns. Some regard them as ancestral to flowering plants. Now extinct, they were common in the coal-forming swamps, forming low-growing plants whose leaves *Neuropteris* and *Alethopteris* are common fossils.

The other major group of coal-forming plants was the Cordaitales. The cordaitaleans, now extinct, were probably ancestral to the conifers, growing to a height of 100 feet or so. *Cordaites* had long, straplike leaves with dichotomous although apparently parallel veining and have been found in a variety of environments, ranging from coal swamps to upland areas. The seeds of *Cordaites* grew up to half an inch in size and have been given the name *Cordiacarpus*.

The conifers (Coniferophyta) include the great variety of living and fossil conifers. Such trees as the pine, spruce, larch, cedar, monkey puzzle (*Araucaria*), sequoia, and cypress are familiar examples. Most members of the group have an evergreen foliage consisting of needle- or straplike leaves, and the naked seeds are generally borne in cones. Many of these forms have a long fossil history.

By Mid-Permian times—some 270–260 million years ago—the giant lycopods of the coal swamps were almost completely replaced by conifers, which seem to have been better suited to drier climatic conditions. Ginkgoes and cycads also emerged during Permian times.

We have already seen that, in Devonian times, plants were worldwide in their distribution and exhibited no obvious regional variation. This implies an absence of sharply zoned climates, with the present polar and tropical extremes subdued. By Late Carboniferous times, however, things were quite different. The great coal forest trees we have described were characteristic of western Europe and eastern North America (the Euroamerican flora), but their eastward equivalent in Southeast Asia (and possibly western North America) was represented by a rather different flora, the Cathaysian, dominated by the plant *Gigantopteris,* which occupied a tropical area before extending north in Mesozoic times. The Euroamerican flora developed in what were then tropical regions, but extended north and south into drier climatic zones.

To the north and south of this central floral belt, two quite distinct floral zones were present—the Angaran in northern Asia and northeastern Russia, and the Gondwanan, characterized by *Glossopteris*, in the Southern Hemisphere. The broad similarities between these two floras are probably the result of similar ecological adaptation, rather than genetic affinity, and the form of their plants suggests growth in temperate conditions, the growth rings implying seasonal changes. This is in striking contrast to the more uniform warm or subtropical forests of the central European-American-Cathaysian zones. Indeed, the southern Gondwana province showed much less diversity than the northern Angaran flora, and this lack of variety is also shared by its insect fauna. This lack of diversity may well be a reflection of the harsh glacial conditions that had already begun in the southern parts of Gondwanaland.

Such were the mighty forests through which the sunlight of distant Carboniferous skies filtered. They were sultry perpetual swamps, with gigantic insects creeping through the tangled undergrowth and filling the heavy air. They were not to survive for long, but before we follow their decline, we must examine one final aspect of life on the land—and to trace its origins we return for a while to the water.

8

THE AMPHIBIAN FOOTHOLD

LIFE ON THE LAND

The vertebrate "conquest of the land," as some authors have described it, may have been less a triumphant expansion than an ecological necessity or a dietary opportunity. Both points of view have been suggested. It was once argued, for example, that the drying up of ponds and streams during seasonal drought may have encouraged fish to invade the land to exploit its opportunities. Others have suggested that food supplies on land provided an attractive supplement, or that isolated marshy pools provided protein for newly laid eggs or larvae, although this involved crossing land to lay the eggs in the first place. One such explanation does not, of course, exclude another.

What is clear, however, is that plants, vertebrates, and arthropods all spread onto the land in late Silurian and Devonian times (425–360 million years ago), and, in each case, the move both followed and involved major adaptations. All three groups, for example, had to compensate for the lack of support that had sustained them in the buoyancy provided by water; all faced the hazards of desiccation, great temperature variation, and new sources of food and the problems of aerial respiration.

Exactly when plants first established themselves in coastal or estuarine environments is still unknown, but it seems probable that they provided food for the early arthropods, whose footprints have been preserved in rocks of Late Silurian age. Fossil arthropods (mites, scorpions, springtails, and spiders) are known from Early Devonian rocks. It was against this background that the early tetrapods emerged in Late Devonian times. And that emergence may at first have involved little more than fishlike splashing in the mud at the water's edge, hauling themselves up with their strong pectoral "limbs," much as crocodiles do today.

But exactly how and when did tetrapods emerge?

CRAWLING ASHORE

In the summer of 1933, a young Swedish geologist, Günar Säve-Söderbergh, worked his way across the ice-bound mountains of Celsius Bjerk in East Greenland, climbing slowly across a steep cliff face. He was inspecting the dark red rocks, bed by bed, layer by layer, searching for fishes—fossil fishes—that had lived there in Late Devonian times, about 360 million years ago. He was also searching for fossil amphibia, the oldest of which he had discovered and described (*Ichthyostega*) in 1932 (fig. 8.1). The stratified rocks he studied were formed from the sand and silt that had accumulated in an ancient river system once covering what was then an ice-free tropical or subtropical region. In fact the region had then supported a lush cover of lowly vegetation—mostly creeping, low-growing club mosses and other simple plants. At that distant time, land plants, even simple, low-growing ones, were still a novelty on the planet: there were no forests, no trees, no grasses, no flowering plants. Even the amphibians and fish for which Säve-Söderbergh was searching were still relative newcomers in those ancient days. They were the only vertebrates in a world of invertebrates—scorpions, insects, and snails—slowly making a laborious transition from the turbulent stability of the waters of the oceans, to the turbulent instability of the waters and shores of the land.

Säve-Söderbergh died of tuberculosis in 1948 at the age of only thirty-eight, and the task of completing the fossil descriptions was undertaken by Erik Jarvik, who had been a student member of the Greenland expedition. The fossils for which both Säve-Söderbergh and Jarvik hunted were scarcely spectacular: small fishlike creatures entombed and thus preserved in what were once the soft muds and sands at the bottom of the water that had once been their home. But the other vertebrates for which they searched were altogether more substantial. Up to 3 feet in length, they were fishlike creatures but had four short, stubby legs and feet: they were part fish—part amphibian, tied to the water but capable of clambering onto the land. These specimens—creatures of lazy river meanders and warm swamplands, appearing in the rocks of what are now the frigid Arctic wastes—included a second genus, *Acanthostega*, described by Jarvik in 1952, which created a sensation when descriptions were published, for the fossil remains were widely recognized as typifying the ancestors of all the later legions of air-breathing amphibia, reptiles, and mammals. They were not fully described until 1996, by which time Jarvik was eighty-nine years old.

The more recent discovery by Neil Shubin, Ted Deschler, and others of another fossil, the 6-foot-long *Tiktaalik* (see figs. 8.2, 8.4) from 375-million-year-old rocks from Ellesmere Island in the Canadian Arctic, about 600 miles south of the North Pole, suggests it may well have been a truly intermediate form between fish and amphibians. *Tiktaalik*, though still clearly a lobe-finned fish, is so transitional in form that it blurs all the conventional distinctions between fish and tetrapods (four-footed vertebrates). It has the flattened, triangular skull, with the close-set eyes on top, characteristic of amphibians, for example, but it also has fishlike fins and scales. It has amphibian-type ribs strong enough to support it on land, but its fins have bony components that resemble our own bones. It seems probable that, in view of the structure of its pectoral fins, it used fins both to paddle and to raise itself up, and perhaps to scramble ashore, though not, necessarily, to "walk." Specimens of this genus are well preserved and represent a variety of growth stages. The fossils are found in association with several fossil fishes.

Another possible transitional form was *Panderichthys*—a large predatory fish whose stout, paddlelike pectoral fins were more strongly developed than its pelvic fins (see fig. 6.8B). It had a spearlike tail and no dorsal fins. It seems to have lived in rivers and shallow coastal swamps and may have been able to scramble on to land by dragging itself by its pectoral fins, much as modern catfish do. Jennifer Clack has suggested that this sunbathing may have assisted in raising its body temperature.

But back to those remarkable fossils—*Ichthyostega* and *Acanthostega*. Where do they fit into this picture? They are not themselves our earliest ancestors: they are too late in time to be that. But they must be broadly similar to what our distant ancestors looked like. They were a jumble of fish and amphibian characteristics, with fishlike bodies, but well-formed walking limbs; skulls that were fish-patterned, but had amphibian-like proportions and eye placings; fishlike tails, but distinct limbs and fingers; part-fish, part-amphibian; part-aquatic, part-terrestrial. This intricate mixture of ancestral-descendant features has been described as "mosaic evolution." And there were significant differences between the two forms found by Säve-Söderbergh and Jarvik. Jenny Clack and Michael Coates redescribed the original fossil material and also described additional material in 1987 and showed that

Figure 8.1. *Ichthyostega.*

(Illustration © Stephanie E. Pierce. Reprinted with permission of the artist. Adapted from S. E. Pierce, J. A. Clack, and J. R. Hutchinson, "Three-Dimensional Limb Joint Mobility in the Early Tetrapod *Ichthyostega*," *Nature* 486 [2012]: 523–26.)

Acanthostega (fig. 8.3) was a fish with legs. It retained its internal fish-like gills and had such a weakly constructed skeleton that it probably never left the water—in spite of its having four limbs (with eight digits on the forelimbs and seven on the hind limbs) and a strong pelvic girdle. *Ichthyostega,* in contrast, seems to have had sufficiently strong forelimbs to lift the front part of its body. It appears to have been "more terrestrial" than *Acanthostega,* but still capable only of crawling or dragging across land surfaces.

One puzzling feature in the search for early tetrapods is the existence of 395-million-year-old Middle Devonian fossil tracks in rocks from the Holy Cross Mountains of Poland, made by four-legged vertebrates. These are the oldest known footprints; they are significantly older than the oldest known fossil tetrapods and also older than *Tiktaalik* and *Panderichthys.* They were made in muddy salt tidal flats by tetrapods estimated to have been about 8 feet long. This does not necessarily mean that these two genera were not somewhere on the ancestral line; it may imply, rather, that they were long-lasting survivors of that ancestral stock.

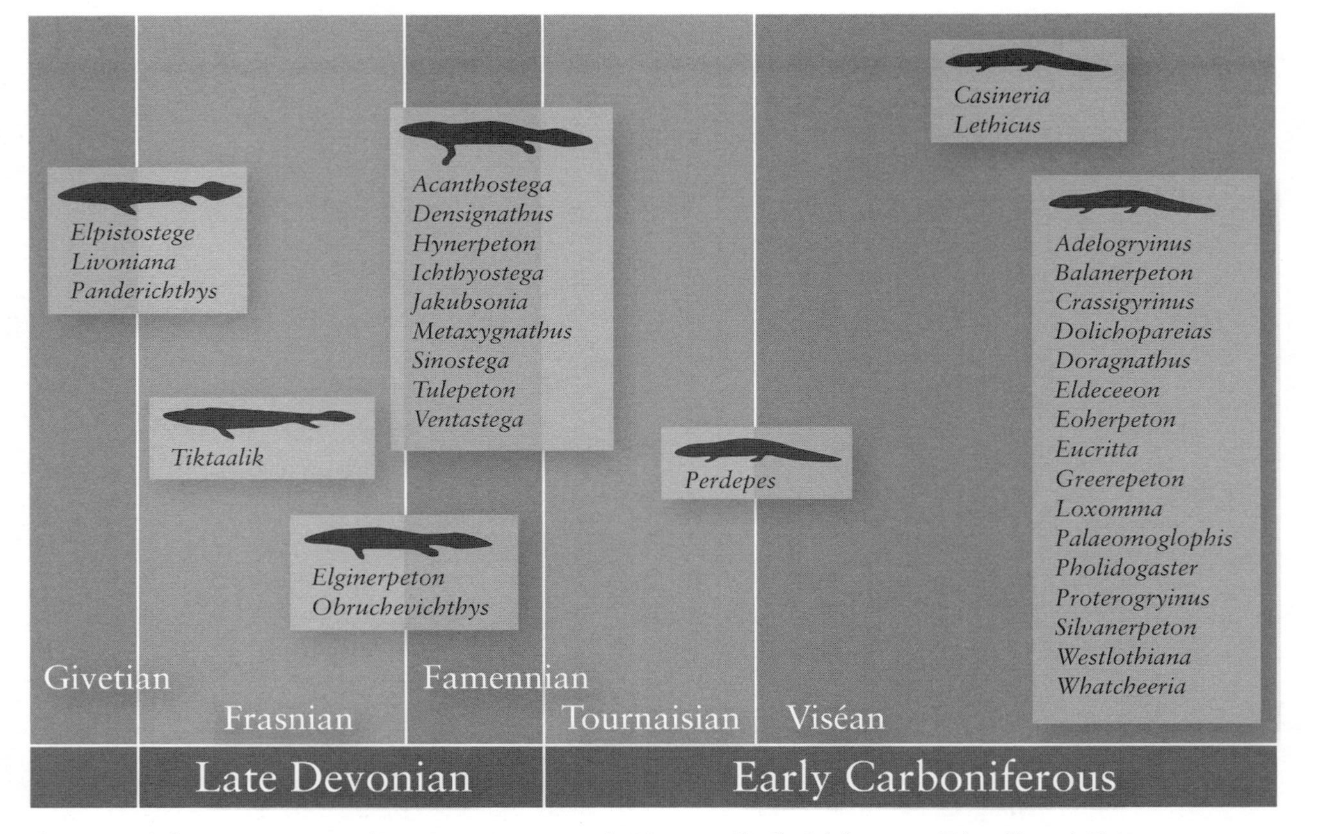

Figure 8.2. A diverse assortment of Late Devonian tetrapods. Note specifically *Ichthyostega*, *Tiktaalik*, and *Elginerpeton*. (Illustration by Dennis C. Murphy, www.devoniantimes.org. Used with permission.)

THE "FISHAPODS"

The oldest known tetrapod is *Elginerpeton* (see fig. 8.4), from 368-million-year-old Late Devonian rocks of Scotland. Though still incompletely known, it has proved to be a treasure. When first found, it was classified as an unidentified fish, but it is so transitional in its characteristics that it shares a rich jumble of distinctive features of both fish and amphibians and is now regarded as an early tetrapod. *Tulerpeton* (fig. 8.4), from estuarine rocks of the Tula area of the USSR, shares a comparable mosaic of fish-amphibian features.

This remarkable array of new "fishapod" discoveries over the last decade includes three genera of transitional fishes, as well as a dozen tetrapod genera that provide the broad outline of the fish-amphibian transition. This took place in Late Devonian times, some 385–359 million years ago, and is represented over an extensive geographic area of Europe and North America. One ironic feature is that most of these early tetrapods, for all their structural novelty, appear to have been fully aquatic in habit: "fish with legs," as one writer has described them. This is interesting because characteristics we associate with life on land, such as limbs and aerial respiration, evolved as adaptations to life as fishes in water.

Ichthyostega lived about 363 million years before humans first appeared on Earth, but it had one essential feature linking it with every member of the human race: it was an air breather, capable of living outside the water from which it emerged, but to which it returned to lay its eggs (see fig. 8.1).

Ever since the days of *Ichthyostega* and its various recently discovered relatives, Earth's lands have been inhabited by vertebrates. But animal life on the land was the direct result of the earlier growth and spread of lowly algae and other plants that transformed Earth's early atmosphere by photosynthesis, in which chlorophyll is transmuted into carbon and oxygen in the presence of sunlight. It was these early plants that made the lands habitable for animals.

The transition from life in the water to life on the land involved profound changes in all organisms—not only plants, but also animals—that moved ashore. The development and mutual interdependence of these profound changes within such different animal groups as vertebrates and arthropods is an extraordinary process. Remarkable as such change is, we can trace at least some of the exquisite transitional changes in fishes by which it came about, in part, it seems, as a response to life in growingly harsh freshwater environments, perhaps in areas of seasonal climate change, or perhaps as newly developing terrestrial opportunities with the added enhancements of oxygen and food.

Living amphibians are confined to moist and relatively warm environments; the fact that their earliest ancestors are found in Greenland reminds us of how little the climate and geography of the present resemble those of the past. In Devonian times, Greenland was situated much nearer the equator than it is today.

The early amphibians we have just described—the ichthyostegids and all the other broadly related forms—appeared in Late Devonian rocks deposited about 375–360 million years ago. A dozen different genera are now known from Scotland, Latvia, Russia, Australia, Brazil, the United States, Greenland, and China. Few, if any, of these seem to have been land dwellers, their strong limbs assisting an active carnivorous life in the water. Some were clearly freshwater dwellers, but others seem to have inhabited coastal wetlands and brackish deltas, all of which supported a lush growth of vegetation by Late Devonian times. They were, presumably, like their present-day descendants—the frogs, toads, and salamanders—confined to moist and warm environments, where their distinctive reproductive structure of fishlike eggs, external fertilization, and lack of protection against drying required them to be present in the water to breed. Unlike later tetrapods, the ichthyostegids had seven or eight toes on each limb, but in detailed skull pattern and fundamental limb structure, they show many similarities to the crossopterygian fish from which they arose.

The pioneer work of the early Swedish workers, Günar Säve-Söderbergh and Erik Jarvik, done under conditions of great hardship, has been followed by more recent studies built on the foundation they provided. The person who has done much to further and integrate the study of these early amphibians is Jennifer Clack, professor and curator of vertebrate paleontology at Cambridge University. Jenny, as she prefers to be called, took her first degree at the University of Newcastle upon Tyne, where she developed an interest in vertebrate paleontology. She spent several years working as a technician and later in education in museums in central England before returning to Newcastle as a PhD student. She later became an assistant curator at the Museum of Zoology at Cambridge. It was there that she came across the fossils of *Acanthostega*, which, as we have seen (see fig. 8.3), was a Late Devonian tetrapod, with eight fingers and toes, transitional in form between fishes and tetrapods.

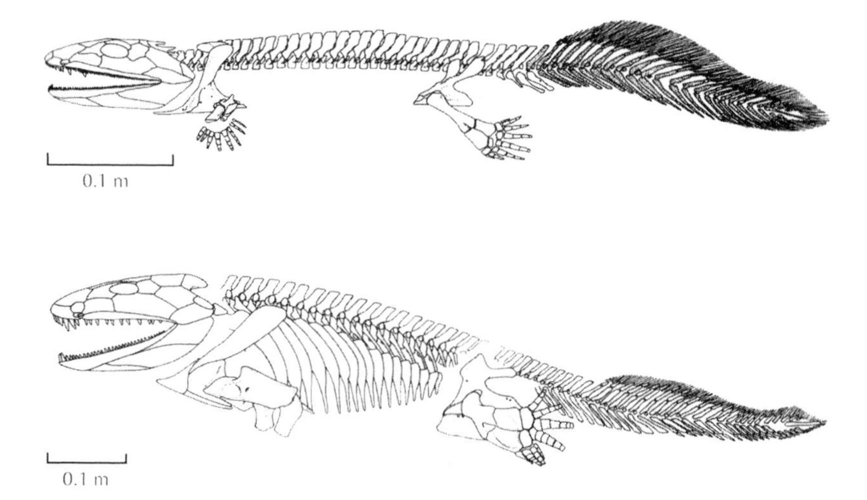

Figure 8.3. Sketch of the skeletons of *Acanthostega* (top) and *Ichthyostega* (bottom), showing the mixture of fishlike features (tail fins, lateral line systems, gill slits) and tetrapod features (robust limbs and shoulder and hip bones, reduced back of skull, expansion of snout).

(Drawing courtesy of M. Coates. From Donald R. Prothero, *Evolution: What the Fossils Say and Why It Matters* [New York: Columbia University Press, 2007], fig. 10.8, p. 227. Copyright Columbia University Press. Reprinted with permission of the publisher.)

Jenny Clack led field expeditions to East Greenland in 1987 and 1998 that have provided critical fossils, as well as leading a series of collecting field trips in the UK that have yielded an enticing collection of early tetrapods (Clack, 2012).

THE AMPHIBIAN HEYDAY

Living amphibians are typically small animals, tending to be soft-bodied and relatively inconspicuous in appearance. In contrast, a great variety of tetrapods appeared in Early Carboniferous times (about 359–201 million years ago) and survived into Triassic times that were anything but small and inconspicuous. Among the most distinctive groups were the Temnospondyls or labyrinthodonts, a name based on the infolded wall structure of their hollow teeth: yet another feature they share with the crossopterygian fish. This group also included some small terrestrial creatures, but the most distinctive members were

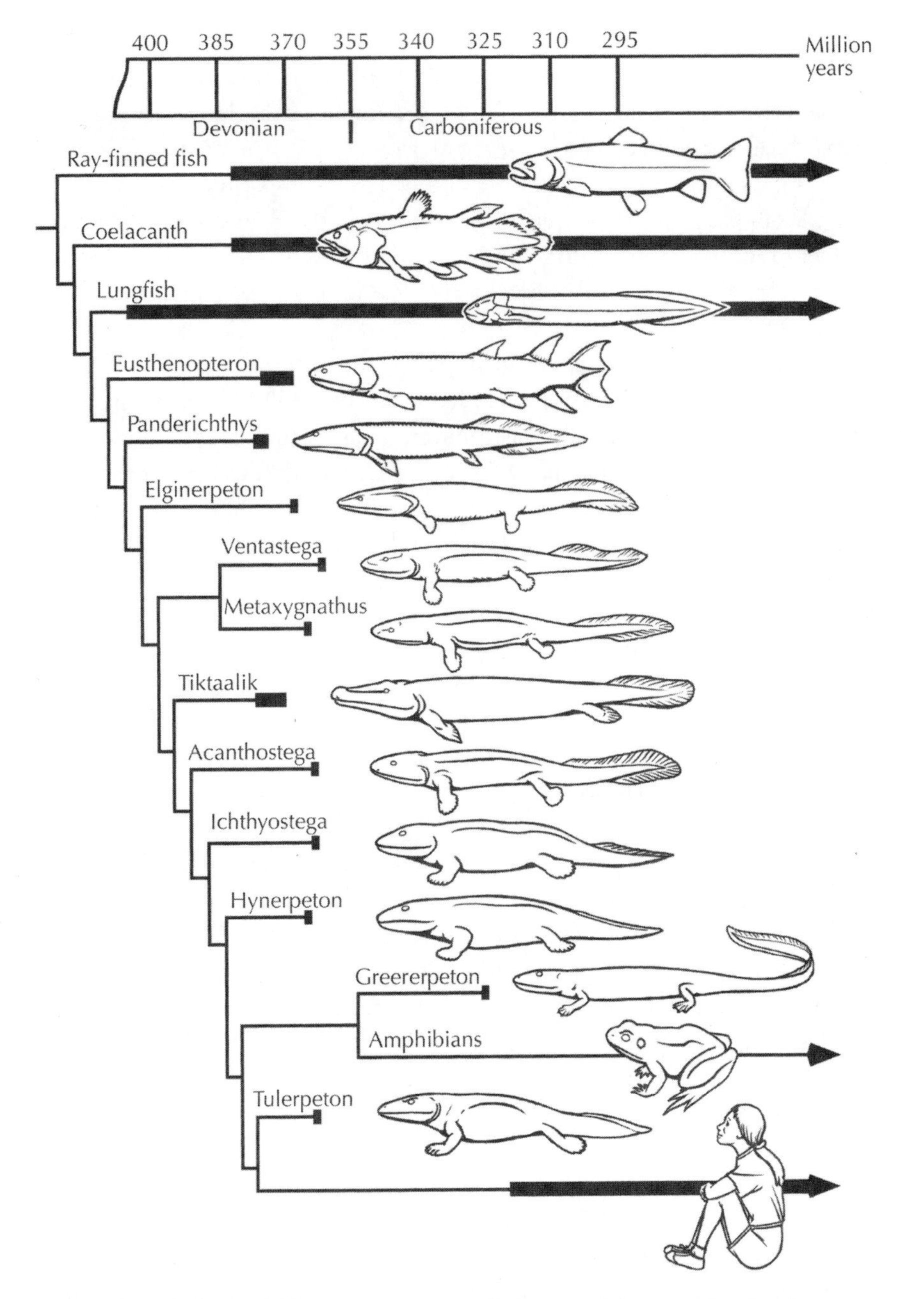

Figure 8.4. Early amphibian representatives. Phylogeny of the transitional series from "rhipidistians" through primitive tetrapods.

(Drawing by Carl Buell. From Donald R. Prothero, *Evolution: What the Fossils Say and Why It Matters* [New York: Columbia University Press, 2007], fig. 10.6, p. 226. Reproduced with permission of the publisher.)

large, sprawling animals, such as *Eryops* (fig. 8.5A), from the Permian of Texas, which had a massive, flattened skull and reached 6 feet in length. More than 170 different genera of these labyrinthodonts have been recognized, most of them from the Permian and Triassic, although some persisted until the end of Cretaceous times. *Seymouria* (fig. 8.5B–C), another powerful, sprawling carnivorous creature, about 2 feet long, from the Permian of Texas and Europe, is such a mixture of characteristics that it is difficult to decide whether it was an amphibian or a reptile.

Lepospondyls were a diversified group, characterized by their spool-shaped vertebrae. The group included small lizardlike forms, elongated snakelike creatures, and the nectridians, such as the bizarre *Diplocaulus*, an elongated aquatic form from the Carboniferous to Late Permian, reaching more than 3 feet in length. Its boomerang-shaped head and slender body with weakly developed limbs gave it a distinctive kite-like appearance, and its structure makes it clear that it was essentially aquatic, rather than terrestrial, in habit.

Microsaurs ("small lizards") were a varied group of short-tailed, long-legged creatures about whose affinities there is little agreement.

LIVING AMPHIBIANS

It was from such distant forms that living amphibians—salamanders, newts, frogs, toads, and caecelians—originated about 200–150 million years ago. They tend to live in damp temperate or tropical areas, their active lives (reproduction, growth, and feeding) being confined to the warmer season, with some hibernating in the cooler areas.

Their eggs, laid in the water, have a relatively small food supply and rapidly hatch into larvae or tadpoles. The remarkable metamorphosis of these large-headed, long-tailed creatures with external gills into four-legged, air-breathing adults is one of the most wondrous transformations in the world of living things. It also emphasizes both the biological unity and the limited terrestrial adaptation of the amphibians. Pioneers though they were in clambering ashore, their reproduction tied them to the water. And, dominant though the labyrinthodonts were in Late Paleozoic times, their range of adaptability to life on the land was limited by their reproductive dependence on the water.

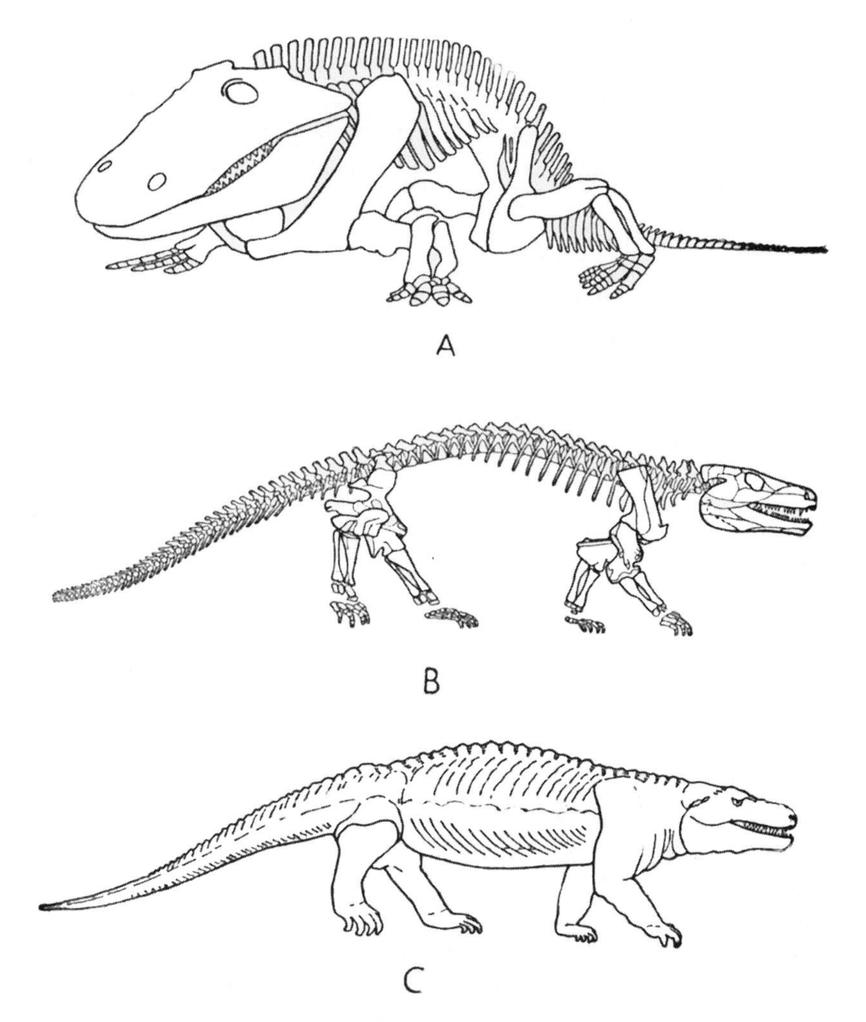

Figure 8.5. The relationships of the four-legged animals, or tetrapods. (A) *Eryops*, a Permian labyrinthodont amphibian, about 5 feet in length. (B–C) *Seymouria*, a Permian "progressive amphibian or primitive reptile" about 2.5 feet in length: (B) Skeleton; (C) Restoration.

([A] after Colbert; [B] after T. E. White; [C] after Gregory. From Frank Rhodes, *Evolution of Life* [New York: Penguin, 1976], fig. 29, p. 183. Reproduced with permission.)

Because they were tied to the water to reproduce, they were—as their name implies—amphibian. While some developed into formidable terrestrial carnivores, others—seemingly perversely—returned to life in the water from which their ancestors had slowly and recently emerged.

Fossils of frogs, toads, and other amphibians are rare, largely, one supposes, because of their fragile skeletal structures. The oldest are found in rocks of Early Triassic age from Madagascar (252–247 million years old); the oldest salamanders in rocks of Late Jurassic age (164–145 million years old). Living amphibians are represented by some 4,000 different species.

9

THE REIGN OF THE REPTILES

WHAT IS A REPTILE?

We all know, of course. Or rather, we know one when we see one: scaly, slithering, snapping, sliding, cold-blooded, crawling creatures, and all the other uncomplimentary adjectives sometimes applied to them. Even the adjective *reptilian*, applied to human behavior, has the implication of a rather shady character. But reptiles are wonderful creatures, the first major group to become adapted to life on the land. Their dry, scaly skin protects them against abrasion and loss of moisture; almost all of them reproduce by laying eggs on land, from which the young hatch as adults; they breathe by means of lungs, and they are cold-blooded; that is, they do not generate their own heat. And, though that cold-bloodedness tends to concentrate their distribution in temperate and tropical regions, it also means that they do not need to burn calories to regulate their body temperature.

There are some 8,000 living species of reptiles, descendants of a race that once ruled the Earth—land, sky, and seas—and it is from them that we—the mammals—ultimately developed.

THE AMNIOTIC EGG

The secret of the reptiles' success has been the development of the amniotic egg. Just as the amphibians first established the transition from the water to the land, so also the development of the amniotic egg freed the reptiles from the water-dependency of their amphibian forebears. The reptilian egg is internally fertilized and then laid on the land, enclosed in a protective shell, where it remains until the animal hatches at an advanced stage of development. Within the egg, the growing embryo is nourished by a large yolk enclosed in a fluid-filled sac, and it is also attached to the allantois, which disposes of waste products from the embryo, via the protective but porous shell.

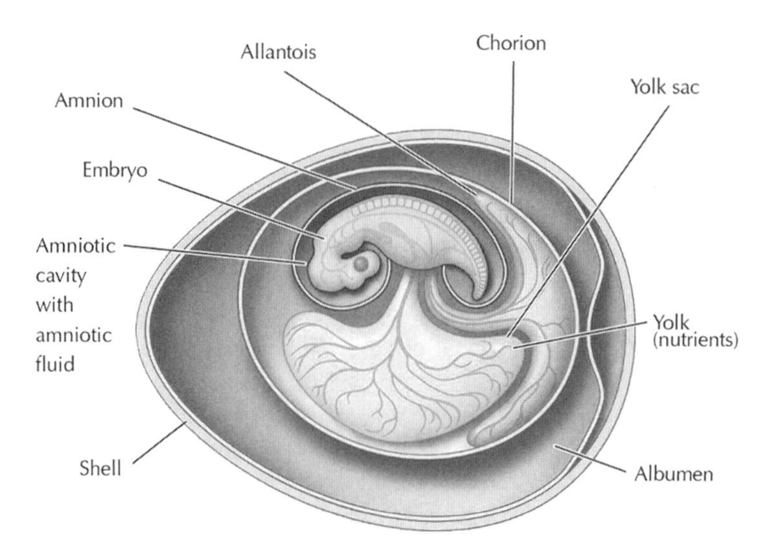

Figure 9.1. Structure of an amniote egg.

(From Neil A. Campbell, Jane B. Reece, Martha R. Taylor, Eric J. Simon, and Jean L. Dickey, *Biology: Concepts and Connections*, 6th ed. [New York, Pearson Education, 2009]. Reprinted by permission of Pearson Education, Inc., New York.)

This sheltered existence allows the reptilian young to be born at a far more advanced stage of development than the young of amphibians, so increasing the prospects for their survival. The oldest known fossil amniote egg comes from the Early Permian of Texas, but it is probable that the structure developed earlier than this. The conventional explanation for this amniote reproduction is that it provided reptiles with a means of exploiting the rich food resources of the land. Alfred Sherwood Romer argued, however, that—so meager were the sources of food on the land in Carboniferous times—it was more likely that the amniote egg developed as a defensive means of getting the eggs out of the hazardous waters and onto the comparative safety of land. Whichever explanation proves correct, the development of the amniote egg marked a major step in the evolution of the vertebrates. Without it, we should still be living gill-breathing, finned, aquatic lives.

ORIGIN AND EARLY EVOLUTION OF REPTILES

It seems likely that the reptiles arose from a large group of amphibians called anthracosaurs. These were sturdy, long-bodied, four-legged carnivores, with a superficial resemblance to living crocodiles. *Seymouria* (see fig. 8.5B–C) is a typical representative, and it is generally accepted that it was probably broadly similar in form to the ancestral reptiles. *Seymouria* itself is, however, of early Permian age and so is probably too late in time to be itself ancestral. Earlier potential ancestors come from Late Carboniferous (Pennsylvanian) siliceous deposits of East Kirkton, in the Midland Valley of Scotland, which have yielded the remains of three creatures: *Westlothiana*, *Cassineria*, and *Eucritta*. *Westlothiana*, affectionately known as "Lizzie," was only about 8 inches long, and it, as well as the associated but headless fossil *Cassineria*, could well be close to the ancestral reptilian stock, though this is not certain. Whether they laid amniotic eggs is also unclear. The small size of these reptile-like creatures may have favored them in their need for temperature regulation and food capture in the terrestrial environment to which they became adapted.

The remains of some other small early reptiles have also been found inside the hollow fossil tree stumps of a fossil forest found at Joggins, Nova Scotia.

These early reptiles and reptile-like forms seem to have been carnivores, and the land onto which they clambered must have offered at least two sources of food: one, of course, was other vertebrates—amphibians, other early reptiles, and perhaps occasional fish from ponds and lakes. The other potential source of food was insects, which exploded in variety and in size in Carboniferous times. Huge dragonflies had a wingspan of up to 2 feet. Some 150 genera of arthropods have been described from the famous Mazon Creek Formation of Illinois, including spiders and giant millipedes, reaching up to 6 feet in length. For the members of the reptilian vanguard, these insects must have offered a substantial food supply.

THE REPTILIAN EXPLOSION

The development of the amniote egg (see fig. 9.1) opened a wide new range of environments to the reptiles. The oldest known reptiles, as we have seen, appeared in Late Carboniferous (Pennsylvanian) times (about 323–299 million years ago) and the reptiles then exploded into a vast variety of forms in Permian and later Mesozoic times.

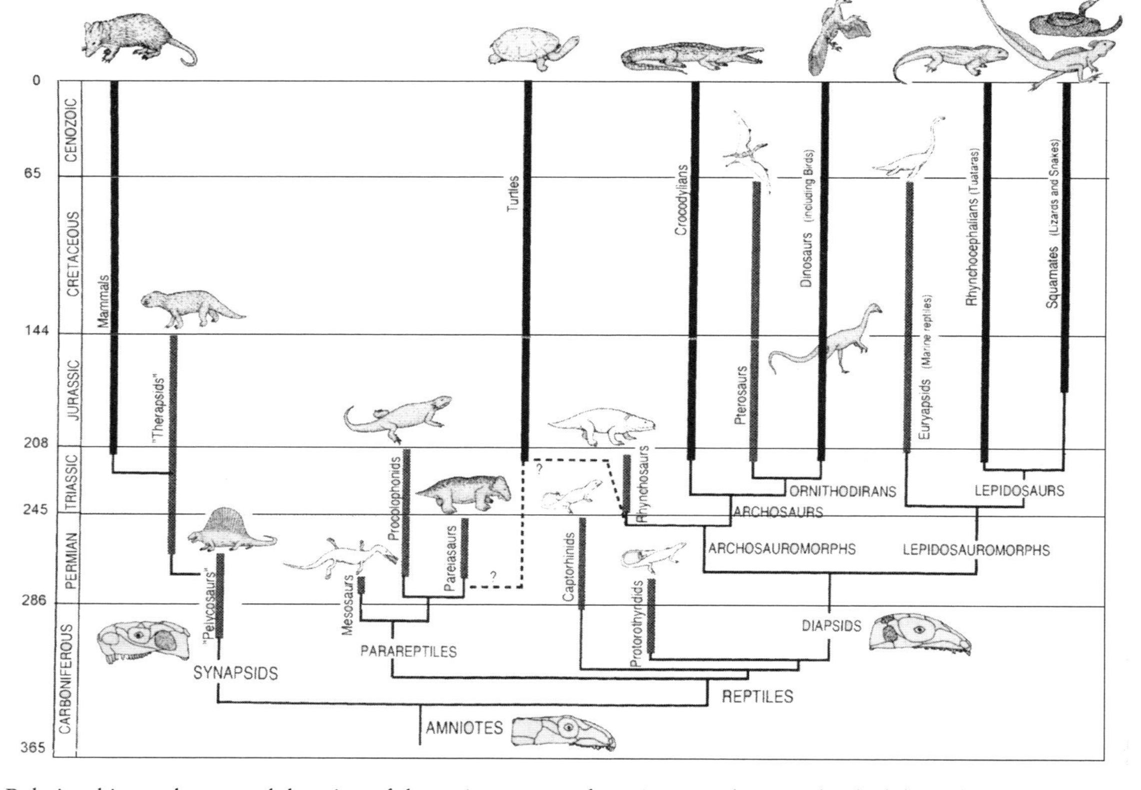

Figure 9.2. Relationships and temporal duration of the major groups of amniote vertebrates. The thick lines depict the known fossil duration of each group, excluding contentious finds (e.g., the Triassic "bird" *Protoavis* and the Cenozoic "therapsid" *Chronoperates*); dashed lines indicate uncertain relationships. Examples from each lineage are illustrated. The skull diagrams show the three major skull types found in amniotes; synapsid (found in synapsids), diapsid (found in diapsid reptiles), and anapsid (found in turtles, parareptiles, captorhinids, and protorothyridids). Note that the synapsid and diapsid skulls characterize discrete lineages but the anapsid skull does not.

(From Joel Cracraft and Michael J. Donoghue, *Assembling the Tree of Life* [New York: Oxford University Press, 2004], fig. 26.1, p. 452. Reproduced with permission of Oxford University Press, USA.)

Three major groups of reptiles have been established on the basis of their skull structure.

Anapsids have no opening in the skull behind the eye socket. Turtles and a small group of the earliest reptiles are included in this group.

Synapsids have one opening in the skull behind the eye socket. The pelycosaurs ("sail-backed lizards"), the therapsids, and the mammals, which later arose from them, are included in this group. We will review the mammals in chapters 12 and 13.

Diapsids have two openings in the skull behind the eye socket. This large group includes the archosaurs, from which several subsequent groups developed, including snakes, lizards, crocodiles, various sea-going reptiles (ichthyosaurs, plesiosaurs, and placodonts), the pterosaurs (flying reptiles), and the dinosaurs and their descendants, the birds.

These three skull types, all appearing early in reptile history, may perhaps be characterized as "evolutionary experiments" in ways to anchor a powerful jaw-closing musculature. Far from being the dry essentials of comparative anatomy, they are now coming to be understood in terms of the lives of animals involved and the history of the groups they represent. It is to the history of these groups of synapsid and diapsid reptiles, in all their wonderful variety, that we now turn.

THE RULING REPTILES

The Mesozoic—that great era of "Middle Life"—was above all else the Age of the Dinosaurs: the "kings" of the Ruling Reptiles. Dinosaurs appeared, as did the mammals, in Triassic times, about 230 million years ago, and for some 140 million years they dominated life on the land. Great as was the history of their contemporary reptilian relatives—lizards, turtles, crocodiles, snakes, and an armada of aquatic reptiles, as well as their reptilian descendants, the birds—it was the dinosaurs that dominated the life of the Mesozoic lands. There were two great groups of dinosaurs: the Ornithischians and the Saurischians.

Ornithischians (order Ornithischia) were herbivorous dinosaurs; their name is based on the "birdlike" structures of their hip bones (see fig. 9.3). Birds themselves, however, were not Ornithischians and their similarity of hip structure is the result of parallel but independent evolution. Birds are now regarded as theropod dinosaurs, and there are striking resemblances between such dinosaurs as *Struthiomimus* and modern ostriches,

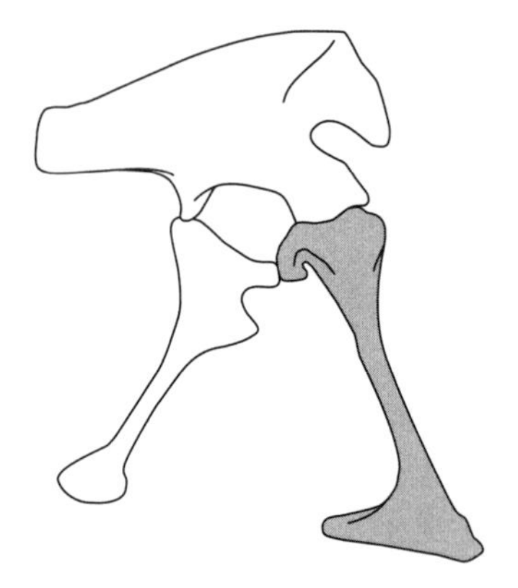

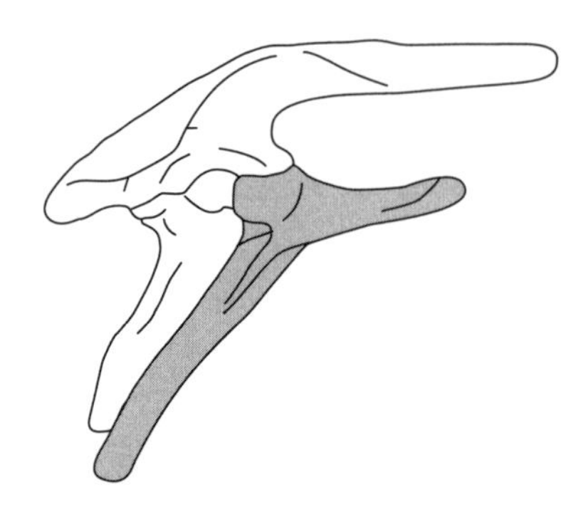

Figure 9.3. Ornithischian and saurischian hip structure. The pelvis in the two dinosaurian orders. *Left*: The saurischian pelvis of *Allosaurus*, with pubis directed forward; *right*: the ornithischian pelvis of *Stegasaurus*, with the pubis parallel to the ischium. Both animals are facing to the right.

(From Edwin H. Colbert, *Colbert's Evolution of the Vertebrates* [New York: Wiley-Liss, 2001], fig. 14-1, p. 188. Reproduced with permission of the publisher. Permission conveyed through Copyright Clearance Center, Inc.)

for example, and between small carnivorous dinosaurs (carnosaurs) and large, carnivorous fossil birds such as the 6-foot *Diatryma* and the *Phororhacos*. It may be that these different hip structures were alternative outcomes in hip musculature in the transition to an erect posture. The lack of a clear-cut distinction between the two groups is seen in small, diminutive genera such as *Staurikosaurus* from the late Triassic of Brazil, some 225 million years ago, when "experiments" were still in progress.

They included an extraordinary variety of forms, including the armored dinosaurs (stegosaurs and ankylosaurs), the ornithopods (iguanodonts and the duck-billed hadrosaurs) and the horned dinosaurs (the ceratopsians).

Saurischia, the other great order of dinosaurs (fig. 9.4), had a "lizardlike" pelvic bone arrangement, with a pubis that pointed forward, away from the ischium. The structure of the hip, though not the only diagnostic difference between the two great divisions of the dinosaurs, was fundamental to their posture and thus their overall structure and lifestyle. The saurischians included two great groups.

The theropods were bipedal carnivores that included not only the oldest known dinosaurs (the Late Triassic *Eoraptor* and *Coelophysis)*, but also the great bipedal carnivores (allosaurs and tyrannosaurs) of Jurassic and Cretaceous times.

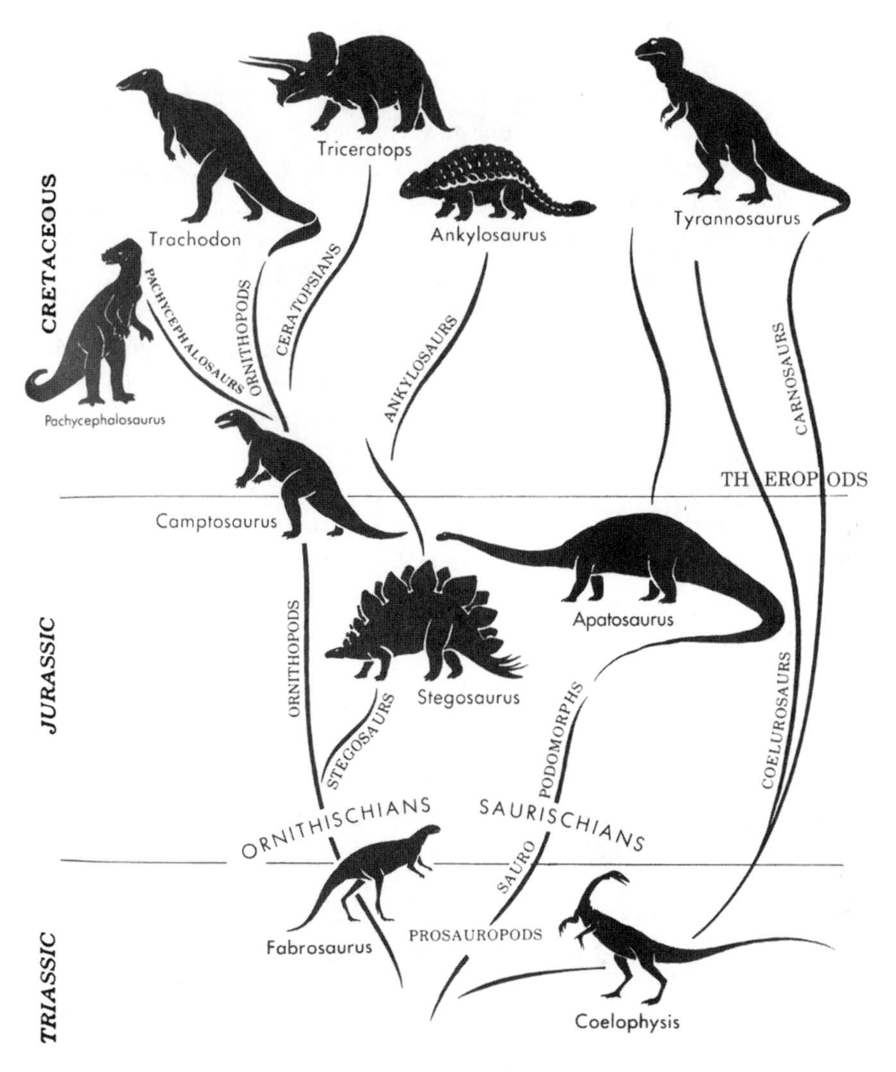

Figure 9.4. Family tree of dinosaurs and related forms.

(Prepared by Lois M. Darling. From Edwin H. Colbert, *Colbert's Evolution of the Vertebrates* [New York: Wiley-Liss, 2001], fig. 14-3, p. 191. Reproduced with permission of the publisher. Permission conveyed through Copyright Clearance Center, Inc.)

Sauropods, the other group of saurischian dinosaurs, were mostly large, herbivorous quadrupeds that flourished from late Triassic until the end of Cretaceous times.

Let us start our review of the dinosaurs by looking at one or two representative saurischians, which happen to represent the earliest known dinosaurs.

The best known of the early carnivores comes from Late Triassic rocks (about 237–201 million years old) of New Mexico, where a rich assortment of fossils have been preserved, including scores of specimens of *Coelophysis,* a delicately built but powerful bipedal carnivore, some 8–9 feet long, with a long tail, small forelimbs with sharp claws, and an elongate skull with sharp serrated teeth. A *Coelophysis* skull, incidentally, was the second of only two dinosaur fossils to have been carried into space, where it was briefly taken onto a space station.

Broadly similar, small, bipedal carnivorous dinosaurs of this general size and type persisted until late Cretaceous times. Though showing variation in detailed structure and size that has led to their recognition as distinct genera, they maintained an overall similarity in general form and structure to ancestral forms, such as *Coelophysis.*

A related group of theropods—the carnosaurs—were, however, far more substantial creatures and included the largest known terrestrial carnivores. *Allosaurus,* from the Late Jurassic (about 155–50 million years ago) and *Tyrannosaurus* from the Cretaceous (about 166–45 million years ago) were huge creatures, reaching up to 20 feet in height and 40 feet in length, having reduced forelimbs—but enormously powerful hind limbs and huge heads with sharp, daggerlike teeth. This group also included *Deinonychus,* a slender creature about 11 feet in length, that had a greatly enlarged, talonlike claw on its hind feet and extended forelimbs with long claws. The long, slender tail was reinforced and stiffened by elongated bony extensions of the distal vertebrae. *Velociraptor* was also a typical member of this group. One other remarkable feature of these reptiles is that some of them developed feathers, and they are now widely regarded as being ancestral to the birds. Few ancestral linkages are more surprising. In spite of their "lizardlike" hips, it was these carnivorous saurischians—not the ornithischians—that gave rise to the birds. It has also been suggested that they may have been warm-blooded.

But the theropods make up only one major division of the saurischian dinosaurs; the other, no less remarkable, was the giant sauropods, the largest of all dinosaurs. Every young admirer will recognize their names: *Apatosaurus* ("*Brontosaurus*"), *Diplodocus, Brachiosaurus,* and others. Huge creatures of this general type existed through Jurassic and Cretaceous times. Some of them were the largest land animals ever to have lived, reaching up to 150 feet or so in length and weighing, perhaps, some 80 tons. There is perennial competition to discover the latest, biggest, heaviest, or longest dinosaur. One current contender is *Argentinosaurus,*

Figure 9.5. Representative saurischian theropods: (A) *Coelophysis;* (B) *Allosaurus;* (C) *Tyrannosaurus;* (D) *Velociraptor;* (E) *Deinonychus.*

(Illustrations by Joe Tucciarone. Reproduced with permission.)

Figure 9.5. Continued

about 120 feet long (the length of three school buses) and 70 feet high (about the height of a six-story building). These herbivorous quadrupeds had long necks, small heads with peglike teeth, massive pillarlike limbs, and long tails. Although it was long assumed that such gigantic creatures must have been aquatic in habit, recent studies indicate that they were more likely terrestrial, though probably preferring swamps and coastal areas. In fact, there is evidence from fossil tracks that they could move through the shallow water of these areas, propelled chiefly by the front feet. Fossil footprints from rocks of Jurassic and Cretaceous age in Texas, Montana, and elsewhere suggest they might have traveled in herds and that at least some of these herds consisted only of juveniles segregated by age.

One lingering but still unanswered question is why these animals should have developed such extraordinary size. What possible benefit could have overcome the immense labor of dragging an 80-plus-ton body around, especially on the land? The answer may well lie in the need to digest the bulk and quantity of vegetation needed to survive. In living herbivores, large size—and thus larger digestive systems—facilitates the process of digestion; thus larger size may possibly have been of great value to creatures living in semiarid areas.

The other great tribe of dinosaurs—the ornithischians—were all herbivores, and the number, form, and placement of their teeth reflect their dietary habits. They blossomed into a wide array of both bipedal and quadrupedal forms in Jurassic and Cretaceous times. One of the major divisions—the ornithopods—the "duck-billed dinosaurs"—were bipedal, semiaquatic in habit, with the jaws developed into a flat, ducklike bill. Appearing in the Late Triassic, they flourished in Jurassic and Cretaceous times.

The Jurassic stegosaurs were heavy, armored, herbivorous quadrupeds. In *Stegosaurus*, two alternating rows of bony, triangular plates extended along the back, culminating in four spikes on the tail. The function of these striking plates is unclear; temperature regulation, protection, and sexual display have all been suggested. Stegosaurs reached a length of 20 feet or so and, though dominantly Jurassic, also survived into the Early Cretaceous. The standard reconstruction of *Stegosaurus* with erect plates is probably inaccurate. It seems more likely that the "armored" plates lay parallel to the body surface, giving the animals a broad resemblance to ankylosaurs (see below).

Another related group of armored ornithopods were the tanklike ankylosaurs, which lived from mid-Jurassic to Cretaceous times. Squat in form and reaching 20 feet in length, the whole body, including the head, was covered by bony scales, while the sides were protected by spikes. The tail terminated in a heavy, macelike, bony mass that must have provided a powerful offensive weapon. The lumbering form of the ankylosaurs has a broad resemblance to the quite unrelated and much later but recently extinct giant South American armadillos.

The remaining group of ornithischians were the ceratopsians, a truly extraordinary group of rhinoceros-like, late Cretaceous horned

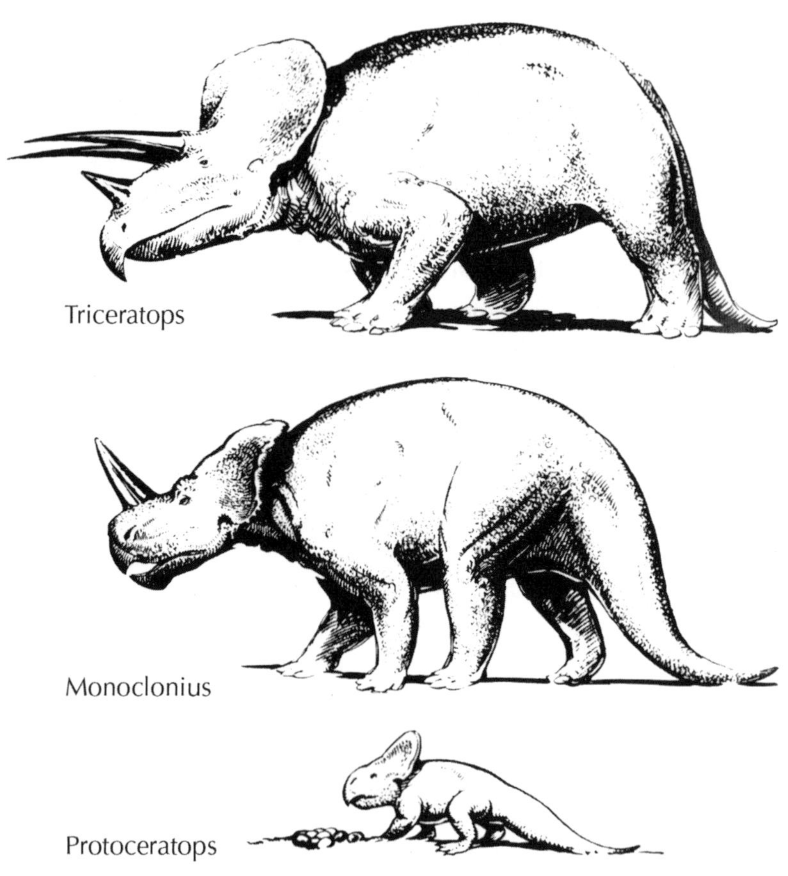

Figure 9.6. Evolution of ceratopsian reptiles. Three ceratopsians or horned dinosaurs, drawn to the same scale. *Protoceratops*, shown guarding its eggs, was about 9.4 feet long and one of the earliest and most primitive of ceratopsians. It occurs in the Bain Dzak (or Djadochta) formation of Mongolia. *Monoclonius*, about 24 feet long, is known from the Oldman formation of the Belly River series in North America. *Triceratops*, about 25 feet long, one of the last of the ceratopsians, occurs in the Lance formation of North America.

(Prepared by Lois M. Darling. From Edwin H. Colbert, *Colbert's Evolution of the Vertebrates* [New York: Wiley-Liss, 2001], fig. 14-13, p. 212. Reproduced with permission of the publisher. Permission conveyed through Copyright Clearance Center, Inc.)

dinosaurs. Their rise can probably be traced to *Yinglong* (from western China), a poorly known late Jurassic dinosaur from Xinjang, and *Protoceratops* from Mongolia, a creature some 10 feet long, with a large bony collared head shield, ending in a beaklike structure. Nests of eggs, some with preserved bones of hatchlings, have been found in association with,

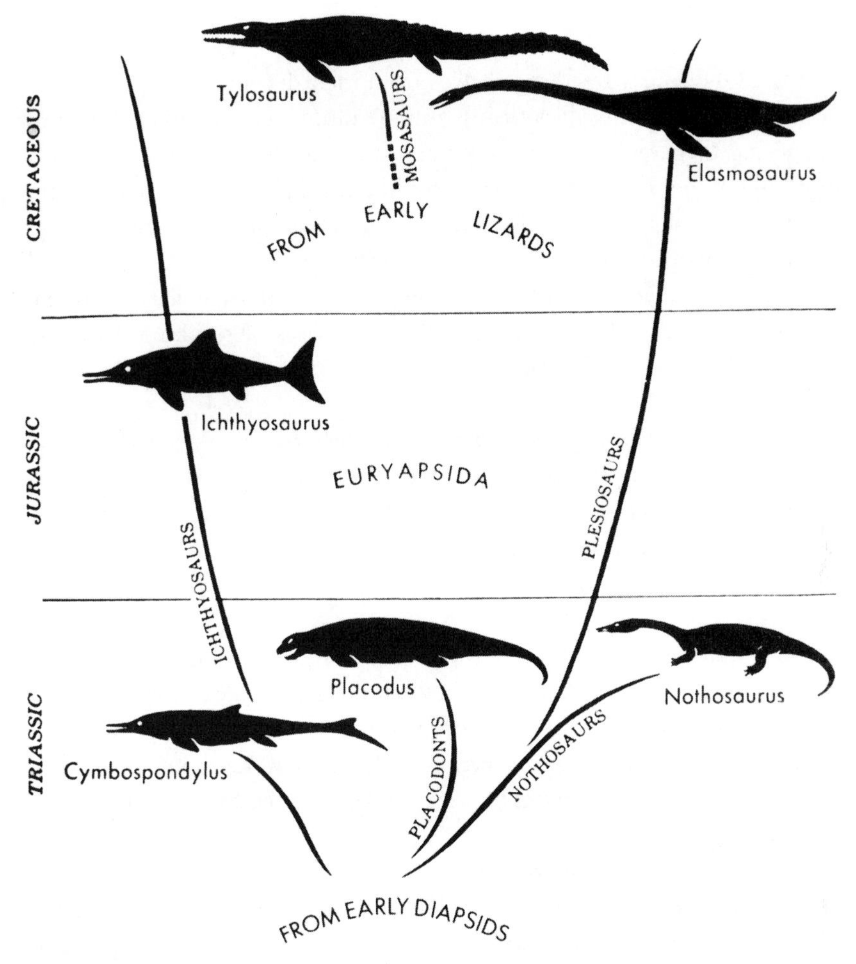

Figure 9.7. Aquatic reptiles. A family tree of Mesozoic marine reptiles.
(Modified after an earlier drawing prepared by Lois M. Darling. From Edwin H. Colbert, *Colbert's Evolution of the Vertebrates* [New York: Wiley-Liss, 2001], fig. 12-7, p. 169. Reproduced with permission of the publisher. Permission conveyed through Copyright Clearance Center, Inc.)

but not belonging to (see p. 14), this dinosaur. The geologically younger *Monoclonius* was larger, with a single horn developed on the head shield, while the well-known *Triceratops* had three such horns and a parrotlike, beaked nose. These are only three of more than thirty other named ceratopsian genera that show great variety in head shield form and armor. All had the rather massive build of those we have described and were vegetarians.

AQUATIC REPTILES

Having laboriously freed themselves from their dependence on the water and established themselves firmly on the land, several significant groups displayed their reptilian perversity by returning to an aquatic existence, not only in the freshwaters of the land but also in the oceans. This meant another series of wholesale changes, but the reptiles returned to an aquatic existence not as reconstituted amphibians, but as reengineered reptiles. This involved transformation, either by replacement—walking limbs became fins, lungs became gills as a means of respiration—or by modification: live birth was developed, with fewer but more fully developed offspring produced.

One of the earliest aquatic reptiles was *Mesosaurus*, a slender, long-tailed, long-necked, sharp-toothed, web-footed creature about 18 inches long from Early Permian rocks of both South Africa and Brazil. Its presence in these two now widely separated continents reflects their earlier proximity as part of the ancient supercontinent of Gondwanaland (see fig. 1.4).

Four other groups of marine reptiles emerged during early Triassic times. The placodonts (*Placodus*) had a heavyset body of walruslike proportions, but their skulls bore sharp incisors, with flattened, mollusk-crunching, platform teeth on their palates. They were about 5 feet long and some had armored plates on their backs. They became extinct in Late Triassic times (fig. 9.7).

The nothosaurs were streamlined reptiles, about 12 feet long, with long necks, long tails, long limbs, and squat triangular heads that carried sharp teeth. They were probably fish eaters but may have been capable of limited land life. They are confined to rocks of Triassic age.

The third group of marine reptiles—the plesiosaurs—though also first appearing in Triassic times, became worldwide in their distribution and flourished in the Jurassic and Cretaceous. Plesiosaurs resembled the nothosaurs in general form but grew up to 40 feet in length. Some plesiosaurs had long necks and strong paddlelike limbs, but others had short necks with much longer skulls.

Ichthyosaurs ("fish-lizards") became perhaps the most superbly adapted of all reptiles to an aquatic existence. Beautifully streamlined, similar in general appearance to both fish and dolphins, they were widely distributed in Triassic, Jurassic, and Cretaceous times, though their origin is unclear. With long, sharply toothed jaws, large eyes, long fins, and forked tails, they are known from fossils to have given birth to live young. The largest known ichthyosaur—the deep-bodied *Shonisaurus* from the Triassic of Nevada—was 50 feet long.

Of these four extinct groups of marine reptiles, three bore a strong external resemblance to superficially similar living mammalian forms. Thus the placodonts resembled manatees, the ichthyosaurs resembled porpoises, and the mosasaurs, the archaeocete whales. In contrast, the plesiosaurs were an "engineering solution," not represented among living forms.

Our knowledge of the evolution of dinosaurs and other fossil reptiles owes much to the fierce rivalry between two of the leading nineteenth-century paleontologists: Othniel Charles Marsh (1831–1899) and Edward Drinker Cope (1840–1897). Both came from wealthy families: Marsh's uncle was George Peabody, while Cope's father was a wealthy Philadelphia banker. Marsh was educated at Yale and Berlin, while Cope spent two years at the University of Pennsylvania. They met in Germany in 1864 and became such close friends that each named a fossil after the other. Four years later, two events led to a lifelong feud that was to have a lasting impact. Cope invited Marsh to inspect a quarry in Haddonfield, New Jersey, from which he had recently obtained fossils. Marsh, behind Cope's back, then secretly paid the quarry owner to ship any future fossil finds to him. Cope regarded that as a betrayal, and the feud deepened when Cope showed Marsh the fossil skeleton of a new 35-foot-long plesiosaur, *Elasmosaurus platyurus,* which he had named and illustrated in a recent publication. Marsh remarked—rightly as it turned out—that Cope had mistakenly mounted the head on the tail, and the humiliation aggravated the tension between them.

From then on each used his substantial wealth to corner the dinosaur market of the largely unexplored western states. In a race that did credit to neither, each hired scores of collectors, shipping vast quantities of bones in what became known as the "Bone Wars." Marsh even paid spies to locate Cope's collecting sites. He also used his political connections to have himself appointed chief paleontologist to the newly formed U.S. Geological Survey, subsequently using his appointment to isolate Cope.

Cope, struggling to keep up the search, lost most of his fortune, which he had invested in a speculative silver mine in New Mexico. He then leaked stories of corruption and misuse of government funds by Marsh to the *New York Herald*, and a prolonged press battle led to a congressional inquiry and the elimination of Marsh's Survey position. The two adversaries died within two years of one another, each made almost penniless by their long and bitter struggle. The only beneficiary was the science of paleontology: between them they discovered more than 100 new species of dinosaurs. And, despite their rush to publish and the torrent of publications each produced—1,400 in the case of Cope, for example—between them they left behind thousands of important specimens, as well as some studies of enduring value and significance. Charles Darwin described Marsh's fossils as "the best support of the theory of evolution."[1]

Three other groups made up the remainder of this remarkable reptilian armada: marine turtles, lizards, and crocodiles.

Most turtles live either in fresh water or on land (tortoises) but Cretaceous turtles included one giant marine form, *Archelon*, that had a huge shell of about eight feet in length.

The mosasaurs, such as *Tylosaurus*, were modified lizards, adapted to a marine life. They appeared in Cretaceous times and were fast, powerfully built carnivores, reaching a length of 25 feet or more.

Crocodiles, too, were represented by seagoing forms—the geosaurs—in Jurassic and Cretaceous times.

This vast Mesozoic reptilian armada represents an event every bit as significant as the rise and dominance of the dinosaurs on the land. From Triassic through Cretaceous times, a period of some 179 million years, six major reptilian groups became the top carnivores of the oceans, with a range of diets and a great variety of body form. So total was their readaptation to an aquatic existence that they dominated the life of the oceans, just as their relatives dominated life on the land.

FLYING REPTILES

Extraordinary as it was for reptiles to return to the oceans, it was perhaps even more extraordinary for them to take to the air. It is a measure of the extent of their Mesozoic radiation that, as early as Permian or Triassic times, gliding reptiles of exquisite form are known from three continents, and other reptiles adapted to sustained flight are known throughout the Mesozoic. We will discuss reptilian aviators more fully in chapter 10, but they deserve a brief word of recognition here. Pterosaurs ("winged reptiles") showed significant variation in both size and structure. Some of the smaller ones, such as *Rhamphorhyncus* from the Jurassic, had a wingspan of about 2 feet, with sharp-toothed, pointed jaws, a long neck, an elongated tail ending in a rudderlike structure, and the fourth finger greatly extended to carry the leading edge of the wing. *Pteranodon*, a "hammer-headed" Cretaceous form, had a toothless skull that was elongated in both the front and rear to an extraordinary degree, virtually no tail, and a wingspan of some 20 feet. The largest known pterosaur, the giant *Quetzalocoatlus*, had a 40-foot wingspan. All these pterosaurs had wing structures that suggest they were efficient flyers, though the largest ones may have also been soarers.

More than 100 species of pterosaurs have been described. Though many seem to have been fish eaters, others seem to have teeth adapted for other diets, including crushing of insects and straining of plankton. The toothless pterosaurs may have fed much as living pelicans do.

LIFE ON THE LAND

The emergence of the early reptiles in the coal swamps of late Carboniferous times was followed by their growing range of adaptation and widespread radiation during the succeeding Permian Period and the long Mesozoic Era that followed. This expansion may have been facilitated by the geography of those distant times, when all the continents were joined into a single landmass—Pangaea—made up of the assembled northern continents (Laurasia) and the southern (Gondwanaland). In Middle Permian times the fusion of these continental masses led to extensive climatic change, with the spread of drier conditions at the expense of the earlier warm, wet conditions favoring the spread of the lush vegetation of the coal swamps. Continental ice sheets spread across the southern continents of Africa, India, Antarctica, South America, and Australia; as the ice sheets gradually retreated some 307–290 million years ago, the therapsids, the ancestors of the reptiles, spread into the southern continents. Their fossil remains, widely distributed across Antarctica, southern Africa, and South America, offer mute testimony of the former conjunction of these now-separated continents, a conjunction confirmed by the widespread presence of the distinctive fossil plant *Glossopteris* and the presence of striated rock surfaces covered with glacial debris (tillites).

CRETACEOUS EXTINCTION

The reptilian dominance, not only on land, but also in the oceans and in the air, extended from the Permian to the end of Cretaceous times, some 250 million years. It produced a dazzling array of forms that have been described by a rich assortment of superlatives: the largest land animals ever, the most fearsome carnivores of all time, the largest flying animal ever to have existed, and so on.

And then, in what amounts to the blink of a geological eye, they were gone. It was not just one group that disappeared or one environment that was affected. Marine reptiles—plesiosaurs and mosasaurs—dinosaurs of all kinds, flying reptiles, occupying every major environmental niche on land, sea, and air and pursuing almost every imaginable way of life, vanished. Equally tantalizing, several contemporary groups—crocodiles, turtles, lizards, and snakes—survived, as did the flowering plants, which had become widespread in Cretaceous times and continued, after a very brief interval, undiminished to the Cenozoic. So too did most mammals and some birds, as well as most insects. Most fish were unaffected. Nor was it only the vertebrates that suffered: some once-dominant invertebrate groups, such as the ammonites and belemnites, as well as many planktonic microorganisms, such as foraminifera, vanished from the scene. Even as we catalog some of the details, the varying impact of extinction remains tantalizingly puzzling. Thus ichthyosaurs, once widespread and abundant, died out 30 million years before the Cretaceous-Tertiary, or K-T extinction, even though their habitat and way of life so resembled those of the plesiosaurs, which vanished in the K-T extinction. Nine of the twelve families of birds became extinct, as did three out of four marsupial families and almost 50 percent of the families of tetrapods and amniotes (see table 1.1)

In seeking to identify the cause of this widespread extinction, it would be helpful to know just how fast it occurred. Was it, for example, extinction over a year or so, or did it take much longer—say, 2 or 3 million years? Unfortunately, though the "iridium spike" marking the K-T or K-Pg (for Paleogene) boundary, is precisely located, too little is yet known about the detailed distribution of the various biological groups in the rocks above and below this precise boundary to allow us to describe the extinction as either catastrophically fast or more gradually occurring.

Those who argue that it was a gradual event, in which the extinction points of particular groups were not identical, point, for example, to the fact that, as we have seen, ichthyosaurs disappeared 30 million years before the plesiosaurs and mosasaurs, which shared both their lifestyle and their marine environment. They also point out that some three-quarters of all then-living families *survived* the K-T extinction. Gradualists also tend to assert that extinction is a fact of life (forgive the phrase) for all groups and that well-documented climatic and geographic changes in late Cretaceous times could have merely accelerated background extinction rates.

Those who regard the K-T extinction as far more rapid tend to invoke a single unusual mechanism. Two of these proposed mechanisms are now popular. The less popular relies on the vast outpouring of volcanic lavas—the Deccan Traps of Western India and the Indian Ocean—which were produced at the end of Cretaecous times, when the moving Indian plate passed over a rising volcanic plume. The Hawaiian Islands were formed by a more recent event involving the same mechanism, though it is argued that the Deccan explosion was at least thirty times more powerful. Huge volcanic lava flows, some 1.2 miles in thickness, now cover an area of more than a million square miles. Proponents of this mechanism argue that this gigantic eruption, which occurred at about the K-T boundary, could have produced a cooling effect, a greenhouse effect, acid rain, atmospheric changes, or some similar effect, or even some combination of these various effects. The best analogy, perhaps, is the eruption of the volcano Tambora in Indonesia in 1815, which produced enough ash and atmospheric dust to make 1816 the "year without a summer," leading to widespread crop failures over a vast area. The Deccan eruptions were many, many times more powerful than Tambora and, proponents argue, could have depleted the food sources of the herbivorous dinosaurs and thus, in turn, impacted the food chain of the carnivores. But whether this Deccan Trap plateau volcanism would produce such dire climatic effects is debatable.

The alternative, and currently more popular, hypothesis involves the impact of an asteroid or comet that struck the Earth at the time represented by the K-T boundary. The evidence for this was discovered by Walter Alvarez and others in a well-preserved rock sequence in Gubbio, Italy, that spans this interval. A sharp increase of about 100 times, a "spike," in the iridium content of rocks marks the boundary, and this spike has since been detected in well over a hundred different localities at precisely the same point in the stratigraphic sequence and in the rocks representing many different depositional environments. Now, iridium is present only in trace amounts in sediments deposited on Earth, but it is present in much higher concentrations in meteorites. Walter Alvarez and his colleagues—his late father, Nobel Laureate Luis Alvarez, and two geochemists, F. Asono and H. V. Michel—argued that two other features of the boundary rocks provide further evidence of a meteorite impact: shocked quartz crystals showing two intersecting sets of fractures, and tiny glass spherules, both known to be characteristic products of the impacts of an asteroid, some 6.2 miles (10 km) in diameter.

But if such an impact had taken place, where was the crater? After all, if an asteroid of the size of the one suggested to produce such debris and such effects was 6.2 miles or so in diameter, and *if* such a body *had* struck the Earth, it should have left a substantial crater at the point of impact. Such a crater has now been discovered at Chicxulub in the Yucatan peninsula of Mexico. The structure is about 112 miles (180 km) in diameter, and it has all the marks of impact, including igneous rocks rich in iridium and dated at 65 million years. Other associated features, both igneous and sedimentary, confirm the impact.

Few, if any, would now doubt that an asteroid of substantial size impacted Earth at the K-T boundary. The difficult question is whether such an impact could have produced the dire worldwide climatic effects that have been postulated to account for such widespread extinction: aerosols, dust, smoke, greenhouse atmospheric changes, acid raid, and all the rest. And the same question must be raised about the "volcanic" hypothesis.

There is yet one more ambiguous aspect to this K-T extinction event, not unlike the Sherlock Holmes story of the dog who shared a stall with a stolen racehorse, "Silver Blaze." In a story of that name by Conan Doyle, Inspector Gregory of Scotland Yard asks Holmes, "Is there any other point to which you wish to draw my attention?" Holmes replies, "To the curious incident of the dog in the night-time." Gregory comments, "The dog did *nothing* in the night-time." And Holmes replies, "*That* was the curious incident."

The curious incident in the K-T extinction is the number of groups of animals and plants that were *not* profoundly affected by the wave of extinction. Thus mammals survived, as did birds (in spite of the extinction of some groups), planktonic diatoms survived but planktonic foraminifera were decimated, dinosaurs were wiped out, but crocodiles survived, as did lizards, snakes, and turtles; belemnite and ammonite cephalopods, abundantly worldwide in their distribution, were wiped out, but the closely related nautiloids were not; the flourishing angiosperms were replaced for a brief interval by fern spores, before springing back in early Paleocene times; the bivalved trigonid inoceramids became extinct, but other bivalves were unaffected. These remain puzzling features.

And there is still one more puzzle. The K-T extinction is but one of a dozen or so major extinction events in the fossil record. Do we invoke the same mechanisms for each of them? The Permo-Triassic (P-T_R) extinction, 251–250 million years ago, was of far greater extent: 81 percent of amphibian families, 75 percent of reptilian families, one-half of all marine families, and four-fifths of all genera became extinct in that sweeping event. One estimate is that only 5 percent of species survived.[2] In the much smaller end Ordovician extinction, one-fifth of all families died out.

It is tempting, but surely unwise, to seek a single mechanism to explain all such mass extinctions. Changes in sea level, oceanic circulation, continental form and aggregation, climate, glaciations, forest fires, volcanism, and changes in incoming solar radiation have all been suggested as gradual, non extraterrestrial causes, in addition to the possible extraterrestrial impact we have already described. It has also been suggested that mass extinction events show periodic cycles, of the order of 26 million years, perhaps associated with cyclicity of a comet, asteroid, star, or companion planet.

Our discussion of K-T extinction has carried us far. Fascinating as the topic is, we simply do not yet have enough information to reach a convincing solution. That will involve more collecting, more observation, more detailed information, and almost certainly, more surprises.

We will review the more general problem of extinction in chapter 16.

10

THE AIR

THE BENEFITS OF FLIGHT

Earth's earliest atmosphere was very different from that of the present. Its oldest atmosphere probably consisted largely of hydrogen, and perhaps helium, from the solar nebula from which Earth formed. Subsequent volcanism produced Earth's second atmosphere, consisting largely of nitrogen, carbon dioxide, and inert gases. Its third atmosphere developed about 2.4 billion years ago, during the Great Oxygenation Event, when, after the deposition of the banded iron formations, cyanobacteria began to synthesize oxygen. The modern atmosphere, then, is essentially life's creation, and a breathable atmosphere was a prerequisite for flight. But why, if you had made your laborious way onto the land, would you ever subsequently be tempted to fly? After all, that's what the insects, a few amphibians, many reptiles, and their prolific offspring, the birds, as well as a few mammals, have all done. What an extraordinary thing it is, in retrospect, that so many different groups should have established themselves ashore, only subsequently to take to the air. Flying, after all, requires the expenditure of enormous energy, as well as facing great hazards. Even the seeds of some plants are passive gliders, depending, as they do, on airborne dispersal for their successful reproduction.

Of course, not all these groups are involved in active flight. Plant spores, pollen, and seeds (such as cottonwood and dandelion) float downward on the wind. "Flying" frogs, "flying" geckos, "flying" fish, and "flying" squirrels all tend to parachute or glide, rather than fly. But the extinct flying reptiles (pterosaurs), and some fossil birds were true aviators, while living birds are capable of extraordinary and sustained flight, sometimes over distances of thousands of miles. What are the benefits of life in the air?

For plants, the benefits are most conspicuous. Seed dispersal by the wind increases the distribution of the parent plants or trees. For animals, flight can also extend the range of a species, as well as providing added sources of food. It also provides a powerful means of hunting for some birds, such as raptors, and it can provide an important means of escape from predators. Extended migration flights can take advantage of the changing seasons, opening up favorable breeding and feeding areas that would otherwise be unavailable.

INSECTS: THE EARLIEST AVIATORS

The earliest known aviators were flying insects, whose remains come from rocks of Carboniferous age, deposited some 304–313 million years ago. These included giant dragonfly-like creatures, some of which had a wingspan of about 30 inches. Mayflies also appeared at about this time. Neither dragonflies nor mayflies are able to close their wings over the abdomen when they are at rest; their mobility on trees, and thus their habitat, is therefore limited. All other insect groups are able to fold back their forewings, which in some groups—such as grasshoppers and cockroaches—are modified to provide a protective cover.

But the success of the insects as the 'first aviators' is only a partial measure of their achievement. They are the most abundant of all living groups. In numbers of species they outnumber all other living animals combined by about three to one. In terms of abundance of individuals they are equally prolific. They multiply with astonishing rapidity under favorable conditions. It has been suggested that the offspring of a single aphid could cover the earth in one season if all survived! They are highly adapted to life in all environments except the oceans (although some are littoral and one species lives on the surface of the sea), and they have a wider distribution than any other group of land animals. Some forms are parasitic, others live in ponds and streams; some are known to migrate to new feeding areas over distances of several hundred miles. Many insects are solitary, but others live a social existence in highly organized societies. Insects play a significant role in the life of humankind. Some play an important role in food production, including not only bees that produce honey but also other insects that play an essential part in the cross-fertilization of many fruits and crops. Others are pests and carriers of disease, against which the human race is obliged to fight a continuous battle. Indeed, in many ways the insects are our greatest competitors on Earth.

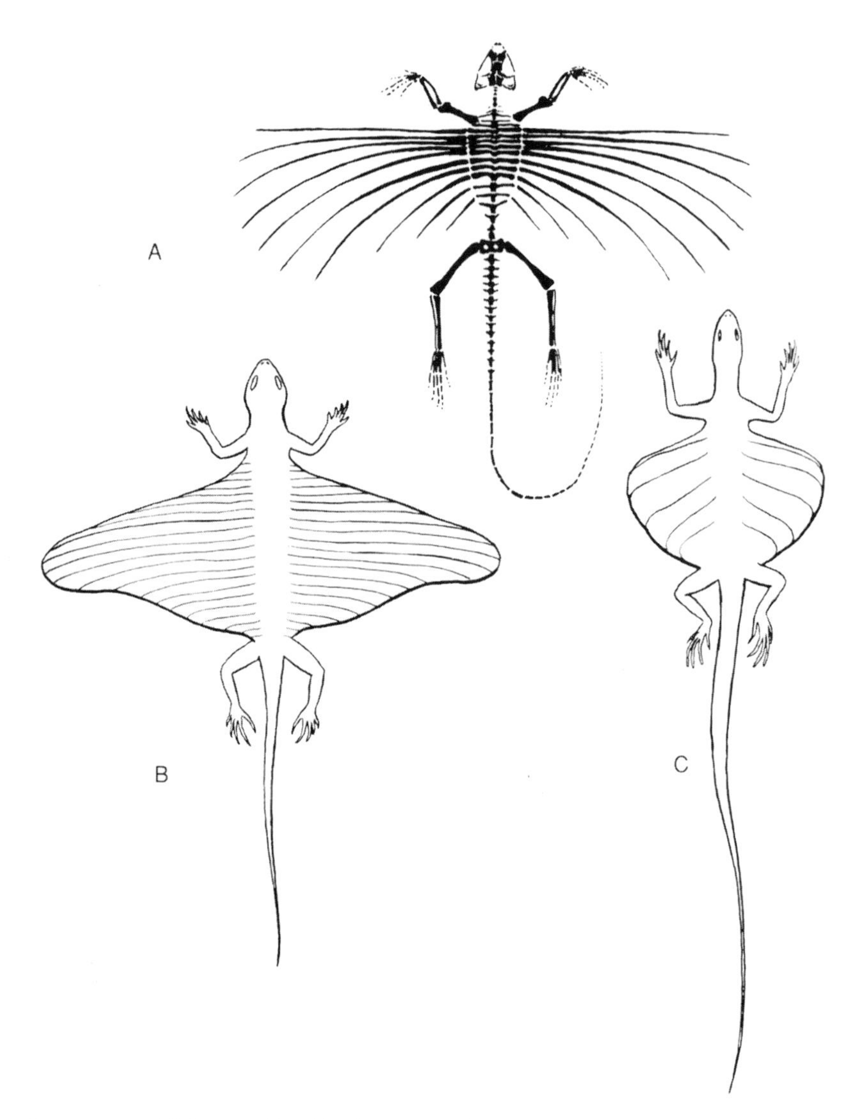

Figure 10.1. Gliding reptiles. (A) Skeleton of the early lepidosauromorph diapsid *Icarosaurus* from the late Triassic of eastern North America, maximum width about 7 inches; (B) Body outline of the primitive diapsid *Coelurosauravus* from the later Permian of Madagascar and Europe, maximum width about 12 inches; (C) Body outline of the living gliding lizard *Draco*, maximum width about 3 inches, total length about 10 inches.

(From Edwin H. Colbert, *Colbert's Evolution of the Vertebrates* [New York: Wiley-Liss, 2001], fig. 15-1, p. 219. Reproduced with permission of the publisher. Permission conveyed through Copyright Clearance Center, Inc.)

Beetles—the largest group of living insects—are first known from the Early Permian, though not all beetles can fly.

The "flying" vertebrates include the gliders, the pterosaurs, the bats, and the birds. Among the oldest known gliders is *Coelurosauravus*, from late Permian rocks of the UK, Germany, and Madagascar, which was a gliding reptile, a foot or so in length, that had extended rib bones supporting a winglike structure (fig. 10.1). Two later genera, of Late Triassic age, also show delicate extended ribs that formed an airfoil.

These structures are essentially similar to those of the living gliding lizard *Draco*. In none of these forms was sustained "flapping" flying possible.

"Flying" fish, incidentally are also gliders or jumpers, flipping themselves out of the water and using their fins to escape predators.

The only true flying mammals are the bats, although both "flying" lemurs and "flying" squirrels have winglike membranes that enable them to glide. We will revisit bats in chapter 13.

FLYING REPTILES

The oldest known true-flying vertebrates come from the Late Triassic. The first of these, the pterosaurs ("winged lizards"), remained reptiles. The second, which developed later, were reptiles with feathers: the ancestral birds.

A great variety of forms developed during Jurassic and Cretaceous times. The pterosaurs showed three essential modifications for life in the air: the presence of wings, a light but strong skeleton, and large cerebral hemispheres controlling the important senses of sight and coordination. Thus many of the bones were hollow and air-filled, the skeleton was given added rigidity by the fusion of certain parts, and some structures (such as the sternum or breast bone) were enlarged.

Other pterosaur structures were modified, such as the fourth finger, which formed an elongated support for the broad membranous wings, the remaining three fingers protruding as claws from the leading edge (fig. 10.2). The Jurassic pterosaurs were rather small creatures (no larger than a sparrow) with, in some cases, toothed jaws and long spadelike tails, but the Cretaceous forms were much larger, some, such as *Pteranodon,* having a wingspan of 25 feet and long pointed "hammer heads."

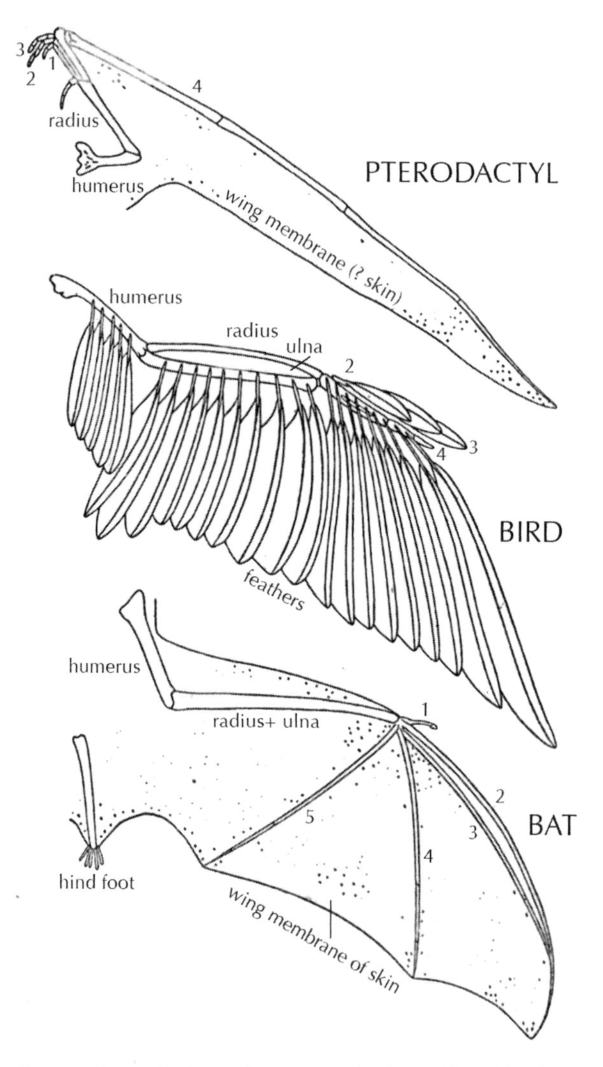

Figure 10.2. Comparison of wings of pterosaur, bird, and bat. The fingers are numbered 1–5. In the reptile the membranous wing was supported by the elongated fourth finger, the other fingers being hooklike claws. In birds, the fingers are combined to give added strength, and the wing is formed of stout primary feathers. In bats, four fingers are extended to support the wing membrane.

(After Storer. From Frank Rhodes, *Evolution of Life* [New York: Penguin, 1976], fig. 34, p. 215. Reproduced with permission.)

The skull was nearly twice as long as the vertebral column and bore pointed teeth.

The general construction of pterosaurs suggests that they were less active fliers than most birds, and they probably supported themselves by extended soaring and gliding. Although fossil walking tracks have been found, pterosaurs may well have been largely arboreal in habit, for the claws of some forms were well adapted to clinging to rocks and trees. Their wing structure was less robust and less efficient than that of birds or bats, and, if they shared the cold-blooded existence of other reptiles—though this is far from certain—they must have been severely limited in their level of activity. Some recent studies suggest, however, that they were warm-blooded and more active fliers than previously thought.

Most of their remains are found in marine strata, suggesting that these creatures were probably fish-eating sea fliers, though some others are found in fresh water deposits and apparently lived on land, perhaps feeding on insects or fruit.

The pterosaurs included the largest fliers ever known. Their bones were lightly built, with air spaces adding to their lightness. The greatly elongated fourth finger which supported the wing was over 10 feet long in the larger types (fig. 10.2). These huge creatures are found almost exclusively in marine sediments and seem to have been chiefly fish eaters.

Two main groups of pterosaurs have been recognized: the rhamphorhynchoids, which flourished from Late Triassic to Early Cretaceous times (some 209–140 million years ago) and had greatly elongated tails culminating in a rudderlike structure, and the pterodactyloids that existed from Late Jurassic to Late Cretaceous times and had greatly reduced tails. *Pteranodon* was a typical representative of this latter group, the largest member of which, the giant *Quetzalocoatlus* from the Late Cretaceous of Texas, had a wingspan of some 40 feet, about the size of a small plane. Whether these huge creatures were cold- or warm-blooded we do not know. Some have suggested they may have been too large for sustained flight.

More than 100 species of pterosaurs have been described, but for all their variety, they shared the fate of their reptilian cousins; by the close of Cretaceous times, they were gone. Why they should have become extinct when birds prospered and expanded in numbers is not yet clear.

One of the surprising and most remarkable developments of the last decade has been the discovery of feathers in many genera of small theropod dinosaurs from Early Cretaceous (125–120-million-year-old) rocks of China and Mongolia. These feathers range in form from small filament-like tufts to birdlike contour feathers. It seems unlikely, however, that these dinosaurs were capable of flight; the feathers may perhaps have served as a mating display or filled some other function, such as insulation. The presence of feathers provides further support for the idea that dinosaurs were warm blooded. One of these feathered dinosaurs, *Caudipteryx*, even has gastroliths preserved within its stomach, thus displaying still another birdlike characteristic.

The occurrence of feathers on creatures whose forelimbs are clearly not developed as wings is the prime evidence that birds are dinosaurs, developed from maniraptoran ("seizing hands") theropods, a group of dinosaurs that includes such forms as *Deinonychus* and *Oviraptor,* even though the function of the feathers is still unclear. Maniraptorans were the only dinosaurs known to have breastbones.

THE BIRDS: AVIAN DINOSAURS

The second major group of flying vertebrates are the birds. The birds (class Aves*)* include some of the most familiar and distinctive members of the animal kingdom. However hazy our notions of the characteristics of other groups, none of us needs reminding that birds are covered with feathers, that they possess wings, that their skeleton is of delicate but strong construction, that they have beaks but no teeth, and that they lay eggs. Clearly, in many of these respects they stand out from other animal groups. What is rather less obvious, however, is that the level of their body structure is "higher" than that of all other animals except the mammals. Thus the feathers provide an insulating covering, their body temperature is regulated (they are warm-blooded), and their circulatory system highly specialized, their ability to fly has allowed them to occupy unique environments, their senses of hearing and vision are highly developed, and their song has been a source of delight down the ages. But perhaps the most important feature of all is the relationship between the parent birds and their offspring. Many birds are born in a naked, helpless condition, others are born fully feathered and able to "stand on their own feet," yet in both cases the fledglings are nurtured and cared for by their parents. Just as reptiles' reproduction marked the consolidation of life on the land, so also this distinctive feature of avian and mammalian reproduction marked the initiation of family and ultimately societal relationships that have played so fundamental a role in the life of our own and other species. The oldest remains of the other major group in which this parental care has developed, the mammals, appear a little earlier in time (Middle Triassic) than the birds.

The oldest known birds come from a remarkable Late Jurassic (150-million-year-old) deposit known as the Solnhofen Limestone of Bavaria, Germany. The limestone of Solnhofen is so fine grained that it is known as lithographic, being widely used in lithographic printing in the past, and it represents the deposits of a tropical lagoon into which the remains of nearby creatures were carried, buried, and preserved. These fossil remains are exquisitely preserved and include more than 600 different species. Even jellyfish—soft-bodied as they are—and insects are preserved, as are scores of perfect pterosaurs, representing eight different genera.

The greatest treasure of all among the creatures of Solnhofen is *Archaeopteryx*, which is known from six different skeletons and a single isolated feather. *Archaeopteryx* was discovered and first described within a couple of years of the publication of *The Origin of Species* and provided dramatic confirmation of Darwin's theory of evolution. *Archaeopteryx* (fig. 10.3) was a small creature, no larger than a modern crow—about 2 feet long—whose skeleton, while reptilian, shows beautifully preserved imprints of feathers on the wings and tail. Its jaws had sharp teeth, its three fingers were clawed, and it had a long bony tail. Another fossil bird, *Confuciusornis* of Early Cretaceous age, has recently been described from northeastern China; unlike *Archaeopteryx*, it lacked teeth. The genus is represented by six pigeon-sized species and many hundreds of specimens.

Archaeopteryx does not itself solve the puzzle of the origin of feathers, but it does provide a classic example of the transitional pattern of evolution between two major groups. Critics of evolution have argued that it was inconceivable that random mutations could have transformed arms into wings, provided a breastbone, aerated bones and feathers, and caused all the other modifications that would have converted a reptile into a bird. But *Archaeopteryx* indicates that the transition from reptiles to birds was a piecemeal process, rather than one in which the development of all features was closely integrated. It is, in fact, another example of mosaic evolution, which we encountered in the early amphibians. The general evidence suggests *Archaeopteryx* was a tree climber, using the claws of both its wings and its toes to claw its way up trees or rocks from which it may have then glided or flown. It seems entirely possible that powerful flight developed later from such gliding ability.

Figure 10.3. *Archaeopteryx.*
(Illustration by Joe Tucciarone. Used with permission.)

Bird skeletons are of such light construction that they are poor candidates for fossilization, but after *Archaeopteryx* emerged, the evolution of birds was very rapid. Early Cretaceous birds are known from Asia, Europe, North America, and Australia and reveal a variety of smaller birds. By Late Cretaceous times, various seabirds are known. One, a strong, ternlike flier, *Ichthyornis,* with toothed beak, was probably a fish eater. Another, *Hesperornis*, was a large diving bird, about 6 feet long, but flightless.

So, after the long ascent to the air, some birds soon abandoned the power of flight, just as some amphibia cumbersomely adapted themselves to life on the land, only to return to the water, and some reptiles, having strengthened their hold on the land, returned to the seas. Such is the diversity and range of adaptation—and such also its irony!

Half a dozen familiar groups of modern birds are now known from the very late Cretaceous, including gallinaceous birds, ducks, and sandpipers.

Interestingly, perching birds (Passeriformes), now the most species-rich group of land birds, are not definitely known until the Miocene, when their diversification began.

Unlike the flying reptiles, the birds survived the end of the Mesozoic, but their subsequent history is rather patchily recorded in the fossil record. Such birds as are known have a distinctively modern appearance. Teeth are no longer present, and, unlike the pterosaurs, the hind legs are usually well developed. Among the most spectacular Cenozoic birds were the giant flightless birds, the forebears of the living ostriches, cassowaries, and rheas. Some extinct forms were very like these tall wingless creatures, and their structure was apparently a response to conditions of plentiful food supply and absence of predatory enemies. Even in early Cenozoic times such forms existed; some seem to have been savage carnivores, and they may well have been formidable local rivals to the mammals.

Other Cenozoic birds became adapted to widely different kinds of life, such as the penguins of Antarctic regions, waders of the shores and estuaries, seabirds of the open oceans, songbirds of woodland and meadow, game birds of upland areas, and powerful birds of prey of mountains and other areas. Living birds vary enormously in size, from the recently extinct 10-foot-high elephant bird of Madagascar to some tiny hummingbirds that weigh only 0.1 ounce. Some birds (pelicans and cormorants) are voiceless, others have characteristic songs of great complexity; many birds are solitary, others live in flocks of up to 100,000; some birds—swifts and falcons—fly at speeds of more than 150 miles per hour, others are flightless; some are restricted to one locality, others undertake short seasonal migrations, while others travel several thousand miles.

The largest known bird is the Pleistocene teratorn vulture, *Teratornis incrediblis*, with a wingspan of 17 feet, while the largest living birds, the California and Andean condors and the Wandering Albatross have wingspans of about 12 feet. The smallest bird, the Bee Hummingbird of Cuba, weighs only 0.056–0.071 ounces and is 2.0–2.4 inches long. Its name reflects that it is about the size of a bee.

Birds are also highly intelligent. Alex, a celebrated African Grey Parrot, was the subject of an extended experimental study over a period of some thirty years in which he was taught to count, distinguish colors and shapes, solve problems, and hold simple "conversations."

Judged by any standards, the birds are not only the champions of the air, but also one of the most successful of all living groups of animals.

11

THE BLOSSOMING EARTH

THE INFLUENCE OF PLANTS

The history of life is almost always viewed in terms of the history of animal life. Even the names of the great eras of geologic time—the Proterozoic, the Paleozoic, the Mesozoic, and the Cenozoic—are animal-based. Indeed, if the eras had been established on the basis of plant life, the Paleophytic and other eras would almost certainly have had quite different boundaries. The Paleophytic, for example, would extend from the Silurian to the Permian, and the Cenophytic would have begun in the Late Cretaceous. But the developments of animals and plants, though not coincident, are of course intimately intertwined. The changes in Cenozoic mammals, for example, can be understood only in terms of changing vegetation that provided their diet (and ours) and especially in the profound changes involved in the spread of grasses, fruits, and flowering plants. And changes in both plants and animals have been influenced by changing climate and geography. So also the development of pollinating insects has had a huge impact on the evolution of plants. Some indication of the changing balance between the major plant groups is given in figure 11.1.

We have seen that the dominant plants of the "Paleophytic" were club mosses, ferns, seed ferns, and cordaitaleans. By Late Carboniferous times, it has been estimated that club mosses accounted for half the known plant species. It was these ancient trees—so much mightier, more varied, and more abundant than their living descendants—that formed a major part of the coal forests. This assemblage of trees declined, perhaps in a complex response to growing aridity, and was gradually replaced in Permo-Triassic times by conifers, cycadeoids, and cycads. That this period was one of considerable aridity is reflected in widespread, thick deposits of red beds and evaporites, which are indicative of these conditions.

In the Southern Hemisphere, a distinctive (*Dicroidium*) flora persisted. In northern lands a number of familiar Paleozoic forms were still present;

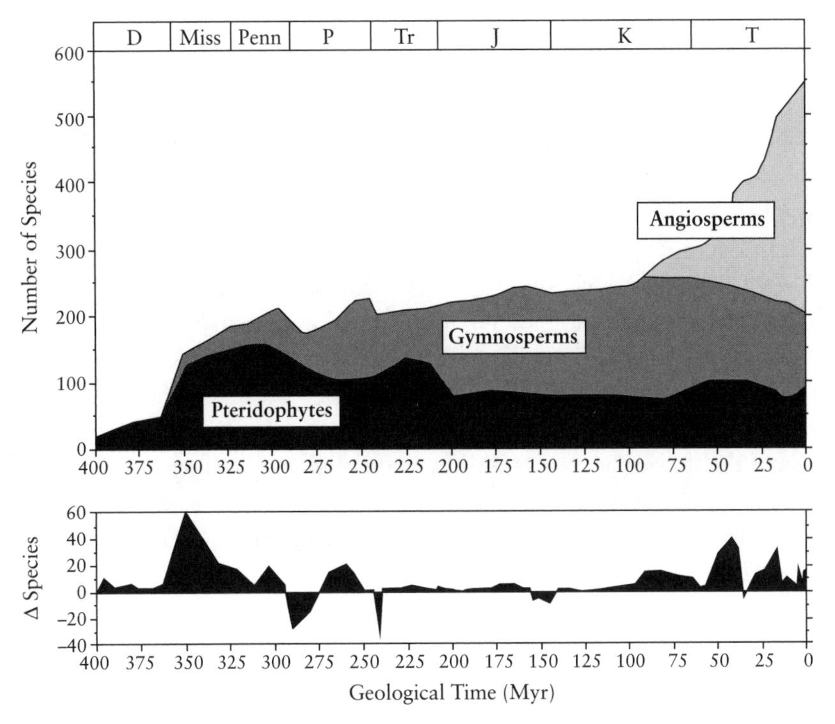

Figure 11.1. Diversity of major plant groups through geologic time. Taxonomic turnover during the Phanerozoic. Paleofloras were dominated by pteridophyte species during much of the early Paleozoic. These floras were replaced by ones dominated by gymnosperm species during the Mesozoic. The most recent floras are dominated by angiosperm species. The trend of increasing species numbers throughout the history of the vascular land plants is the result of a slight surplus of species accumulated over hundreds of millions of years (lower graph).

(Adapted from Niklas, Tiffney, and Knoll 1980. From Karl J. Niklas, *The Evolutionary Biology of Plants* [Chicago: University of Chicago Press, 1997], fig 8.15, p. 397. Reproduced with permission.)

some cordaitaleans, seed ferns, and scale trees survived for a short time, there were scouring rushes, and conifers were a major component of the flora. Some of these conifers grew to be of extraordinary size and are represented by the beautifully preserved silicified fossils of the Triassic petrified forest of Arizona, which included trees up to 10 feet in diameter and 200 feet in height.

The ferns were equally important. They formed the lower-growing vegetation. The main groups represented had first appeared in the Carboniferous, though many genera were new. And some unrelated groups made their first appearance. The most striking newcomers were the now extinct cycadeoids (or Bennettiales). These forms were broadly similar to the living cycads; they had rather short, stumpy, generally unbranched trunks crowned by clusters of palmlike fronds, with hermaphroditic cones growing from leaf bases. They differed from living cycads, however, in their method of reproduction, for in most species both the male and female reproductive organs were combined to form a flowerlike structure. But in spite of their superficial similarity, these structures were really pseudoflowers and do not appear to have given rise to the true flowers of later plants. The cycadeoids, which appeared in the Triassic and persisted through the Cretaceous, are among the most widespread and abundant and, with the ferns, cycads, and conifers, the most characteristic Mesozoic plants. Unlike the two latter groups, however, the cycadeoids became extinct before the end of Cretaceous times. Their decline shows some correlation with the rise of the flowering plants, and the two events may well be connected.

The cycads still survive, represented by about 300 species in widely scattered warm areas, and in them the male and female cells are produced in separate conelike structures. The cycads appeared at roughly the same time as the cycadeoids, and because both were common, the Mesozoic is often referred to as the "age of the cycads."

The gradual transition in composition of plant communities can be traced in the geologic succession of fossil plants. We have already seen that by Late Carboniferous (Pennsylvanian) times, no fewer than half of all the known plants were club mosses, although this group constitutes substantially less than 1 percent of living plant species. This difference gives emphasis to the need to consider, not only the total stratigraphic range, but also the relative abundance of plant groups in attempting to reconstruct the vegetation of the past.

All the plants we have so far described—seed ferns, conifers, cycads, cycadeoids, and ginkgoes—which between them dominated the Triassic, Jurassic, and Early Cretaceous forests—were gymnosperms, "naked seeds" in which the fertilized embryos are protected and nourished but not enclosed in an ovary. Some indication of their variety is given by the fact that the Middle Jurassic rocks of northeast England, which have been studied in detail at 600 fossil-bearing levels, were found to contain no fewer than 260 species of these fossil gymnosperm plants.

The rise in gymnosperm variety and abundance was marked by a decline in the proportion of the ferns and giant club mosses that had dominated the flora of the Pennsylvanian coal-forming swamps. It is worth noting, here, the difference between patterns of extinction in animals and plants. In animals, extinction events are clear-cut: dinosaurs, large marine reptiles, and ammonites, for example, all disappeared in a "catastrophically" sudden moment in time. Plants seem less affected by such episodes of mass mortality. In plants, it seems competition and more gradual geographic and climatic change are the more significant influences.

Living ginkgoes and dawn redwoods are a reminder of the earlier dominance of both groups in those ancient forests. Both were once thought to be extinct in the wild, although each one had been preserved by its careful cultivation as an ornamental tree. In 1691, the German Engelbert Kaempfer discovered the ginkgo growing in the wild in Japan. Ginkgoes are deciduous trees, reaching up to 90 feet in height and branching repeatedly. They have distinctive, fan-shaped leaves, with individual trees being either male or female. Similarly, *Metasequoia*, the dawn redwood, was known only from fossils until living specimens were discovered in the wild in Szechwan Province, China, in 1946. Both these existing genera are living fossils, surviving remnants of once widespread groups.

THE RISE OF THE ANGIOSPERMS

By mid-Cretaceous times, the once dominant cycads and ginkgoes were declining in numbers, though the conifers still dominated the forests, especially in the drier upland areas. It was against this background that the earliest flowering plants (angiosperms) appeared, about 140 million years ago. Unlike the naked seeds of their gymnosperm predecessors, the seeds of angiosperms are covered and protected within a carpel or ovary, providing protection against pests and desiccation, as well as providing a vehicle for seed dissemination, as carpels evolved into edible fruits in certain groups of flowering plants. The endoderm tissue within the seed provides nutrition for the growing embryo. Flowers, and the fruits that contain the seeds, are the most visible expression of these reproductive changes, which became widespread during mid-Cretaceous times.

But where did angiosperms come from, with all the complicated arrangements involved in their pollination and development?[1]

Charles Darwin, in a letter written in July 1879 to his close friend Joseph Hooker, described the first appearance of fossil angiosperms as an "abominable mystery."[2] Darwin was troubled by what he regarded as the abrupt origin and extraordinarily rapid diversification of flowering plants. He worried that this might imply a rapid developmental jump or a saltation, rather than the more gradual evolution for which he was the advocate. Recent discoveries make these mysteries "less abominable," though they have not yet identified the ancestral group from which angiosperms arose. What frustrated earlier researchers was the apparent absence of early flowers in the fossil record, although fossil pollen was known from rocks that are some 135 million years old. But recent research has uncovered microscopic flower structures from 70–130-million-year-old Cretaceous rocks from southern Europe and New Jersey, while a 125-million-year-old flower, and the aquatic plant that bore it, have been discovered in northeastern China. The plant—*Archaefructus*—lacked both petals and sepals but did have an enclosed carpel.

The search for early fossil angiosperms, supplemented by molecular studies, continues, but one aspect of Darwin's abominable mystery remains.[3] Gymnosperms dominated plant life for 250 million years, yet in a brief interval of perhaps 10 million years they were displaced by angiosperms, which now make up more than 90 percent of living land plants. Just how this extraordinary change took place, and with it the associated changes in pollinating insects, birds, bats, and other mammals, remains a tantalizing question.

By the close of the Cretaceous, several hundred species of flowering plants were in existence and had replaced the conifers as the most common trees. Some of these were forms now long extinct. Others were the earliest representatives of familiar genera that still survive—beech, birch, maple, oak, walnut, fig, magnolia, poplar, and willow, for example—as well as holly, ivy, laurel, vines, and other plants and shrubs. Temperate conditions apparently existed far to the north of their present limits, for a large number of typical temperate genera occur in Cretaceous strata of Arctic regions.

It may well be that the development and spread of the angiosperms was favored both by increasing aridity in some areas in Late Jurassic to Early Cretaceous times and by the great mid-Cretaceous (Cenomanian) marine transgression that brought extinction to many flora. When the seas retreated toward the close of the period, it was the angiosperms that colonized the empty lands.

Living angiosperms have become the most successful of all plants. Some are minute, only a fraction of an inch in height; others such as *Eucalyptus* can be more than 200 feet in height. They are found throughout the Earth, in every type of terrestrial environment, from the tropics to the polar regions and in every climatic zone, from deserts to monsoon jungles. They far outnumber all other plants combined in their total numbers of both species and individuals, and their level of organization is more complex than that of any other group. Almost all plants play a part in the life of humans, but we are particularly dependent on the angiosperms. They make up most of our vegetable diet (fruits, vegetables, cereals, and so on) and are the sources of many of our most important drugs, while most of the animals bred for meat also exist on them. Angiosperm trees supply a large share of our timber and their flowers provide an endless source of joy.

Angiosperms can be divided into two groups. The monocotyledons develop with a single embryonic leaf, have stems that are not differentiated into pith, wood, or bark, and have no growth rings. The lilies, palms, onion, bananas, and grasses are familiar monocots. In these the roots are fibrous, the leaves generally parallel veined, and the flower parts usually developed in groups of three.

The dicotyledon seedling has two leaflets, the stem is differentiated into pith, wood, and bark, and in woody dicots growing in seasonal climates, annual growth rings are formed. Almost all the familiar garden plants (except tulips, orchids, and lilies), the deciduous trees and shrubs, and fruits and vegetables fall within this group, in which the leaf veining is usually reticulate, and the flower parts have a four- or fivefold grouping.

The rise of the angiosperms and the transformations they brought about in the plant world were of the greatest importance, not only for the vegetable kingdom but also for the animal kingdom, for the appearance of plentiful supplies of edible nuts, fruit, seeds, and leaves was a major factor in the rapid contemporary evolution of birds and mammals, and also possibly the flying reptiles. The development of flowers implies the spread of pollinating insects, though little is known from the fossil record about their earliest history.

The conifers remained dependent on the wind as the chief agent for seed dispersal, but many angiosperms became adapted, as we have seen, to dispersal by other means, chiefly birds, insects, and mammals. This adaptation has probably been an important factor in their success, and it is yet another example of evolutionary interdependence.

What a contrast these changes must have brought to the face of the Earth. When the Mesozoic dawned, the Earth was clothed with conifers, ferns, cycads, and a few relics of the trees of the coal forests: a somber landscape of greens and browns. Reptiles spread across the land, and only the archaic insects rose through the heavy air. One gains the sense that the changes of mid-Mesozoic times came as a worldwide spring in the long life of the Earth. Flowers, butterflies, birds—these were tokens of a new age.

The cycadeoids became extinct during the Cretaceous. The cycads and conifers still survive, but both groups are very much reduced. The cycads, although widespread, are confined to ten genera in warm or subtropical areas. The conifers, for all their mighty forests, are also a limited group. In their natural state, they tend to be confined to the less hospitable, cold, upland environments of the Earth and are represented by comparatively few species. They were dominant plants for 75 million years, but their decline began with the blossoming of the angiosperms. For in plants, no less than in animals, the principle of natural selection applies—it is the better adapted that survive.

We have seen something of the far-reaching implications of the rise of the angiosperms, not only in the general change they brought to the face of the Earth, but also in the related changes that took place in insects and mammals, as well as in other plants. For one group of animals, the angiosperms might have been a major factor in their decline. The dinosaurs had spread and multiplied throughout the Mesozoic, in a balanced economy that was at once both successful and precarious. In Late Cretaceous times they vanished from the scene. We have already speculated about the factors involved in their sudden extinction and have seen the difficulty of any simple explanation. Yet it has been suggested that the great plant changes of Cretaceous times were influential factors in the decline of the giant herbivores and thus, indirectly, in that of the carnivores. Whether or not this is so is problematic, because some herbivorous dinosaurs underwent a considerable expansion in Cretaceous times. It may be true, however, of the huge sauropods, which must have required enormous quantities of plant food for subsistence.

CENOZOIC PLANTS

To write of Cenozoic plants is to risk confusion, for the Cenozoic, the era of modern life, is based on characteristic fossil animals. Had it been based on fossil plants (the Cenophytic), it would, as we have seen, presumably have begun with the Cretaceous. Nevertheless, it remains true that Cretaceous and later plants were different in some respects from those of the Mesozoic. Most of these differences are rather minor, however, and are represented chiefly by the incoming of modern species. Many hundreds of species of Tertiary plants are known, and members of almost all genera show an increasing similarity to forms still living. Because of this similarity, the abundance of fossils, and the relatively short period of time involved, these fossil flora provide an unusually good opportunity to compare the distribution of present floral zones with those of the past. If we assume that the fossil species occupied broadly similar environments to those of their direct descendants, it becomes possible to reconstruct ancient floral zones. The general result is one of permanence, with change; permanence because floral zones undoubtedly existed in the Tertiary and appear to have been parallel to those still existing today; change because with changing climates the plants constituting the zones have constantly migrated back and forth. London, for example, was the site of great tropical forests in early Tertiary times, and the remains of these ancient plants are preserved in countless numbers in the London Clay. In the Pacific northwestern states of Oregon and Washington, the Tertiary Eocene strata yield abundant palms and other plants that today grow only in the wet and warm areas of Central America and southern Mexico. Beds of the same age in Alaska contain fossil plants more typical of present northern plants, such as maples, sequoias, poplars, willows, and birches. In the overlying Miocene rocks of Washington and Oregon, these drier, colder forms have replaced those of the Eocene, the more luxuriant fauna having moved southward.

Perhaps nowhere is this floral migration better shown than in the Pleistocene glaciation, when ice sheets of continental proportions advanced and retreated across the Northern Hemisphere in a series of climatic fluctuations. Very clear evidence is provided of major glacial advances gradually giving rise to interglacial periods in which varying groups of plants re-established themselves.

So far we have traced two aspects of Cenozoic flora—their increasing modernization and their changing distribution. One further factor remains: the spread of the grasses. Grasses are unlikely candidates for preservation as fossils. Occasional and authentic records exist of megafossil grasses in the Eocene, including flowers with pollen in the anthers. The pollen is very distinctive in grasses. It has a complex circular aperture with an extended inclusion called an operculum, with an annulus surrounding the opening; the wall ultrastructure is also distinguished by narrow channels.[4] Mostly unauthenticated records of grasses exist in

Late Cretaceous and Early Tertiary rocks, but they are generally absent. Recent studies suggest that they first became abundant in Miocene times, for their seeds are known in great numbers from the Miocene strata of the High Plains of Kansas. Their common appearance has a double interest. First, it is an indication of widespread physical changes that brought about a general transition from forest conditions to open prairies. Second, their appearance coincides with very rapid and profound changes in the mammals. These great changes are reflected in many mammalian groups and seem to point clearly to a general change from browsing to grazing habits. There can be little doubt that the changes are very closely related. To describe them as cause and effect may perhaps be an oversimplification—but it is not far wide of the mark.

The spectacular success of plants in clothing the land and thus changing the chemistry of both the atmosphere and the rocky surface of the planet was the result of their earlier emergence from lowly aquatic forebears in Silurian or earlier times. As we recognize the evolutionary success of land plants, it is worth pausing for a moment to describe the simpler plants that remained in the water. These plants, often grouped together as the Thallophytes, include the algae, bacteria, and fungi, and are all soft "bodied," with no differentiation into the characteristic roots, stems, and leaves of "higher" plants. They reproduce mainly by asexual single cells—spores—that develop into mature plants, although some reproduce sexually. Algae are essentially aquatic plants and include not only the seaweeds, but also the phytoplankton, which populate the surface waters of the oceans in countless numbers and form the base of the food pyramid for many of its creatures. Three particular groups of phytoplankton have had a significant impact on the geologic record. The calcareous *cocospheres* are green algae with minute calcareous disks—coccoliths—contained in their walls. These elaborately patterned structures are known as nanofossils—visible only with the electron microscope—and they exist in such abundance that they have been important rock-formers in the geologic past. Much of the Cretaceous chalk that forms the white cliffs of Dover, for example, is made of coccoliths. They represent the largest single constituent of deep-sea sediments and are known in rocks ranging from Triassic to Recent in age.

A second group of abundant phytoplankton—diatoms—have siliceous (SiO_2) "skeletons," of exquisite intricacy, formed of two platelike forms and known as a frustule. They vary greatly in form and are found, not only in the oceans—where they are major contributors to the food chain—but also in fresh waters and soil. Diatoms are known in both marine and freshwater rocks, the oldest dating back to the Early Jurassic (some 185 million years ago). So abundant were they that in some places their remains cover vast areas of the ocean floor and form rock deposits known as diatomite, or diatomaceous earth, a rock mined commercially for its value in filtration devices, paints, and as an abrasive mixture. Their fossil remains have also been used in establishing stratigraphic zones. There are reported to be 200 genera and many thousands of species of diatoms.

Another group of green algae can synthesize oil, and there is now considerable commercial interest in their potential value as a source of fuel.

This long history of plants—from these "simple" algae to giant redwoods—is the foundation on which the whole pyramid of life has developed. The ability of plants to synthesize biomass from carbon dioxide—photosynthesis—developed at least 2.5 billion years ago. Since those ancient days, this "great web of life" has sustained all living things, providing the basis for our daily existence.

12

THE RISE OF THE MAMMALS

WHAT IS A MAMMAL?

Until about ten years ago, it was possible to describe the evolutionary history and classification of mammals in relatively simple terms. Mammals were air-breathing vertebrates with mammary glands, hair, and a distinctive middle-ear structure. They still are, but there are two new developments in their classification. First, the recent adoption of cladistics—a system of biological classification based on the evolutionary history and inferred relationships of a group, rather than on the traditional basis of structural similarities—has modified and profoundly disturbed long-established groupings and classification. Since some 1 million species of animals and about half a million species of plants have already been named, and several million still unnamed species are also estimated to exist, this is a matter of no small importance. In cladistics, groups are defined on the basis of shared characters derived from a common ancestor and exclusive to a particular group.[1] This new method seeks to reduce the subjectivity of traditional taxonomic classification in favor of some more objective method, less subject to personal judgments of degree of structural similarity and supposed evolutionary relationships. Cladistics bases classification squarely on unique characteristics that are both shared and inherited. If we want to define mammals, for example, we should concentrate on their unique reproductive and "hairy" qualities, rather than the fact that they have a backbone, a feature they share with other groups, such as reptiles and amphibians.

These relationships are expressed by constructing a cladogram, a branching diagram with the nodes of the branches representing the common ancestor of all the groups at the tips of the branches above it.

This new method of classification allows us to retain the use of traditional groups—such as Amphibia, for example—by referring to them as "grades," because they do not include all descendants of that group.

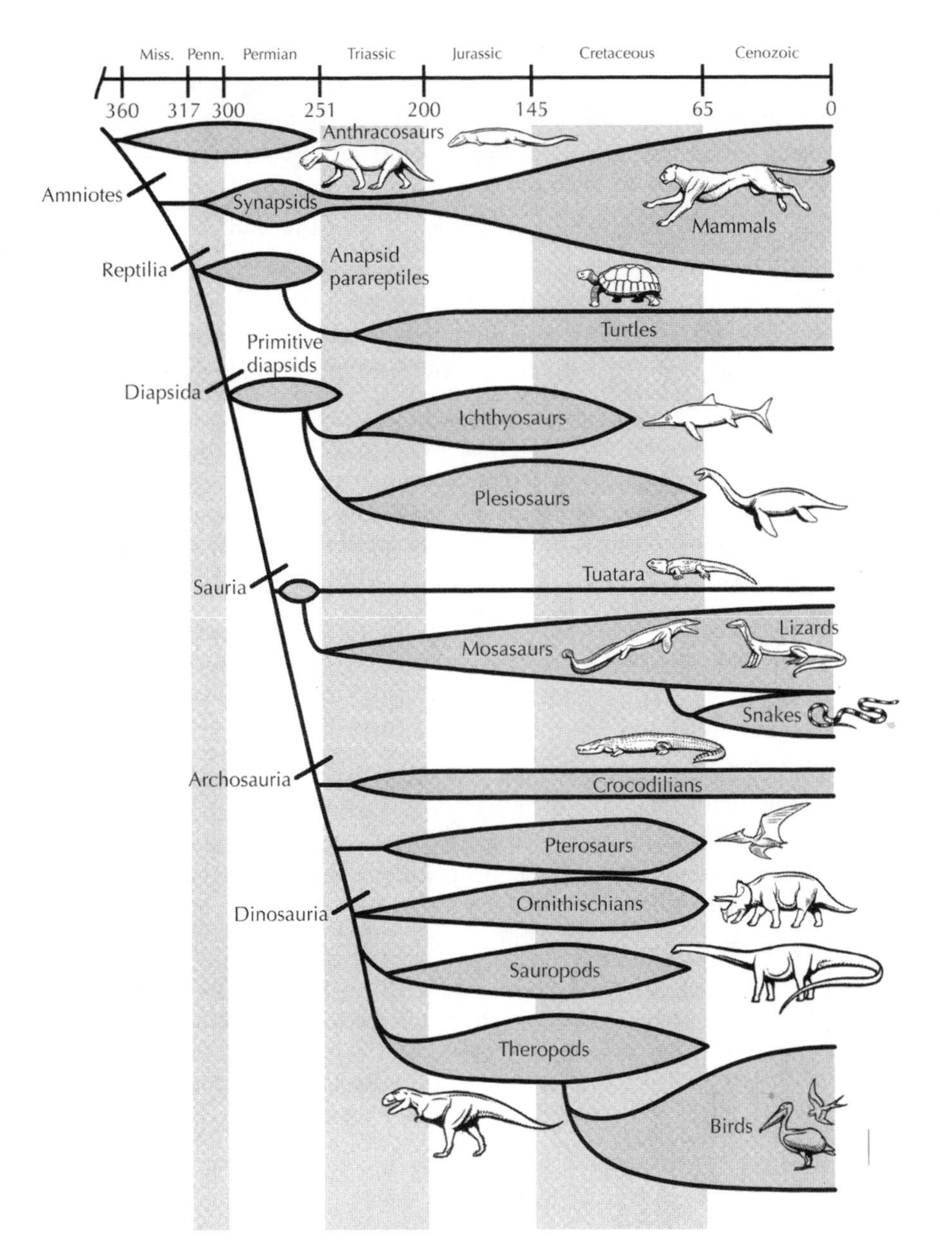

Figure 12.1. Family tree of the amniotes.

(Drawing by Carl Buell. From Donald R. Prothero, *Evolution: What the Fossils Say and Why It Matters* [New York: Columbia University Press, 2007], fig. 11.1, p. 232. Reproduced with permission of the publisher.)

The cladogram is regarded as more inclusive than a traditional evolutionary tree, because it allows us to compare relationships with *any* taxa, rather than only ancestors and descendants.

Cladistics is now the prevailing method of classification because it forces us to distinguish between less significant, "*primitive*" anatomical features, derived from a distant common ancestor, and *advanced* features, derived from recent ancestors. In the case of our own species, for example, our distinctive "humanity" rests on advanced features of our large brains and bipedal locomotion, rather than our vertebral column, hair, mammary glands, and presence of four limbs, which we share with many other mammals. Cladistics, in this respect, reflects phylogeny. Thus reptiles and birds, traditionally recognized as separate classes, are lumped together because birds are derived from theropod dinosaurs. Cladistics classifies both reptiles and birds with mammals as Amniotes, because all these traditional classes share a method of reproduction using an amnion, an enclosing fluid-filled sac that protects the embryo (see fig. 9.1).

The arguments for using cladistics are that it is testable and thus more objective than traditional methods and that it reflects the actual process of evolution. Its critics argue that it bases classification on supposed—but perhaps unproven—ancestry, rather than on readily observable characteristics, that it ignores sensible readily understood groups, such as "reptiles," and that it elevates theoretical ancestry at the expense of biological sense.

This new terminology, though arousing intense enthusiasm among specialists, makes difficult going for the general reader. Long-established groups—fish, amphibians, reptiles, mammals, and so on—are no longer regarded as natural groups because they do not include all their descendants. Thus, as we have seen, we are required to refer to them as "grades" and to regard them as "archaic" terms.

Alongside this development a second, alternative, method of classification has emerged, based on recent molecular DNA genetic biomarker data, which sometimes indicates affinities that differ to some degree from those suggested by cladistics. The result is that taxonomic classification is now in more of a state of flux than at any time in the last century or so. Nor is that all. To compound the problem, not only are these biomarkers undetectable in almost all fossil specimens, but the alternative lines of descent and ancestry they suggest in many cases have not yet been confirmed by the available fossil evidence.

No resolution of this debate is yet in sight, and this ongoing taxonomic turmoil makes it particularly difficult to present a brief, nontechnical, and noncontroversial description of mammalian evolution for the general reader without—forgive the phrase—stepping on one or two sacred cows. But, with these warnings, I will try.[2] What I propose to do in this brief overview is to use traditional terminology ("reptiles," "birds," "snakes," and so on) but to illustrate the relationships of these groups ("grades") to the new cladistics by using a series of excellent illustrations by Donald Prothero (e.g., figs. 8.4, 12.1, 12.3) that show these traditional groups in a cladistic context.

THE VARIETY OF MAMMALS

The mammals (class Mammalia) include the great group of living things that most of us refer to simply as "animals." This is a measure both of their familiarity to us, and of our recognition of them as the "highest" forms of life. The grade ("class") includes most of the common domestic and farm animals, as well as many of the familiar zoo animals, the bats, and the whales. All mammals are more or less covered with fur or hair, are warm-blooded, have highly developed senses, and almost all of them give birth to live young and nourish them with milk secreted from the mammary glands of the mother. This parental care is practiced to varying degrees in various forms, reaching its peak in humans. Mammals also share various distinctive skeletal features, such as their strongly differentiated teeth and the structure of the ear and the jaw. One other mammalian feature, less diagnostic than some others, may be the ultimate key to their present dominance. That is their brain size. In no other living creatures is there such striking mental development.

Although they are relative newcomers, the mammals are one of the most successful of all living groups. They exist in virtually every part of the Earth, from the polar regions to the tropics, from the depths of the oceans to the air, and they range in size from the mouse to the blue whale, the largest animal that has ever existed, which reaches a length of 120 feet and weighs up to 150 tons. Some 5,500 species of living mammals have been described.

Though the number of living species of mammals is substantial, they were represented by many more species in the geologic past. Living elephants,

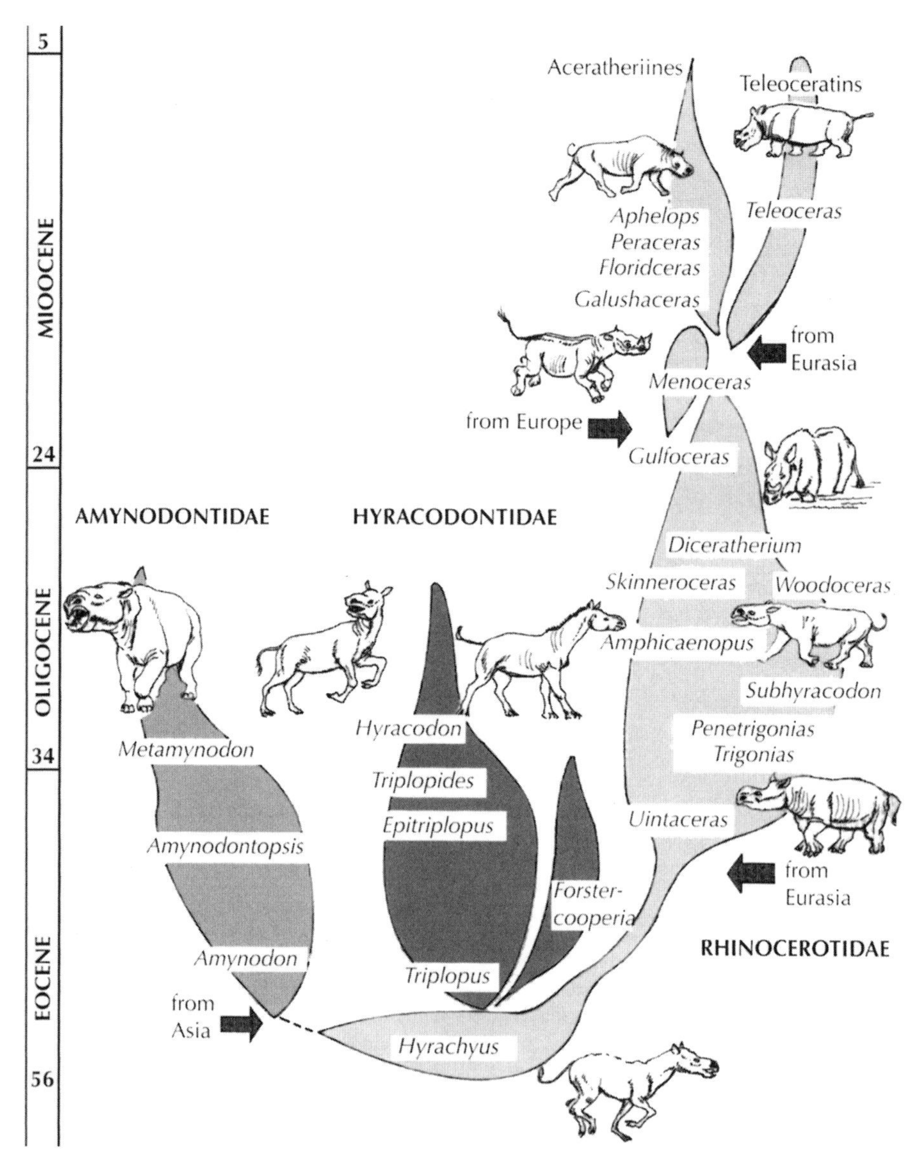

Figure 12.2. The evolutionary history of North American rhinoceroses. In the Eocene, they branched into three families, the hippolike amynodonts, the long-legged running hydrocodonts, and the living family Rhinocerotidae. During their evolution, they varied not only in body size and limb and skeletal proportions but also in the number and position of horns (or lack of horns), the details of their teeth, and many other features.

(Drawing by C. R. Prothero; after Prothero 2005. From Donald R. Prothero, *Evolution: What the Fossils Say and Why It Matters* [New York: Columbia University Press, 2007], fig. 14.6, p. 307. Reproduced with permission of the publisher.)

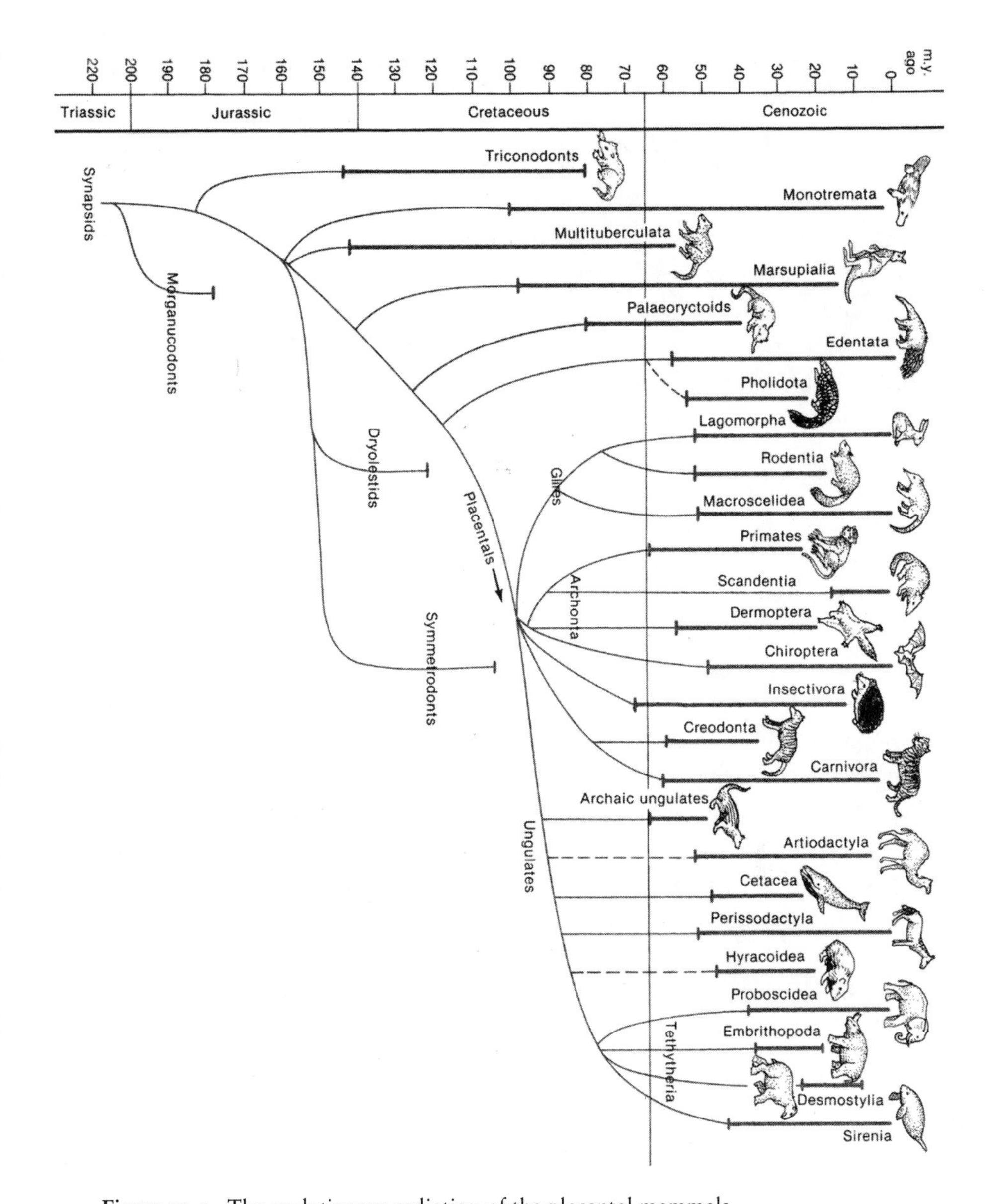

Figure 12.3. The evolutionary radiation of the placental mammals.

(Drawing by Carl Buell. From Donald R. Prothero, *Evolution: What the Fossils Say and Why It Matters* [New York: Columbia University Press, 2007]. Copyright Columbia University Press. Reprinted with permission of the publisher.)

for example, include only two species, the African and Indian elephants, *Loxodonta africana,* and *Elephas maximus,* although a third—*Loxodonta cyclotis*—has recently been proposed. In contrast, named species of prehistoric elephants number at least 150.[3] (See fig. 1.3).

There are five living species of rhinoceroses, but there are more than 200 fossil species, including the biggest land mammal that ever lived (*Paraceratherium,* aka *Baluchitherium*) (fig. 12.2).

In terms of variety of form, environmental adaptation, versatility of diet, and what we—in our mammalian way—describe as intelligence, the mammals have become a supremely successful group. Yet, though they first appeared as fossils in Triassic times, some 208 million years ago, for the rest of Mesozoic time—some 142 million years—they lived in relative obscurity, small shrew-sized creatures, overshadowed by their contemporaries, the dinosaurs.

The modern cladistic classification of mammals, based on their ancestry, rather than their structural resemblances, is illustrated in figure 12.3. This figure illustrates both the complexity of the ancestral relationships involved and the richness and variety of mammalian form and function. We will describe only a few representative members of the 40 or so major mammalian groups that have existed.

Living mammals include three major groups: the placentals or Eutheria, which include the most familiar and by far the largest group, the marsupials or Metatheria, and the monotremes or Monotremata. The marsupials include about 270 living species, such as kangaroos and koalas, while the monotremes include 5 living species, such as the platypus and the spiny echidna. It is these three groups we will now review.

THE MONOTREMES: MONOTREMATA

The monotremes (Monotremata) are the most "primitive" living mammals and include the duck-billed platypus (with a duck bill, webbed feet, and a flat, beaverlike tail) and the echidnas (spiny anteaters) of Australasia. In contrast to most mammals, the monotremes lay eggs, share some "reptilian" skeletal characteristics, and feed their young with milk, secreted from modified sweat glands, which are homologous to the breasts of other mammals. Ogden Nash described this anomalous mixture of characteristics in his poem "The Platypus":

I like the duck-billed platypus
Because it is anomalous.
I like the way it raises its family
Partly birdly, partly mammaly.
I like its independent attitude.
Let no one call it a duck-billed platitude.

The name "monotreme" (derived from the Greek "single opening") describes the single opening—the cloaca—used for reproduction, urination, and defecation. Incidentally, birds and reptiles share this same feature, unlike the placental mammals, which have three: the vagina, the urethra, and the anus.

The oldest known fossil monotremes come from Cretaceous rocks, but it seems likely that they evolved earlier, probably in mid-Jurassic times, 160 million or so years ago.

THE MARSUPIAL MAMMALS: METATHERIA

The second group of mammals are the marsupials (Metatheria), which, along with all other mammals belong to a larger taxonomic group known as Theria or "live-bearing mammals." All give birth to live young by nourishment within the womb from a structure known as the placenta, by which the developing embryo is attached to the uterus of the mother and sustained by oxygen and food from her body.

This structure allows the young to be born at a relatively advanced stage of development. In the larger therian group—the placentals (Eutheria)—the young are highly advanced at birth, but in one small therian group, the marsupials (koalas and kangaroos, for example), the young are born at a relatively early stage and the immature infants carried in a maternal pouch and fed by more or less constant attachment to the mother's teats.

Marsupials were the most abundant mammals of Cretaceous times. About two-thirds of the 330 living marsupials, including the kangaroos, wallabies, wombats, and others, live in Australia and New Guinea, and the rest in South America. Only one species, the opossum, now lives in North America, but fossil marsupials show that the group originated in North America in early Cretaceous times (some 135 million years ago).

Their present scattered distribution reflects both the changing positions of the continental masses over geologic time and their various patterns of migration and subsequent isolation in more recent times. Thus, Australian marsupials have flourished in isolation from most the world except New Guinea. South America was isolated from North America until the elevation of the Isthmus of Panama, some 3.5 million years ago.

Fossil marsupials show a remarkable degree of convergence in outward form to eutherian "placental" mammals, reflecting their adaptation to similar modes of life. Though living marsupials are a relatively small and modest group, their fossil representatives contained some forms that bore an almost uncanny outward resemblance to placental forms. *Thylacosmilus*, for example, was a tiger-sized carnivore, with long, curved, daggerlike teeth, that closely resembled the well-known *Smilodon* saber-toothed cat from the Rancho La Brea tar pits of California. *Borhyaena* was a doglike carnivore from the Miocene of South America that resembled living placental wolves and dogs. We review these creatures in more detail in chapter 13.

Representatives of a second group of marsupials, the edentates (Edentata)—anteaters, sloths, and armadillos—are found chiefly in North and South America. Living forms are animals of modest size, but their fossil representatives include much larger creatures. *Glyptodon* was a giant armadillo as large and as heavy as a Volkswagen Beetle and protected by a thick domelike shell, some 10 feet long, 4 or 5 feet high, and up to 2 inches thick. The animal's head was protected by a strong bony covering and it had a long armored tail that in some species bore a macelike bony knob at the end. It is thought that human hunters preyed on these creatures and may have used their protective domes as shelters. Fossil ground sloths included huge clawed creatures, some 20 feet in length, and weighing up to 8 tons, that developed into many different varieties. These creatures lived in Central and South America from late Pliocene to early Holocene times, from about 5 million years ago, until about 10,000 years ago.

THE PLACENTAL MAMMALS: EUTHERIA

The remaining host of placental mammals are all classified as eutherians, and they give birth to their young at an advanced stage of development. Any brief description of their evolution faces the frustrating problem of how to give even a glimpse of their remarkable diversity and exquisite adaptation. There are some 5,000 species of them. We will review them in chapter 13.

THE ORIGIN OF MAMMALS

Fossil remains of the early mammals are rare, and most consist of fragmentary jaws and teeth. They represent creatures no larger than a shrew, and most were probably nocturnal insectivores, their warm-bloodedness giving them a distinct advantage over the reptiles, whose cold-blooded constitution limited their activities to a largely diurnal existence. It was these lowly creatures that, for most of the long years of the Mesozoic Era, existed inconspicuously alongside the reptilian giants and that ultimately outlived them.

But where did these early mammals originate? Who were *their* ancestors? The most distinctive mammalian characteristics—warm-bloodedness, a covering of hair or fur, the presence of mammary glands, for example—are not preserved in fossil forms. Other distinctive mammalian features, however, include such things as the skull structure, jaw articulation and structure, and the form and fit of the teeth, all of which are preserved as fossils. We find expression of these features in a group of reptiles that were abundant in Triassic times—250–200 million years ago—in the southern continents that made up Gondwanaland. These were the therapsids, which appeared in Late Permian times and included a variety of both carnivorous and herbivorous forms. One particular therapsid group—the cynodonts—seems to have been ancestral to the mammals. These creatures displayed some characteristics so intermediate between reptiles and mammals that they seem to blur the distinction between the two groups. The earliest therapsids show the typical reptilian jaw structure. *Cynognathus*—a typical cynodont—had an upright posture, was about 3 feet in length, and had a secondary palate, allowing it to eat and breathe at one and the same time. *Cynognathus* is essentially a therapsid wolf, and another striking example of evolutionary convergence. *Cynognathus*, whose name, incidentally, means "dog jaw," had a mammal-like skull, with specialized teeth for both biting and chewing. But in many features it remained reptilian, representing an unambiguously transitional form between the two groups. Cynodonts existed from late Permian to Early Cretaceous times (some 260–140 million years ago). Some mammalian features, including teeth, jaw form and articulation, brain size, locomotion, and hearing, appeared in no fewer than six different subgroups of cynodonts (fig. 12.5).

The cynodonts themselves emerged in Permian times from "pelycosaur" stock, which included such creatures as *Dimetrodon* and other sail-backed forms.

The "pelycosaurs"—a polyphyletic group of mixed ancestry (fig. 12.5)—included two distinct suborders: the carnivorous sphenacodonts and the herbivorous edaphosaurs. The sphenacodonts had strongly differentiated, daggerlike teeth, set within powerful, gaping jaws. They reached a length of some 11 feet and had a semiupright posture. Their most distinctive feature was the great elongation of the vertebrae to form a striking sail-like structure from the neck to the hips. This structure is thought to have supported a web of skin; there has been speculation as to its function, with suggestions ranging from sexual display to protection. Perhaps the most persuasive suggestion is that it assisted in temperature regulation, with the animal adjusting its body temperature by orienting its body in relation to the sun.

The other related group were the edaphosaurs, characterized by *Edaphosaurus*, which, in spite of their superficial similarity to the carnivorous sphenacodonts, were herbivorous. The small skull carried uniform teeth, as well as teeth developed on the palate, and the spine—like that of *Dimetrodon*—carried a "sail" that, unlike that of *Dimetrodon*, carried short horizontal extensions, like yards on the masts of a sailing ship. Adding to the enigma, some other members of the edaphosaurs carried no sail at all. Pelycosaurs are found in rocks ranging in age from Late Pennsylvanian to Late Permian (318–254 million years ago) and are especially well represented from the American southwest.

The great extinction of Permo-Triassic times (some 252 million years ago) produced a wholesale turnover of terrestrial vertebrates, with the decline of the therapsid reptiles—the mammals' ancestors—and the rapid expansion of crocodilians, dinosaurs, pterosaurs, and birds. It is soon after this time, in rocks of late Triassic age, that the oldest known true mammals appear, about 230 million years ago, but the group may well have arisen in earlier times from the therapsid reptiles, which, as we saw above, share the same synapsid skull structure. It is noteworthy that the question whether some members of the therapsid reptiles are, or are not, mammals, rather than reptiles is being debated. This blending of characteristics seems, in fact, to confirm the degree of continuity between the two groups.

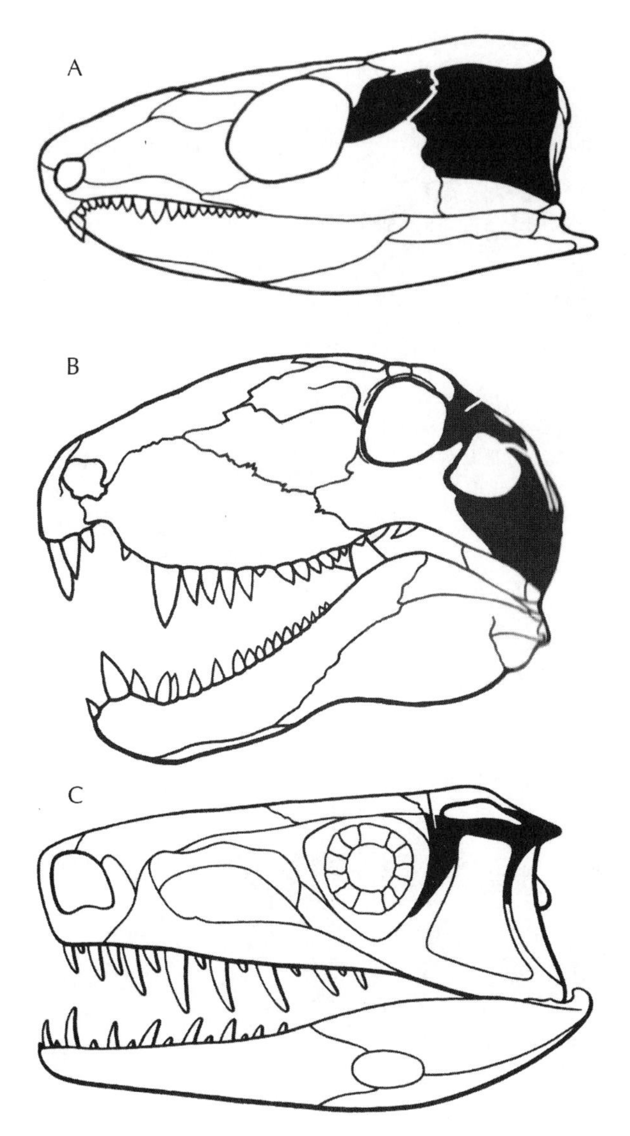

Figure 12.4. Vertebrate skull structure. Relationships of temporal bones and openings in reptiles, illustrated by the skulls of characteristic genera and by schematic diagrams. (A) Anapsid skull with no temporal opening, as in the captorhinid *Captorhinus;* (B) Synapsid skull with a lateral tempora fenestra bounded above by the postorbital and squamosal bones, as in the pelycosaur *Dimetrodon;* (C) Diapsid skull with superior and lateral temporal fenestrae separated by the postorbital and squamosal bones, as in the thecodont *Euparkeria.* Not to scale.

(From Edwin H. Colbert, *Colbert's Evolution of the Vertebrates* [New York: Wiley-Liss, 2001], fig. 9-6, p. 132. Reproduced with permission of the publisher. Permission conveyed through Copyright Clearance Center, Inc.)

Early mammal *(Megazostrodon)*

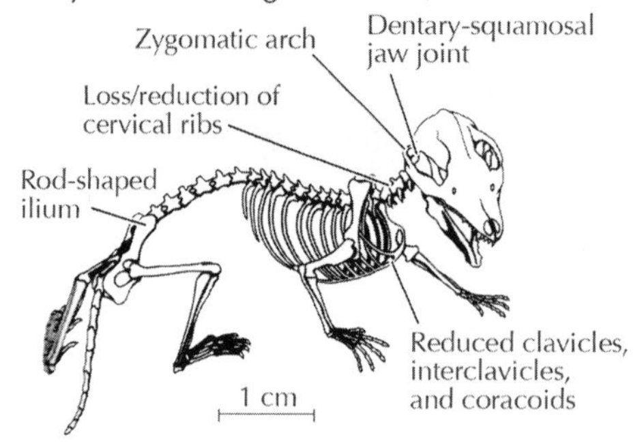

Cynodont therapsid *(Thrinaxodon)*

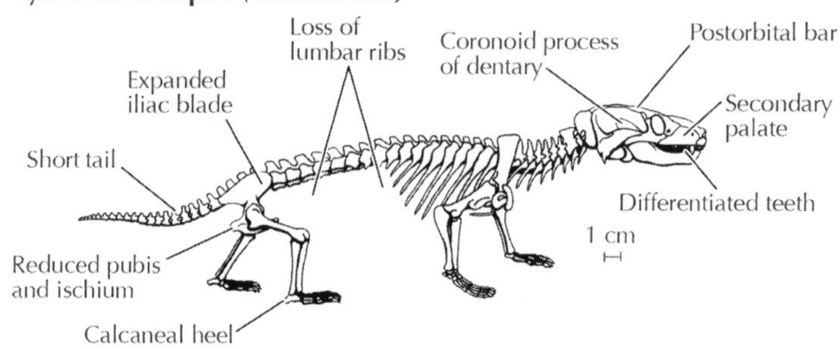

Noncynodont therapsid *(Lycaenops)*

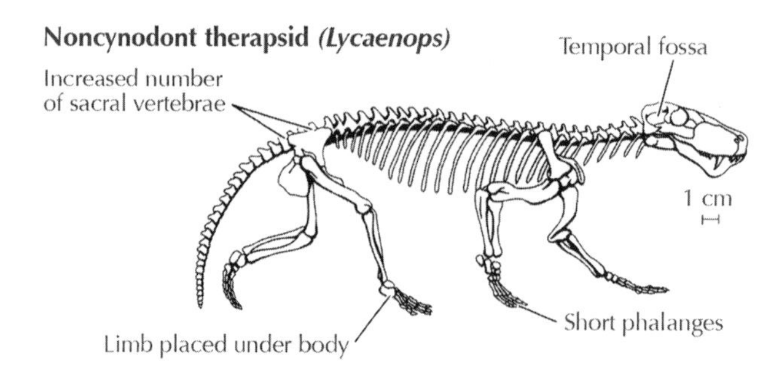

Pelycosaur *(Haptodus)*

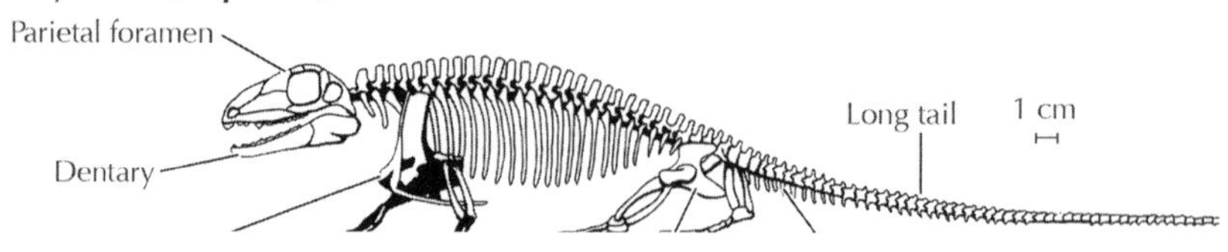

Figure 12.5. Transformation of the synapsid skeleton from "pelycosaurs" through cynodonts to true mammals.

(From Donald R. Prothero, *Evolution: What the Fossils Say and Why It Matters* [New York: Columbia University Press, 2007], fig. 13.4, p. 275. Reproduced with permission of the publisher.)

THE EARLIEST MAMMALS

The oldest known fossil mammals are fragmentary triconodont-like fossils from the Late Triassic (about 208 million years ago). The search for the earliest mammals has recently been complicated by the discovery of two new fossil species reported from China in late 2013[4] that have both expanded and confused the debate. The first, the 160-million-year-old *Arboroharamiya*, was a small arboreal creature with a jaw that was more mammalian than reptilian; it was long-fingered and either herbivorous or omnivorous.

The second, the 164–165-million-year-old *Megaconus*, was probably a dweller on the forest floor and about the size of a chipmunk. Neither of these two new genera seems to have been a "true mammal," but their exact affinities are unclear. More collecting and more analysis are needed to establish the affinities and to reconstruct the earliest history of mammals.

The first reasonably well-represented mammals, however, were the so-called *Morganucodonta* (fig. 12.6), which, not surprisingly, share some reptilian features. One of them, *Hadrocodium,* from 230-million-year-old rocks in southwestern China, was a tiny shrewlike creature, less than an inch and a half in length, but with an exceptionally large brain (*hadro* = large; *codium* = head) and a middle-ear structure that is mammalian. It may well have been nocturnal in its habits. Its diet probably consisted of small insects and worms. There is some debate as to whether this creature—or the larger group it represents—is a "full mammal."

The most abundant of the early fossil mammals were the multituberculates, named for the cusplike structure of their teeth. Teeth are the most readily preserved fossil elements of vertebrate skeletons, and the variety of both form and function of multituberculate teeth suggests the great variety of small rodent or squirrel-like creatures, living from at least early Triassic to late Eocene times, some 250–38 million years ago. Triconodonts were a group of shrew-sized early mammals that may have been ancestral to the multituberculates. Another recently discovered fossil mammal—*Juramaia* ("Jurassic mother") from 160-million-year-old rocks in northeast China—was a shrew-sized creature.

The multituberculates must have been rodentlike in their general appearance, and this similarity is a striking example of both evolutionary convergence and successful ecological replacement by rodents. As G. G. Simpson pointed out, the earliest known rodents barely overlap the latest known multituberculates in the western United States.

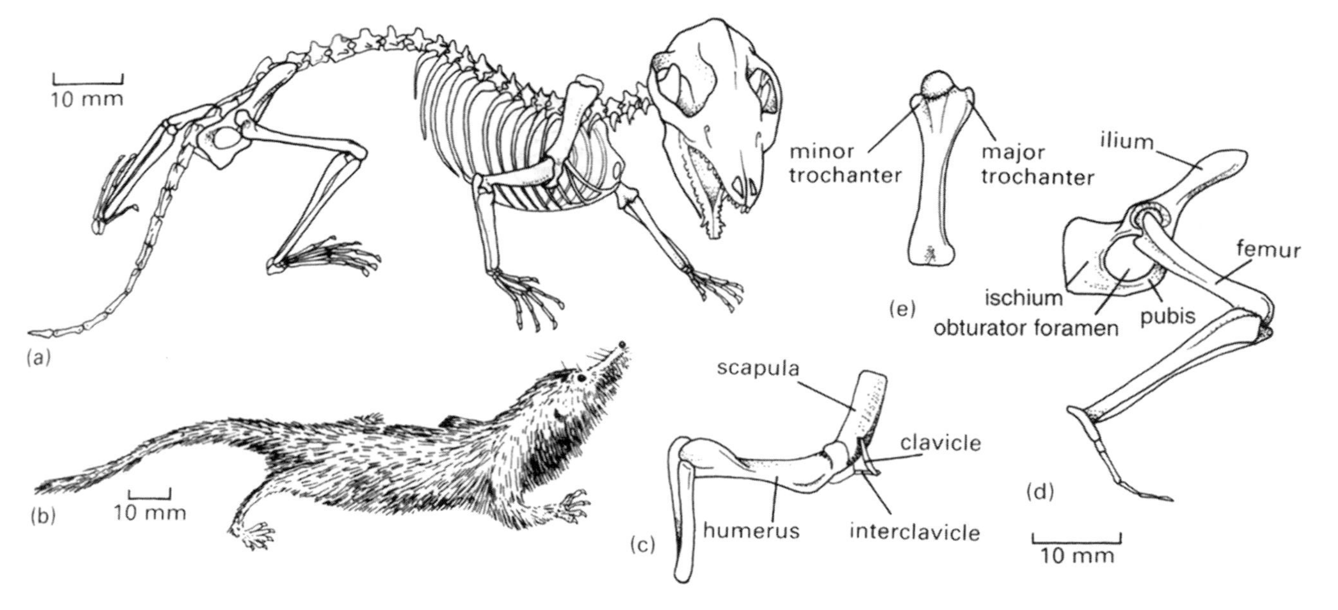

Figure 12.6. Early mammals—morganucodontids. The skeleton of morganucodontids: (a) Skeleton of *Megazostrodon;* (b) Body restoration; (c) Forelimb and pectoral girdle; (d) Hind limb and pelvic girdle of *Morganucodon;* (e) Femur.

(After Jenkins and Parrington, 1976. From Michael Benton, *Vertebrate Paleontology*, 3rd ed. [Oxford, U.K.: Wiley-Blackwell, 2004], fig. 10.7, p. 303. Reproduced with permission.)

Juramaia is the earliest known representative of the eutherian mammals, the group that includes the placental mammals. It has been described as "either a great-grand-aunt or a great-grandmother" of all placental mammals. This fossil from China is remarkably preserved. Until its discovery in 2011, the oldest known eutherian was the 125-million-year-old *Eomaia*, but DNA studies of eutherians suggest that they developed much earlier, about 160 million years ago. *Juramaia* now resolves that dilemma. The forelimbs of this small creature appear to be adapted for climbing, a mode of life that would have provided both food and protection in the tree canopy above the carnivorous competitors of the reptilian world in which *Juramaia* lived.

Throughout the long years of reptilian dominance, the mammals remained rather small, inconspicuous creatures, but with the rapid decline of the ruling reptiles in late Cretaceous times, some 66 million years ago, mammals underwent rapid development into the environmental niches from which they had earlier been excluded. Some mammalian characteristics, such as endothermy ("warm-bloodedness") and a covering of hair, must have increased their ability to colonize these new environments. Their expansion was explosive, and in a brief span of 10,000 years, we find evidence of the first appearance of many of the major mammalian groups, not only those of terrestrial habit, but also whales and bats. Several thousand fossil species are known from this brief interval of time, reflecting a marvelous pattern of adaptive radiation. It is to that explosive radiation that we turn in the next chapter.

13

THE MAMMALIAN EXPLOSION

THE RISE OF THE PLACENTAL MAMMALS

If the fossil record allows the recognition of such things as dynasties, then the overwhelming dominance of placental mammals over the last 80 million years or so would provide an outstanding example. More than 2,500 genera have been recognized, inhabiting every environment from treetops to burrows on the land, in the air, and from the shallow seas to the ocean depths. Manatees and monkeys, camels and cows, rabbits and rodents, dogs and donkeys, whales and walruses, bats and buffalo, elephants and edentates, hedgehogs and humans—all these and countless more have crept, run, swum, flown, and plodded across the world's stage.

The distinctive feature of this extraordinary group is that they bear live young, with the embryo being nourished within the mother's womb by a placenta, which is fixed to the wall of the uterus. Marsupial mammals also have placentas, but in them the young are born at a much less advanced stage of development and are sheltered after birth within the pouch of the mother. In order to distinguish between the two groups, the "true" placental mammals (that is, the nonmarsupial mammals) are frequently described as Eutheria, as described in the previous chapter. In some of the Eutheria, such as the hoofed mammals, the young are born at a very advanced stage, whereas in others, such as humans, they are born without functional teeth and are dependent on a long period of nurturing by the mother. Most placental mammals have an enlarged braincase, and the teeth are differentiated into a typical pattern of three incisors, one canine, four premolars, and three molars on each side of the upper and lower jaws.

The placental mammals are organized into five major groups, and we will select a few of these to illustrate some of the remarkable changes they have undergone over the course of Cenozoic time.

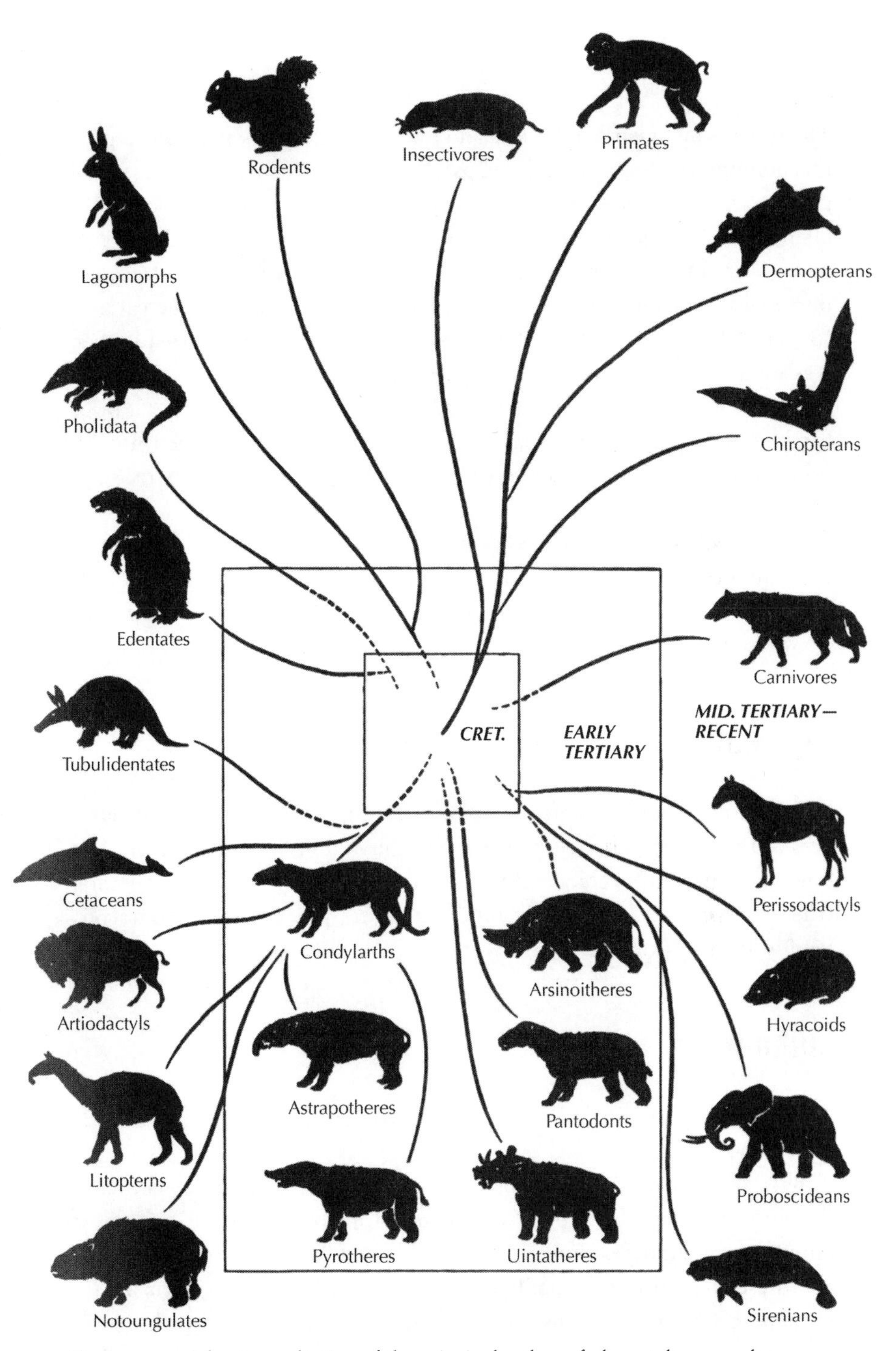

Figure 13.1. Adaptive radiation of the principal orders of placental mammals.

(Modified from an earlier diagram prepared by Lois M. Darling. From Edwin H. Colbert, *Colbert's Evolution of the Vertebrates* [New York: Wiley-Liss, 2001], fig. 21-4, p. 313. Reproduced with permission of the publisher. Permission conveyed through Copyright Clearance Center, Inc.)

RODENTS

The rodents include not only rats and mice, but also squirrels, guinea pigs, and porcupines. Rodents are, perhaps, the most successful of all mammalian groups, with 1,700 species described, representing more than 50 percent of all living mammalian species. They have successfully established themselves in a variety of environments, from aquatic beavers to flying squirrels and burrowing mole rats. Though living forms tend to be rather small creatures, some fossil forms were huge: *Castoroides,* a Pleistocene bear-sized beaver, was some 7 feet long and weighed an estimated 450 pounds.

The evolutionary success of the rodents comes at something of a price for humans. The bubonic plague, for example, which ravaged human populations of Europe in the fourteenth century, was carried by rats and their fleas.

Almost all rodents are herbivorous and have chisel-like incisors in their upper and lower jaws, and these are particularly well seen in beavers. These teeth are well suited to the diet of grubs, seeds, and vegetation on which many rodents depend.

LAGOMORPHS

Rabbits and pikas or conies, as well as the elephant shrews, are members of this group. They seem to have originated in Paleocene times. The long hind limbs of rabbits and hares make them swift and agile runners, whereas the pikas, with short hind limbs, are burrowers. Lagomorphs tend to be prolific in numbers.

CHIROPTERA

This is a group that includes the bats (Chiroptera), tree shrews, and gliding colugos.

Bats are poorly known as fossils, their habits and lightly constructed skeletons making them poor candidates for fossilization. The oldest known representatives come from the Eocene of Wyoming and elsewhere, and these were already broadly similar in structure to living forms. Bats are superbly adapted for sustained flying, their front limbs, together with four fingers, being greatly extended to support the membranous wing (see fig. 10. 2). The claws of the hind feet allow bats to hang vertically. The large fruit-eating bats of the Old World and Pacific, including the flying foxes of Java, with a wingspan of 5 feet, live in trees and are diurnal in habit. In contrast, the generally smaller insectivorous bats, which are worldwide in distribution, live in caves and have a unique system of echolocation that allows them to be nocturnal feeders, capturing insects on the wing, thus giving them a competitive advantage over birds and other groups. Bats are the second largest group of living mammals, after rodents, being represented by more than 1,200 species.

Some alternative classifications group the primates with the bats, although the two could scarcely appear to be more different. But a closer look shows the two groups to share a number of primitive structural features, such as, for example, the five-digit structure of the hands. There are similarities, too, in the structure of the braincase. In the present account we will treat them separately. We will review the history of primates in chapter 14.

CARNIVORA

The carnivorans are a large and successful living group related to the extinct condylarths, which flourished in Paleocene times, and the creodonts, large predators that became extinct at the end of Miocene times (5.3 million years ago). The carnivores include living cats (felids), civets, mongoose and hyenas, dogs (canids), and bears (ursids), as well as raccoons, otters, weasels and badgers, and the related pinnipeds: seals, sea lions, and walruses. Not all carnivores are carnivorous: pandas, for example, are vegetarians, while bears and raccoons are omnivorous.

The oldest members of this group are the creodonts, which existed during early Cenozoic times. Some authors place the creodonts in a separate taxonomic group. Creodonts showed a great diversity of form, ranging from weasel-like creatures to savage beasts as large as lions. Their teeth were broadly similar in form to those of modern carnivores. The decline of the creodonts in late Eocene times (some 40–36.6 million years ago) was marked by the diversification of their successors into two broad groups—the Caniforms (dogs, raccoons, bears, badgers, weasels, and seals) and the Feliforms (cats, mongooses, hyenas, and related forms).

Seals, sea lions, and walrus appear unlikely associates of these groups, but both anatomical and fossil evidence support their affinity with early bearlike forms.

The cats (feliforms) underwent rapid diversification in Miocene times. For all carnivorous mammals, the physical demands of a predatory, hunting existence were substantial, including the need for strong jaws with sharp canine stabbing teeth, as well as sharp claws for attacking prey, well-developed, bladelike slicing teeth for cutting it, well-developed brains, and lithe, muscular limbs for successful hunting, together with a keen sense of smell and acute vision. Living cats and most fossil forms display these characteristic features, but one extinct group—the saber-tooth cats, which existed from Eocene to the end of Pleistocene times—represent a contrasting direction of development. Whereas in most cats, the daggerlike canine teeth were gradually reduced in size, in the saber-tooths they remained very large, probably as an effective means of preying on the large bison, deer, and other animals of late Tertiary times, including, perhaps, young mastodonts. With the decline of these large creatures toward the close of the Pleistocene, the saber-toothed cats also became extinct.

The canid (doglike) carnivores appeared in mid-Eocene times (some 45 million years ago) and underwent rapid diversification into the wolves, dogs, foxes, and their relatives. They developed long limbs, large brains, and sharp teeth for hunting. Hunting has involved, not only individual pursuit, as in foxes, but also the development of family groups and packs and the social structure to support them. Dogs became the first animals to be domesticated by humans and to assist them in various tasks, such as hunting and herding. It seems probable that it was from doglike ancestors that bears, raccoons, and perhaps pandas developed, sometime in the mid-Cenozoic.

The pinniped carnivores include sea lions, walrus, and seals, and they too are members of the carnivore group. Their skeletal structure and fossil history make it clear that they are derived from terrestrial canine ancestors whose modified limbs now serve the pinnipeds as flippers, which they use both for propulsion and balance in the water and for support and ungainly movement on the land. The teeth have been modified in seals for capturing fish and in walrus for crushing shellfish.

Whether judged by their variety and versatility, or by their position at the top of the food chain, the carnivores are one of the more successful living groups.

Ungulates include the hoofed mammals and their descendants. But it is now recognized that other groups shared a common ancestor with the ungulates, and we therefore review the following five related groups separately:

- The *Artiodactyla* include the even-toed (cloven-hoofed) mammals, such as pigs, deer, camels, cattle, and hippopotamuses.
- The *Perissodactyla* include horses, tapirs, and rhinos, whose weight rests on the middle toe.
- The *Cetacea* embrace the whales, dolphins, and porpoises, which are carnivores, whose feet have been transformed into flippers.
- The *Proboscidea* include the elephants and their many prehistoric forebears.
- The *Sirenia* include the manatees or sea cows.

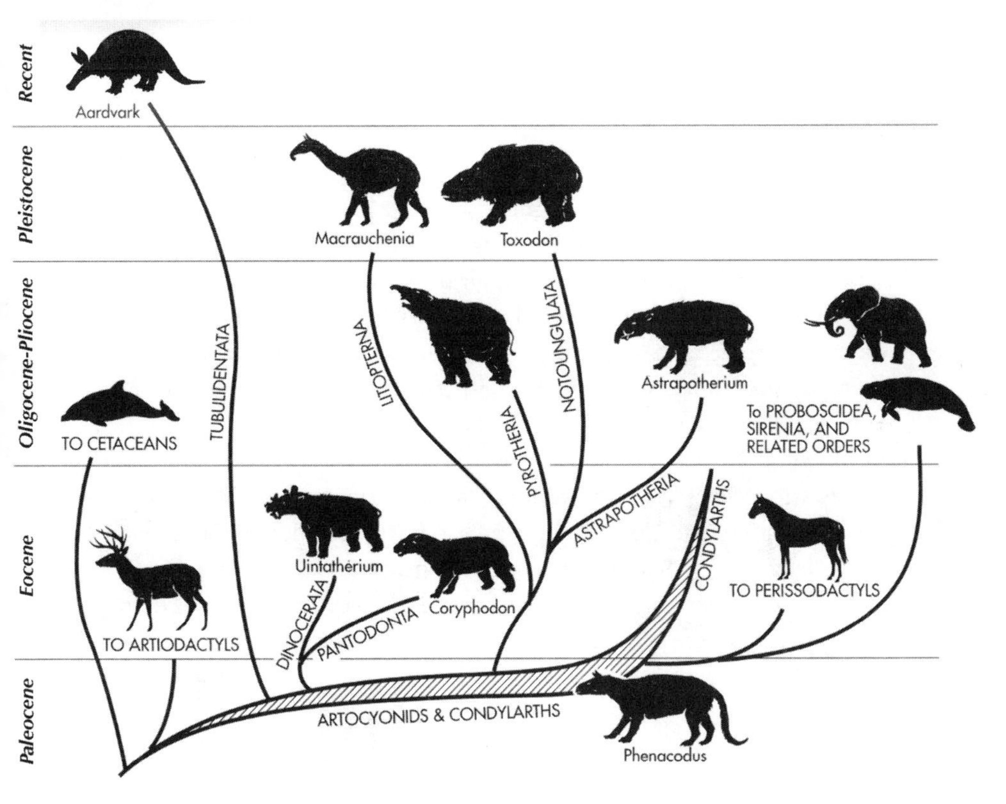

Figure 13.2. The relationships of the primitive ungulate.

(Portions modified from earlier drawings by Lois M. Darling. From Edwin H. Colbert, *Colbert's Evolution of the Vertebrates* [New York: Wiley-Liss, 2001], fig. 27-5, p. 407. Reproduced with permission of the publisher. Permission conveyed through Copyright Clearance Center, Inc.)

Ungulates constitute about a third of all mammalian genera. Most are herbivores, typically of substantial size.

Ungulates emerged in the Cretaceous and began to diversify in early Tertiary times, some 50 million years ago. The early ungulates probably arose from "condylarth" ancestors, though we should note that "condylarth" is also a "catch-all" polyphyletic group. The earliest known condylarths are found in the late Cretaceous, and they were small creatures, known only from their skulls, which had blunt, broad grinding teeth. Later (Paleocene and Eocene) members of the group grew to substantial size. The pantodonts or amblypods ("slow-footed") and dinoceres were more massive creatures, some as large as a rhinoceros, with six horny protuberances on their heads with heavy bodies, pillarlike limbs, and broad feet. They had prominent teeth, including most unherbivore-like enlarged, protruding canines. They became extinct in Oligocene times but included a variety of forms.

ARTIODACTYLA

The *artiodactyls* (even-toed) mammals include many of the most familiar living forms: pigs, cattle, sheep, goats, and deer, for example, as well as camels, llamas, alpacas, hippos, and giraffes. In these, the weight is borne on two toes (the third and fourth) rather than on one, as in horses. They are sometimes called "split-toes" in contrast to even-toed horses. All are herbivores, and they play an important role in providing our everyday needs, including meat, milk, and wool. Their spread between Asia, Europe, and North and South America represents a fascinating pattern of migration, diversification, and extinction. The development by cattle, deer, antelope, pigs, goats, sheep, and giraffes of a ruminant stomach—a very efficient, four-chambered digestive system that allows time to "eat in haste and digest at leisure"—gave them such an evolutionary advantage over other ungulate groups that they became the most successful ungulate group in late Tertiary times.

CETACEA

The *cetaceans* include whales, dolphins, and porpoises. It was long believed that a remarkable series of fossils indicated whales that had developed from a group of long-jawed, wolf-sized, hoofed, carnivorous predators—mesonychids—that flourished in late Paleocene and early Eocene times, some 55 million years ago. The similarity of tooth structure seemed to support this view. On the other hand, recent molecular phylogeny studies suggest the derivation of whales from hippopotamus-like artiodactyls. The recent discovery of a group of terrestrial hoofed mammals—the pakicetids—in the early Eocene of Pakistan, whose teeth show them to be carnivorous, but whose general skull structure links them to whales and whose ear region is intermediate between terrestrial and aquatic mammals, supports this view. From some such creatures as these of Eocene times, the two main modern groups of whales emerged: the predatory toothed whales (including killer whales, sperm whales, porpoises, and dolphins), and the baleen filter-feeders (including the right whales, humpbacks, blue whale, and others) (see fig. 4.2).

The change from predatory life on the land to a predatory existence in the oceans involved every aspect of the creatures' existence, but especially changes in locomotion, reproduction, and respiration. Locomotion involved, not only the transformation of the forelimbs into flippers, but the development of an elongated, torpedo-like body form, and smooth, hairless skin, below which a thick layer of blubber replaced the insulating properties of the hair, and a strong horizontal tail fin (as opposed to the vertical tail fin of fishes), which by greyhound-like vertical undulations of the body provides powerful propulsion. Successful reproduction in the oceans has led to the young being born at a very advanced stage, able both to breathe air at the surface and to swim powerfully, though the young are still nursed by the mother for an extended period. Whales have remained air breathers, and respiration has been assisted by the migration of the nostrils to the back of the skull and the development of a blowhole, which can be closed during diving. Some whales can remain submerged for as long as an hour, often at great depths. The sonar apparatus by which whales navigate is formed by changes in the bones of the ear, and the brain is highly developed. Given the buoyant support provided by the water, many whales have grown to gigantic size. The largest living member, the blue whale, can reach 100 feet in length.

PERISSODACTYLA

The *perissodactyls* include horses, tapirs, rhinos, and their extinct relatives. All have an odd number of toes, and the weight of the body rests on the middle toe, with the inner toe (the thumb or the big toe) being suppressed. Horses have but one toe, tapirs and rhinos have three.

The perissodactyl ungulates also include the extinct brontotheres and chalicotheres, as well as tapirs and rhinoceroses, and the diminutive hyraxes or conies. All these various groups became established in North America, Europe, and Asia by early Eocene times, some 55 million years ago. Brontotheres or titanotheres developed from creatures the size of large pigs to Oligocene forms that reached a height of 8 feet, with large, blunt horns above the nose. Their rapid evolution culminated in their virtual extinction by the end of Eocene times, some 34 million years ago, perhaps because their teeth were less well adapted to the harsh grasses of the prairies that replaced the once widespread forests of early Tertiary times as the Earth began to experience worldwide cooling (fig. 13.3A).

Horses, which we discuss later in this chapter, are, of course, the classic example of evolution.

The related chalicotheres also appeared in Eocene times but persisted until the late Pleistocene, when they and so many other large mammals became extinct. Horselike in size and in general appearance, they differed from horses in having pillarlike limbs, with clawed feet (fig. 13.3B).

Rhinos and tapirs constitute another related perissodactyl group. Rhinoceroses are a large and successful group whose present paucity in numbers of species (two in Africa and three in Asia) and of individuals belies their earlier diversity and success. Fossil forms range all the way from creatures only the size of a dog to others that were huge animals. *Paraceratherium* (*Baluchitherium*), the largest land mammal of all time, was 18 feet high at the shoulder and is estimated to have weighed some 44,000 pounds. Earlier rhinoceroses were hornless, but later forms included horned varieties, one of which—the woolly rhino—was widespread in Europe during the ice ages of Pleistocene times. The horn in rhinoceroses is made of hair and not of bone. Some fossil genera had two horns, situated side by side; others had two horns situated in tandem. Living rhinos hover at the brink of extinction, in large part because of hunting by humans and loss of habitat.

The related tapirs are limited today to four species, distributed in Malaysia and Central and South America. They are hog-sized animals, up to about 7 feet in length, weighing up to 600 pounds, with low-crowned teeth and a distinctive elongated, trunklike nose. Earlier forms are known from Eocene times onward and were a relatively conservative group, even though they were widely distributed throughout North America and Eurasia. The (surprisingly) related hyraxes (conies), now found in Africa and

Figure 13.3A. Comparison of one of the earliest titanotheres with one of the latest. The Oligocene titanothere *Brontotherium* stood about 8 feet high at the shoulders.

(Prepared by Lois M. Darling. From Edwin H. Colbert, *Colbert's Evolution of the Vertebrates* [New York: Wiley-Liss, 2001], fig. 30-11, p. 468. Reproduced with permission of the publisher. Permission conveyed through Copyright Clearance Center, Inc.)

Figure 13.3B A Miocene chalicothere, a clawed perissodactyl about as large as a modern horse.

(Prepared by Lois M. Darling. From Edwin H. Colbert, *Colbert's Evolution of the Vertebrates* [New York: Wiley-Liss, 2001], fig. 30-12, p. 470. Reproduced with permission of the publisher. Permission conveyed through Copyright Clearance Center, Inc.)

the Middle East, are diminutive animals that, though they look like rabbits or rodents, are related to the hoofed mammals. Their feet end in small hooflike nails. They are small herbivores with prominent, chisel-like teeth, though some earlier forms grew to the size of a large dog.

The evolution of some of these groups has been studied in great detail, and that of the horses provides a wonderful example of the evolutionary process, and of the intricate interactions among function, form, and environment.

PROBOSCIDEA

The Proboscideans include the two species of living elephants and the great variety of their fossil antecedents. Henry Fairfield Osborn, the distinguished early twentieth-century head of paleontology at the American Museum of Natural History, identified no fewer than 352 fossil species and subspecies. The earliest members of this group appeared in the Eocene of Egypt and were creatures without tusks or long trunks. *Moeritherium*, a member of this earliest group, was a heavyset, pig-sized creature with stout legs. It may not itself be directly ancestral to later elephants, but it is close to that line. Originating in Africa, proboscideans became worldwide in distribution and developed a great variety of forms, with tusks of various form and length. Prehistoric elephants were widely distributed in North America, Europe, and Asia and were contemporaries of early humans, who hunted them. Living elephants have been domesticated and now perform a variety of human tasks. The evolution of elephants was marked by a general increase in overall size, development of tusks in one or both jaws by transformation of the second incisor teeth, a huge relative increase in size of the head, growth of a trunk by elongation of the nose, and development of pillarlike limbs. A dazzling variety of forms—deinotheres, gomphotheres, trilophodonts, stegodonts, mastodonts, mammoths, and more—are represented. We have already discussed something of prehistoric elephants in chapter 1.

SIRENIA

Another related mammalian group are the living manatees, sea cows or dugongs, which are slow-moving, clumsily built, herbivorous creatures of rivers, estuaries, and coastal waters, in which the limbs have been adapted to an aquatic life by transformation into paddlelike flippers, with loss of the hind limbs and the development of flukelike tails. These large, ungainly, sad-looking creatures (the Sirenia) are thought to be the "sirens," or mermaids, reported in legends to have lured sailors to their doom. Like other mammals, they nurse their young. They are herbivores, eating a huge daily diet of sea grasses. There are four living species, all social animals, each of which grows some 10–12 feet long. A fifth species, Stellar's sea cow, was hunted to extinction in the late eighteenth century.

"INSECTIVORA"

The insectivorans ("insect-eaters") was long used in classification to include moles, hedgehogs, tree shrews, elephant shrews, and other small, insect-eating mammals. Molecular evidence, however, indicates that they do not represent a single order but rather are members of other different groups. We will not review them in detail here.

SOUTH AMERICAN MAMMALS

The mammals with which we are most familiar tend to be those characteristic of North America, Europe, Asia, and Africa, and in general these areas today still exhibit a strong faunal similarity (although we have noted some differences). The explanation lies in their geological history, for they were periodically, though frequently, connected during Tertiary times. There were, of course, and still are local barriers and restricted connections and therefore differences; but there remains a broad similarity between all these continental faunas. Once we look at the Cenozoic faunas of South America, however, we find a completely distinct group of animals. South and Central America (the Neotropical Realm, as zoologists call it) were linked to North America in early Cenozoic times. From then until the late Pliocene, however, 60 million years later, this Neotropical Realm was isolated. It is this isolation that is the clue to its distinctive fauna, just as it is in the case of Australia. Both the marsupials and the placental mammals became established at the dawn of the Cenozoic Era, and in the Northern Hemisphere the marsupials were completely overshadowed by the placental expansion. But in the Southern Hemisphere, they held their own. They became, in fact, the chief carnivores, some of them being wolflike creatures and others growing to become savage predators the size of lions, and bearing a remarkable resemblance to the saber-toothed cats of other areas. These two latter groups were quite unrelated in origin; their similarity was due to similar coincidental adaptation to similar modes of life—a perfect example of convergent evolution (fig. 13.4).

The hoofed marsupial mammals of South America were an equally distinctive group, including five orders of great variety. Some of the creatures were of considerable size: *Toxodon,* a massive Pleistocene form, stood 6 feet high at the shoulder, but many were dog-sized and others still smaller. Many of these South American forms were strikingly similar to contemporary hippopotamuses, horses, camels, elephants, and other creatures in other parts of the world, but here again the resemblance was the result of adaptation to similar environments, rather than immediate genetic affinity. These hoofed mammals flourished throughout the Tertiary, but like the marsupial carnivores, they became extinct in Late Pliocene times, as hordes of North American placental mammals filtered southward over the newly raised Panamanian land bridge between the two continents. It was not a one-way traffic, but the South American mammals had almost no success in their colonization in the opposite direction. They represent a unique and major group of hoofed mammals, and their multiplication stemmed from their isolation. But just as, in one sense, isolation can be productive, so, in another, its end may bring catastrophe, even after some 60 million years of progress. For the isolation of the creatures had brought only "partial" natural selection; with the arrival of North American carnivores and ungulates came competition, for which even the spectacular diversity of an era's cloistered change was no match.

The other distinctive group of South American mammals are the edentates, a group that includes two rather different types of animal. One group includes the living armadillos and the extinct glyptodonts. These latter were creatures some 8 or 9 feet in length and protected by a huge domed armor plate. In some of them the tail ended in a savage clublike structure. The second group includes the herbivorous sloths and the anteaters. In late Cenozoic times, some ground sloths achieved spectacular proportions, one reaching a length of 20 feet—rather larger than a good-sized elephant. Both these creatures and the glyptodonts spread into North America when the land connection was reestablished between

Figure 13.4. Convergent evolution among North and South American mammals. Drawn to same scale.

(After Simpson. From Frank Rhodes, *Evolution of Life* [New York: Penguin, 1976], fig. 45, p. 247. Reproduced with permission.)

the two continents. Their remains have been found associated with stone weapons of early humans, and they probably became extinct in only comparatively recent times.

The other members of the South American fauna, the New World monkeys and some rodents, were later arrivals that migrated into the continent in late Eocene times. Though both are characteristic of the continent, their relatively closer resemblance to forms from other continents is a measure of their shorter period of isolation than other South American forms. More recent arrivals—deer, for example—are even more similar to their North American counterparts, while a few other forms, such as tapirs and llamas, are remnants of a group now extinct in North America.

This distinctive South American fauna of glyptodonts, giant sloths, carnivorous marsupials, and diverse ungulates again illustrates the overriding importance of geographical isolation in the evolutionary process. The same is true of Australia, which has been isolated since Cretaceous times. There, as we have already seen, the marsupials faced no competition from the placentals, and their diversity in that continent seems to be the direct result. Isolation is a factor that operates at all levels—from the level of whole continental faunas, as in this case, to the level of a single species, as in many birds. For wherever populations, of whatever size, are cut off from interbreeding with their neighbors, new species are in the making.

The person to whom we owe an enormous debt for our knowledge of the evolution of South American mammals and, indeed, the whole evolutionary process, is George Gaylord Simpson (1902–1984), described as "the most influential paleontologist of the twentieth century." He was one of the founders of the "synthetic theory of evolution," which combined the Darwinian theory of natural selection with Mendelian genetics—as interpreted by Sewall Wright—and wedded both to paleontology.

Simpson was born in Chicago, the third child of Joseph and Helen Simpson. His father was an attorney who became active in land acquisitions and mining in the west. A precocious, delicate, and rather lonely youth, he was educated at the University of Colorado, Boulder and at Yale.

While a senior at Yale, Simpson secretly married Lydia Pedroja, with whom he had four daughters to whom he was devoted. Simpson continued in graduate school at Yale and, after receiving a PhD, spent time at the British Museum of Natural History, studying fossil mammals. On his return, he was appointed assistant curator of fossil vertebrates at the American Museum of Natural History. With his second wife, Anne Roe, whom he had known since childhood and married in 1938, Simpson wrote a joint book on *Quantitative Zoology*[1] that was published in 1939 and became an influential textbook.

Simpson volunteered for the U.S. Army in 1942 and was commissioned in Military Intelligence, serving in Italy and North Africa before being released because of poor health with the rank of major. Returning to the American Museum of Natural History in 1944, he was appointed chair of the Department of Geology and Paleontology and, soon afterward, professor of vertebrate paleontology at Columbia University. Over the next decade, Simpson undertook extensive fieldwork in the United States and South America. During work at the headwaters of the Amazon in Brazil, he was seriously injured by a falling tree and spent the next two years in and out of hospitals. This accident limited the scope of his future fieldwork. Simpson moved to Harvard in 1959, serving as Alexander Agassiz Professor and remaining there until 1970. He was appointed professor of geosciences at the University of Arizona in 1967 and divided his time between there and Harvard.

Simpson's contributions to the study of evolution were monumental. His book *Tempo and Mode in Evolution*[2] distinguished between the rate of change (tempo) and the pattern of change (mode). He identified three types of evolution: speciation, phyletic, and quantum. He challenged the then-prevailing concept of orthogenesis, for example, by demonstrating that, in contrast to the interpretation of horse evolution as a linear process, it actually involved a far more complex pattern of migration, adaptation, and extinction.[3] He demonstrated that apparent gaps in the fossil record represented rapid evolution in small populations.

Simpson also created a comprehensive classification of mammals that remained the definitive scheme until the recent development of cladistics.[4]

Simpson was a shy, sensitive man who attracted relatively few graduate students but reached the wider world largely through brilliant writing. In all, he wrote some 500 papers and 8 books.

Simpson's work was widely recognized by the honors he received throughout his long career: the Penrose Medal of the Geological Society of America, Fellow of the Royal Society, the Elliot Medal of the National Academy of Sciences, honorary doctorates from a dozen universities, including Harvard, Oxford, and Cambridge, the Darwin Medal of the Royal Society, and the National Medal of Science. These were worthy honors for a man who did so much, not only to advance our understanding of the nature of evolution, and of the evolution of mammals, but also to energize and re-create the science of paleontology.

This long account of mammalian diversification has taken us from the lowly insectivores, living under the shadow and the threat of their dinosaurian competitors of Mesozoic times, to the mammalian exploitation and dominance in every niche of the terrestrial environment during the ensuing Cenozoic. But, paradoxically, just as with the amphibians, reptiles, and birds, this mastery of a new environment was accompanied by return to the old. Just as reptiles—freed from the amphibians' reproductive constraints—established themselves on the land, then to return to dominate the oceans, so also did the mammals, the most fully and perfectly adapted of all animals to life on the land, early in their radiation return to the seas—equally well adapted. Whales and dolphins are the epitome of adaptation to marine life, just as were the ichthyosaurs, plesiosaurs, geosaurs, and mosasaurs before them.

We have now seen something of the great mammalian radiation that took place in Cenozoic times. It carried mammals into every niche and corner of the Earth, into the air, and into the seas. But one other line of mammalian evolution is of particular interest, for we are its children.

14

THE LEAKEYS' LEGACY

OUT OF AFRICA

It was 1976. The working days were long. The African sun was sweltering and the paleontological fieldwork was laboriously slow. In a playful moment, Andrew Hill, David Western, and Kay Behrensmyer were tossing dried elephant dung at one another. Hill dove to avoid an incoming missile and, as he fell, he noticed the hard rock surface below him seemed to be covered with animal footprints. He and Behrensmyer examined the fossil prints, becoming more and more excited about their number and variety. Later counts showed there to be more than 18,400 of them—ranging from insects to birds, antelope, and rhinoceroses.

Mary Leakey, the expedition leader, had been searching for human remains and tools in East Africa for forty years before her team made this discovery. Her husband and fellow paleontologist, Louis, had made his first expedition to East Africa even earlier, in 1926. Realizing the potential of this exquisite preservation, Mary Leakey quietly concentrated the attention of the whole expedition on the search for more footprints, hoping that further search might turn up something even more significant: human footprints. Then, just as the field season was coming to an end, two other members of the group, Peter Jones and Philip Leakey, her son, discovered what they thought might be hominid footprints.

Further careful collecting from the same locality in 1978 ultimately yielded a treasure: an 89-foot-long trail of footprints, made by three individuals of unequal size, walking together across a new fall of rain-soaked volcanic ash that had even retained imprints of rain drops. The wet, cementlike surface was subsequently buried, and thus preserved, by a later eruption from the nearby volcano, Sandinan, some 3.6 million years ago. The individuals who made the prints are generally thought to have been a male, female, and child[1] who stepped into one another's footprints. The length of stride was short, implying these were short-legged creatures, but the form of the footprints is hardly distinguishable from those of modern humans, with the big toe aligned with the other toes, quite unlike the grasping, divergent big toes of chimpanzees. Realizing the significance of this discovery after excavating one particularly well-preserved footprint, Mary Leakey is reported to have "sat up, lit a cigar, exhaled slowly, and announced, 'Now this is really something to put on the mantelpiece.' "[2]

The partnership of Louis and Mary Leakey was of major importance in establishing the African origin and early evolution of prehistoric human populations. Louis Leakey (1903–1972) was born in a tiny mud and thatch hut in the Kikuyu village of Kabete, near Nairobi. The son of British missionary parents, he learned as a young child the language and customs of the Kikuyu people, and such practical skills as how to stalk and track animals. He underwent secret initiation ceremonies of Kikuyu boys to manhood and often said his first and best language was Kikuyu, the language in which he dreamed. He later wrote the definitive grammar of the Kikuyu language and a three-volume history of the tribe.

Louis Leakey was sent to England for more formal education at Weymouth College after World War I, and in 1922 he gained entry into St. John's College, Cambridge, to study modern languages. Injured in a rugby match, he took a year's leave to assist a Canadian paleontologist, W. E. Cutler, on a dinosaur-collecting expedition in Tanganyika Territory. The expedition provided valuable practical experience in handling fossils, as well as the opportunity to learn Swahili.

He returned to Cambridge to take First Class Honors in Languages and then, in May 1926, to gain a First in Archaeology and Anthropology in Part II of the Tripos. The languages that he chose were French and Kikuyu, and in this latter language he was "placed under the care of Mr. W. Crabtree . . . but Mr. Crabtree knew Luganda, a somewhat similar East African Bantu language, so Leakey was instructed to instruct his instructor in the Kikuyu language. And this he duly and solemnly did, teaching Kikuyu to Mr. Crabtree who was subsequently to set most of Louis's examinations."[3] Leakey later led East African Archaeological Research Expeditions in 1926–27, 1928–29, 1931–32, and 1934–35, having become convinced that humans had originated in Africa. On his early expeditions, Louis Leakey was accompanied by his wife, Henrietta Wilfrider ("Frida"). Leakey later scandalized his colleagues and family by leaving Frida and their two young children and marrying Mary Douglas Nicol (1913–1996), who joined him in his quest for human origins.

Mary had a rebellious youth, being expelled from both Catholic schools in which she had been a student, the second after deliberately blowing up the chemistry lab. She was, however, not only a gifted artist but also an enthusiastic amateur archeologist, and it was these qualities that led Louis Leakey to invite her to illustrate his book *Adam's Ancestors*.

In Africa, Mary's skills came into their own. She studied and described Stone Age rock paintings and made some of the most important fossil discoveries that have been recorded from the region. It was Mary who discovered the 18-million-year-old Miocene ape *Proconsul africanus*. It was she who discovered what was first described as *Zinjanthropus boisei*, a 1.8-million-year-old hominid, now assigned to *Australopithecus* (*Paranthropus*). And all this, while bearing four children and caring for her family. She was, concluded the London Times in an obituary, "the scientific anchor without which her husband . . . might have been dismissed as a mere controversialist with an exotic private life. For every vivid claim made by Louis about the origins of man, the supporting evidence tended to come from Mary, whose scrupulous scientific approach contrasted with his taste for publicity and enjoyment of personal battles."[4]

Sadly, Louis and Mary were virtually separated for the last six years of his life, Mary becoming growingly impatient with his showmanship and by his various liaisons. The Leakeys' legacy lives on, their son Richard having made major fossil discoveries, including skulls of *Homo habilis*, *Homo erectus*, and *Australopithecus aethiopicus*.

The Leakey family's achievements have been recognized in a variety of ways, including in some less-than-immortal lines by Peter Suffolk that appeared in *Punch* (June 10, 1964, p. 866):

> When God at First—
> When the first men were fashioned
>   in the good Lord's forge,
> He sent them, it seems, to the Olduvai Gorge,
> There to be tested and kept an eye on
> With the proto-lizard and the proto-lion.
> This-hyphen-pithecus and Homo That,
> With the archaeo-elephant and the palaeo-cat,
> Lived there, and died, and were hidden away
> Till countless millions of years should run
> And Leakey discover them, one by one . . .

While, back in the heavenly forge, the Lord
Went back again to the drawing-board.
I sometimes wonder: suppose that I
Were digging out there, at Olduvai,
And I brought to light a significant bone
Of a kind I could positively call my own;
And under the bone, when I'd worked it free,
I found (let us say) an ignition key-
Should I declare it, as of course I ought,
Or should I just pocket it? Perish the thought![5]

A GORGE CALLED OLDUVAI

Olduvai Gorge, the area most associated with the work of the Leakey family, is a steep-sided valley in the Serengeti Plain near the western margin of the Eastern Rift Valley in northern Tanzania. It was formed over tens of thousands of years by a river, now dry for most of the year, that cut its way down through the underlying rock layers along a 25-mile stretch of the flat, dusty plain to a depth of 300 feet. Olduvai Gorge is notoriously difficult to see from the surrounding plain. It was discovered in 1911 by a German entomologist, Professor Wilhelm Kattwinkel, who was chasing an unusual butterfly across the wide, open grassland of the Serengeti. So single-minded was his pursuit of the butterfly, Kattwinkel failed to notice the gorge opening in front of him and almost fell to his death on the rocks of Olduvai some 300 feet below. After recovering from the shock of his near disaster, Kattwinkel explored the gorge and found many fossils in the exposed rock face.

Kattwinkel's report on Olduvai led to a German expedition two years later under the direction of Hans Reck. Reck found more fossilized animal bones in Olduvai, including those of extinct species and genera, and a human skeleton (Olduvai hominid I), originally thought to be from the Middle Pleistocene. The skeleton later proved to be a relatively recent specimen that had been buried in older deposits.

The outbreak of World War I and the subsequent transfer of Tanganyika from Germany to Britain prevented Reck and his colleagues from exploring and excavating the gorge for several years, but he did participate in Louis Leakey's archaeological expedition to Olduvai Gorge in 1931. Within twenty-four hours of arriving, one of the Kikuyu aides found a hand axe—the forerunner of thousands that were later excavated from the gorge.

Over the next three decades, often using their own funds, Leakey and his wife Mary worked the gorge. Not the least of their hardships was the drinking water. As Leakey phrased it, "we never could get rid of the taste of rhino urine, even after boiling it, filtering it through paper and using it in tea with lemon."

Their efforts in Olduvai yielded thousands of stone tools and fossils of extinct animals, but no additional human fossils were found in Olduvai for nearly three decades.

But back to the fossil footprints that Mary Leakey discovered. Who made these ancient footprints, 3.6 million years ago? Associated in the same stratum with the prints are the fossil remains of *Australopithecus afarensis*, the creature commonly known as "Lucy."

HUMAN ORIGINS

The decades-long search that led to the discovery of these remarkable footprints caught the public interest and stimulated research in the broader field of human origins. How are we related to these ancient bipedal creatures? Do our origins reach back to them?

Any search for our distant *Origins* must begin with ourselves, with our own identity, with our own immediate ancestry and relationships. So how do we define ourselves? Who am I? What is "man"? "A human being, the human race, a person, an individual, one with manly qualities," the dictionary asserts. " [A] husband, a person under one's control" it adds, with an oblique sense of synonymy.

And we define ourselves in other ways, too: by describing or naming ourselves, for example. We categorize ourselves as bipedal mammals, anatomically related to the great apes, but distinguished by our own erect posture, our notable brain size, and our capacity for articulate speech and abstract reasoning. We name ourselves not just *Homo*— "man"—but also *Homo sapiens*, "wise man," so that our taxonomic immodesty is exceeded only by our cultural vanity. Within our species, there is considerable variation in height, in weight, in skin color, and in general appearance. But, for all our differences, our various races all have the capacity to interbreed and produce fertile offspring, and that is the basis of the definition of a species. In fact, the recent mapping of the human genome shows not only our unique and distinctive nature, but also our close affinities with other species. We share 98.8 percent of our genes with our nearest living relative, the bonobo chimpanzee, or "pygmy chimpanzee." This same mapping also provides an indication of the time that has elapsed since our evolving lineage parted company with the chimpanzee, some 6 million years ago. Mitochondrial DNA studies also show that our particular species arose in Africa some 200,000 years ago, from an African woman who has been given the nickname "Eve."

But the chimpanzee is not our only relative. We also share a closer link with no fewer than four earlier but now extinct species of the genus *Homo*. There are, incidentally, some half a dozen other "named" extinct species of *Homo* (such as *H. antecessor, H. ergaster, H. rhodesiensis*, and *H. rudolfensis*) but there is no general agreement that they are distinct and independent species, rather than varieties or races of the species we consider here.

We should also note that one of the most frustrating and perplexing things about describing human evolution is the changing nature of terminology used by specialists. The terms human, hominoid, hominid, and hominin are all in frequent, confusing, and often contradictory use, sometimes to describe the same group. But they are not interchangeable, and they refer to quite different groupings. And, to further complicate matters, their use has changed over the last twenty or so years.

Here's how we use them in the present discussion.

Hominoids (Hominoidea) include living and fossil humans and their ancestors, chimpanzees, gorillas, orangutans, and gibbons.
Hominids include humans, chimpanzees, gorillas, and orangutans.
Homininae include humans, chimpanzees, and gorillas.
Hominins (Hominini) include humans and chimpanzees.
Hominina include humans (*Homo*) and their immediate ancestors, but they are also often referred to as hominids.

This classification is still in a state of flux, and different writers define these categories in different ways. The confusion stems in part from the rarity of fossils of these groups and, in part, from the fiercely competitive taxonomic practices and interpretations of ancestry and affinity adopted by students in the field.

THE HUMAN FAMILY

So let's introduce the family: our closest relatives and recent forebears. Out of the several dozen or so named fossil species of our own genus *Homo* and our ancestral genus *Australopithecus*—some of them debated and their relationships hotly disputed—I have selected nine to illustrate the antiquity, variety, and origin of our family tree (fig. 14.1). Suppose we start with our most ancient family member.

Australopithecus afarensis had a projecting face, a low forehead, prominent brow ridges, jutting jaws with no chin, and a brain size about one-third that of modern humans. Individuals were short (up to about 5 feet for males, and 3.5 feet for females), lightweight (about 100 pounds), and walked upright. Their skulls show a brain comparable in size to that of a chimpanzee. They had an apelike tooth pattern. Their remains have been found near the footprints at Laetoli. "Lucy," a young female, was a member of this species. *A. afarensis* once seemed to be the rootstock from which other hominids—including other *Australopithecus* and *Homo*—developed, but a recent study has cast doubt on this. These creatures lived in East Africa from about 3.6 to 2.9 million years ago.

Australopithecus (Paranthropus) robustus lived from about 2.0 to 1.5 million years ago and differed from *A. afarensis* chiefly in the flatter form of the face, the distinctive crested structure of the skull, with massive protruding jaws, the grinding structure of the large teeth, and its rather smaller body size, ranging from about 4 feet 4 inches in males, to 3 feet 7 inches in females. Broken bone fragments found in caves associated with the fossils suggest this species may have used them as tools.

Australopithecus africanus lived from 3 to 2 million years ago and, while broadly resembling *A. afarensis* in its slender build, rounded jaws, smaller teeth, and brain size, it was more like humans than apes. The first skull of this species—the "Taung Child"— was discovered by workers in a limestone quarry at Taung, South Africa, and given to Raymond Dart in 1924. Dart concluded the skull was intermediate in character between apes and humans, with the attachment of the spinal column to the skull suggesting an upright posture, but his claim was rejected by his professional colleagues. It is now clear, however, that Dart was right and that *A. africanus* lies on or near the branch from which humans developed.

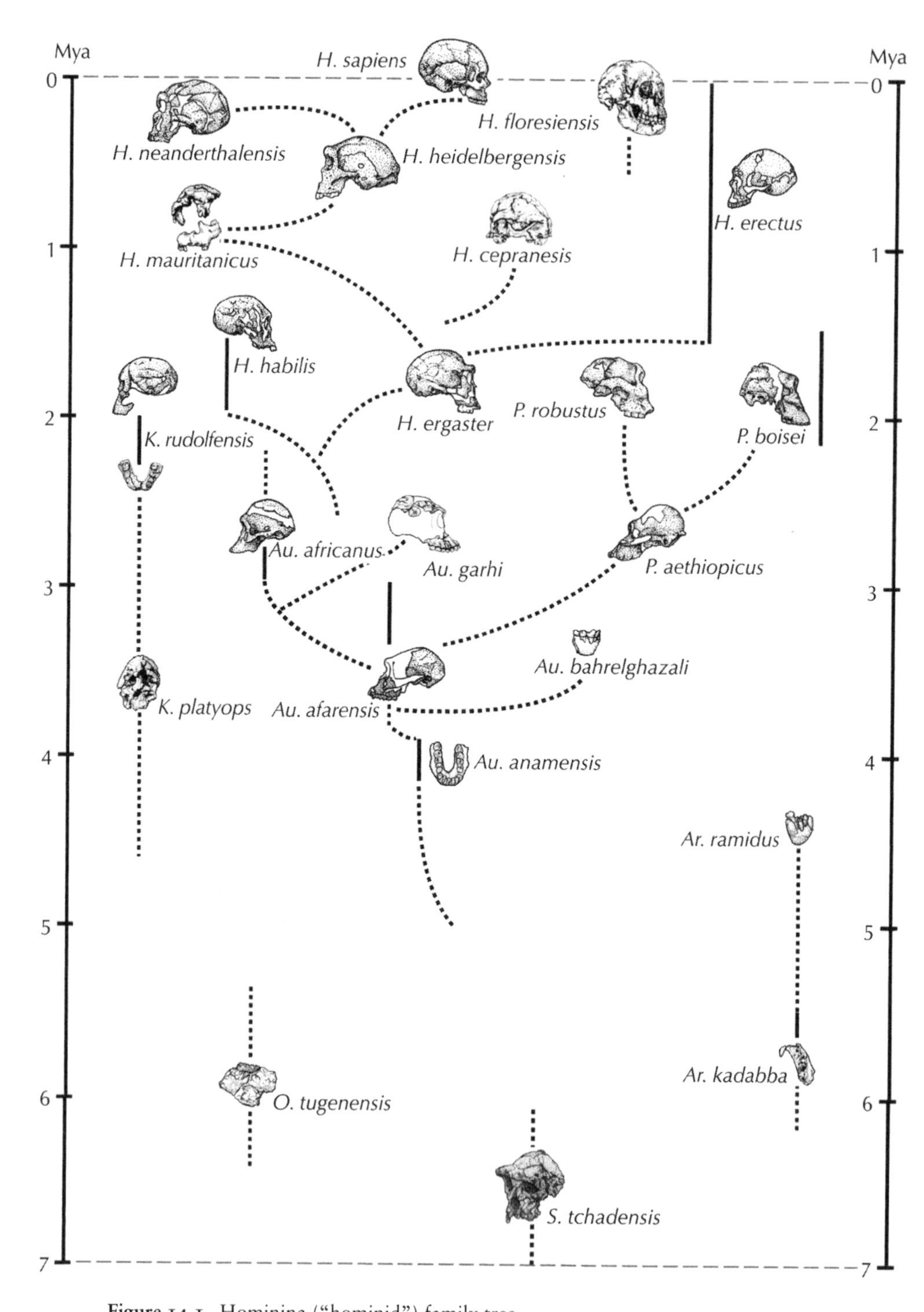

Figure 14.1. Hominina ("hominid") family tree.

(From Ian Tattersall, *The Fossil Trail: How We Know What We Think We Know about Human Evolution,* 2nd ed. [New York: Oxford University Press, 2009], fig. 14.1, p. 280. Used by permission of the author.)

S. L. Washburn, a distinguished anthropologist from Berkeley, has described Raymond Dart's demonstration of the anatomical similarities between human and apelike anatomy and motion:

> Professor Dart would illustrate his lecture on human evolution to his medical class with a graphic display. He would walk to the back of the old medical amphitheater, grab hold of a pipe, then swing back using the pipe as the arboreal route to the stage. When he was over the podium he would call out 'And so he probably came to earth,' let go of the pipe and land right by his lecture notes. I am told no student ever forgot this dramatic proof of human ape-like origins. In his demonstration swing over the medical class, Professor Dart used the anatomy which we share with the contemporary apes. The way humans swing is ape-like and very different from the behavior of monkeys. It depends on the anatomy of hand, wrist, elbow, shoulder and trunk. We stretch to the side as apes do, and this simple action summarizes a great deal of anatomy. Monkeys stretch forward, in the manner of a cat or a dog, summarizing a very different anatomy.[6]

Homo habilis ("handy man") is so named for the rich variety of chipped and flaked stone tools found in association with the fossil remains, which come from South and East Africa. The fossil remains, however, are fragmentary and their precise affinities controversial, even though many regard them as the first member of genus *Homo*. The best-preserved skull shows a larger braincase than the australopithecines and lacks brow ridges. Remains of *H. habilis* are found in deposits ranging from 2.3 to 1.6 million years in age. Their tooth structure suggests these creatures were chiefly vegetarian in their diet. The species was first discovered by Louis Leakey, who regarded it as ancestral to later humans.

Male *H. habilis* were substantially larger than the females. Both males and females had arms considerably longer than those of modern humans, suggesting they may have spent part of each day in the trees, possibly to avoid predators while sleeping.

H. habilis traveled as far as 8 miles from Olduvai to obtain the white quartzite they chipped into rough tools, indicating they were able to plan ahead. Although they were not yet capable of the full range of human speech, changes in their brain and in the position of the larynx were already giving them a far greater vocal repertoire than their ancestors. They probably had the beginnings of language and culture.

Homo erectus is the first hominid known to have migrated beyond the African continent. Fossils have been found, not only in Africa ("Turkana Boy"), but also in Asia ("Java Man" and "Peking Man") and the Republic of Georgia, in rocks ranging from 1.8 to 0.3 million years old. Just when this migration occurred is not known. Individuals had prominent brow ridges, but an inconspicuous chin, rather flattened skull, shortened face, and large molars. They showed increased brain size and greater height than earlier forms, but had sturdier bodies than those of our contemporaries. At Olduvai Gorge, remains of *H. erectus* have been found near the remains of large animals whose long bones appear to have been split for marrow.

H. erectus had a larger brain than *H. habilis* and probably had more complex cultural skills and behavior, suggesting, according to some students, that cultural elaboration and physical evolution were closely linked.

On the other hand, the recent discovery of a particularly well-preserved Early Pleistocene 1.8-million-year-old skull, together with four other specimens from the same site—Dmanisi, in the Republic of Georgia—suggests that the variation that had previously led to the creation of distinct species, such as *Homo habilis, H. erectus,* and *H. rudolfensis,* may actually have existed within a single contemporary interbreeding population. This would lead to a much simplified family tree of our early hominid forebears.

Homo erectus is considered by some to be directly ancestral to *Homo sapiens*. Certainly some of the characteristics of the species suggest qualities, such as skilled toolmaking, the use of fire, and organized hunting, that we associate with our own species.

Skeletons of *Homo heidelbergensis* show them to have been tall, large-brained, and muscular individuals, living in communities. The species—taller and larger-brained than our own—seems to have originated in Africa and subsequently migrated to Europe, where it is known from widespread localities extending from Spain, France, and Germany to England. Individuals of some populations of this species from South Africa are reported to have had an average height of 7 feet. Members of the species lived from about 2.3 million to 1.6 million years ago. *Homo heidelbergensis* was a skilled toolmaker and hunter. Some anthropologists do not regard *H. heidelbergensis* as a valid taxon.

Homo neanderthalensis lived from some 250,000 to 30,000 years ago. Individuals had a protruding jaw and face, brains slightly larger than our own, and a stocky build. Many individuals lived in colder areas of

Europe, but they are also found in the Middle East. Though they overlap both in geographic range and in time with our own species, most students have traditionally regarded them as a distinct species.

The relationship of *Homo neanderthalensis* to our own species remains, however, controversial. The older view was that competition with contemporary *Homo sapiens* led to their extinction. More recently, it has been suggested that they may, in fact, have interbred with members of our own species, into which they became assimilated.

Recent research on "fossil" DNA from Neanderthals has allowed geneticists to reconstruct the entire Neanderthal genome. This is unlike the genome of any living humans, though it has led to the suggestion that humans and Neanderthals shared a common ancestor some 800,000 years ago. This conclusion has not, however, completely ended the debate as to whether Neanderthals and modern humans are, or are not, distinct species.

But if one accepts the definition of a species as a group of actually or potentially interbreeding individuals, reproductively isolated from other such groups, then Neanderthals would be regarded as a distinct species.

"Anatomically archaic" *Homo sapiens* are known from Dali and Mapa in China and the Narmada Valley in India. These fossils range from 250,000 to 300,000 years in age and show no close biological affinities with *Homo erectus,* Neanderthals, or modern humans.

Anatomically "modern" members of our own species, *Homo sapiens*, first appeared about 200,000 years ago near the Omo River in Ethiopia. Large-brained, with prominent forehead and chin and reduced brow ridges, later members of *H. sapiens* used increasingly sophisticated tools crafted from stone and subsequently bone and antlers, and later produced carvings, and clothing of increasing refinement, in which we see the slow emergence of all the qualities we regard as most distinctively human. About 40,000 years ago, tools became more sophisticated (the Cro-Magnon culture) in both materials and use, and cave art later developed.

One more early *hominin* species inhabited the island of Flores in Indonesia between about 95,000 and 17,000 years ago. This is the diminutive *Homo florensis*, which in adulthood stood only about 3.5 feet high, with apelike form and features, and small brains. These "hobbits" shared their island home with now-extinct species of pygmy elephants and were accomplished toolmakers. Dwarfism of isolated island-dwelling species is known in other animal groups.

So what are we to make of these various relatives of ours? Were they human—or just humanlike? Do we refer to individuals as "she" or "he"—or, rather, "it"? It depends, does it not, on what we regard as the distinctive traits that separate humanity from other related groups? W. S. Gilbert (of Gilbert and Sullivan opera fame) had Lady Psyche sing in *Princess Ida*, "While, Darwinian man, though well-behaved, At best is only a monkey-shaved."

Yet few of us would be satisfied with such a definition. When we describe the most despicable behavior, not only as inhumane, but as inhuman, we define humanity by its moral sensitivity and conduct. But such a definition is applicable only to our contemporaries and to our recent forebears, for whose behavior we have adequate historical documentation. The fossil bones of other hominid species offer scant evidence of such morality, one way or another.

If it is not easy to separate modern humans from their ancestors on moral or behavioral characteristics, there are at least four distinctive physical characteristics that combine to make us human:

Bipedalism: Although other primates can run or walk on their hind legs, only humans can stand erect for long periods with straight knees on their hind limbs. Darwin (1871) suggested that bipedalism evolved when human ancestors came to spend less time in trees and more on the ground and needed to see above the savanna grasses. Others have proposed that bipedal walking might have evolved as an energy-efficient way for scavenging humans to cover large distances and to locate sources of meat.

Manufacture and use of tools: Among modern ancestors, the earliest documented tool user is *Homo habilis*. Although *Homo habilis* had a brain roughly half to two-thirds the size of modern humans, "it" seems to have been capable both of a power grip, such as one might use to wield a hammer, and some degree of precision grip, which we use to hold small objects between the thumb and forefinger. The early stone tools, and even some of the more advanced hand axes found in association with *H. habilis* could have been made by individuals who had only power grip, however. Analyses of the way the stone tools were chipped suggest that most *H. habilis* were right handed, indicating that the brain had already evolved into right and left hemispheres, each with its own specialized functions.

Mary Leakey recovered 164 tools of 12 different types from the "Zinj floor" at Olduvai, which seem to be the work of *H. habilis*. Called Oldowan, after Olduvai, they include pebble choppers—smooth, tennis ball-sized cobblestones that were given a rough cutting edge by knocking flakes of stone from both edges—as well as scrapers, burins, anvils, and hammerstones. (fig. 14 .2)

The Oldowan tools allowed *H. habilis* to obtain a varied supply of plant and animal food. Two butchery sites of large animals are known at

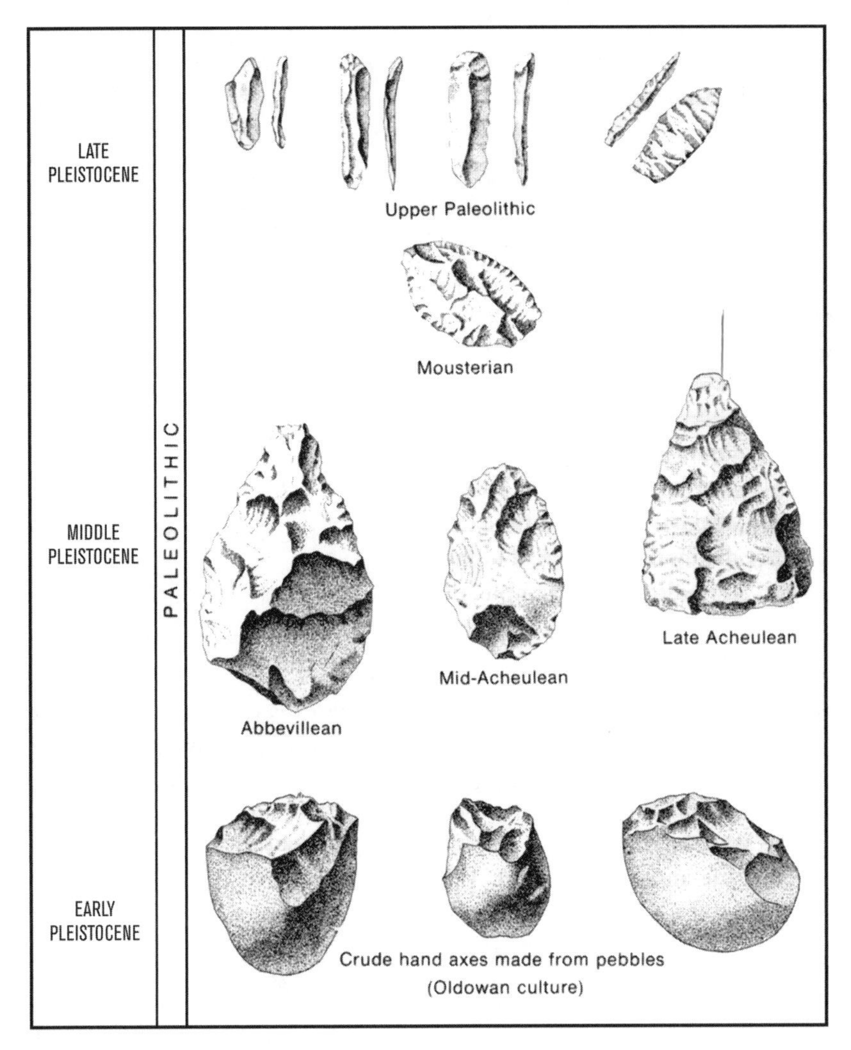

Figure 14.2. Progressive improvement in quality of tools made from stone during the Pleistocene.

(From Harold L. Levin, *The Earth through Time*, 9th ed. [Hoboken, NJ: Wiley, 2009], fig. 17-25, p. 554. Used with the permission of the publisher. Permission conveyed through Copyright Clearance Center, Inc.)

the Olduvai Gorge, and "Oldowan" tool kits have been found at nineteen Olduvai Gorge sites.

Mary Leakey argued that the accumulation of tools at the various Olduvai sites supports the idea that *H. habilis* had a lifestyle similar to modern hunter-gatherer societies. More recently, however, other researchers have suggested that the Olduvai tool sites are simply caches in an area used for foraging by *H. habilis*.

As *H. habilis* evolved into *Homo erectus* some 2.3 to 2.1 million years ago, the Oldowan tool culture was augmented by the Acheulian tool culture. Acheulian tools first appeared about 1.8 million years ago and were made from a wider range of materials—bone and wood as well as stone—and included a range of tools designed for cutting, scraping, pounding, chopping, and piercing. Makers of Acheulian tools had apparently discovered a new way of flaking stone using softer materials such as wood, bone, or antler. This "soft hammer" technique allowed toolmakers to be more precise, something especially evident in the flattened, tear-shaped hand axes that are characteristic of Acheulian culture.

The Acheulian tradition spread far beyond the Rift Valley, and over centuries it appears to have enabled early humans to adapt to a variety of tropical, temperate, and even cold environments. Except for Oldowan, it is the longest-lasting toolmaking style in the archaeological record.

Language: It has been suggested that the development of language may have been a partial response to the need to communicate effectively about the increasing complexities of life—how to make tools, where to find food, and the social traditions of the group—that would be passed on from mother to infant. Fully articulate speech, however, had to await changes in the position of the larynx, which differentiates humans from other mammals.

In almost all mammals, the larynx is positioned high in the neck, lying roughly opposite the first and third cervical vertebrae. This arrangement allows the animal to breathe and swallow at the same time, but it severely limits the range of sounds the animal can produce.

Human infants begin life with their larynxes positioned high in the neck like other mammals, but at about age two, the larynx begins to drop so that in adulthood it is opposite the fourth to almost the seventh cervical vertebrae. Although human adults are more likely to choke when they eat or drink—because they cannot breathe and swallow simultaneously—the lower position of the larynx produces a greatly enlarged pharyngeal chamber, which seems to be the key to speech.

From studies of the interior of fossil skulls, we now know that australopithecines had a larynx similar to that of monkeys or apes, and consequently they had a very restricted vocal repertoire compared with modern humans. The descent of the larynx may have begun in *Homo habilis* and continued in *Homo erectus,* but modern positioning was not complete until the evolution of early *Homo sapiens* some 200,000 years ago.

The development of culture: The brains of modern humans are three times larger than those of early protohumans. This increase in physical size probably increased the brain's memory storage capacity and ability to learn, making it easier to learn complex cultural skills and behavior. For that reason, many authorities believe that the physical evolution and cultural evolution of humankind have gone hand in hand.

Perhaps we are left only with our own species, *Homo sapiens*, as the one we can truly identify as human. But even here, we face uncertainties, for—whatever the skeletal identities of our own ancestors—we know as little of the conduct and morality of those who lived, say, 200,000 years ago, as we do the behavior of other species of *Homo*.

Our own species, like every other, shows all the marks of its affinity and evolution. But the topic of human evolution continues to be marked by lively debate and sometimes bitter controversy. This arises partly from the fact that—for whatever reasons—about 45 percent of the U.S. population is reported not to believe in human evolution. But the controversy also arises partly from intense competition and debate among professional paleontologists concerning human origins, as well as from the fragmentary nature of the human fossil record. Numbers of named hominid species have been based on only a single fossil fragment. Other fossils that have been identified as new species or genera, even by respected leaders in the field, have subsequently been shown to belong to existing taxa. In all, dozens of human and prehuman species have been named. We have reviewed only nine or so of the best known and most representative of these, but even about these species, lively debate frequently continues about their affinities and evolutionary relationships.

This general discussion has concerned the question of human ancestry and the origin of human and prehuman australopithecine species. But where did these come from? What is the origin of the larger group of hominoids—the great apes and the lesser apes—with which we share so many features?

EARLY HUMAN ANCESTORS

As reptiles become less numerous at the start of the Cenozoic (the geological era that began 66 million years ago and continues to the present), various mammalian groups began to compete for the habitat made

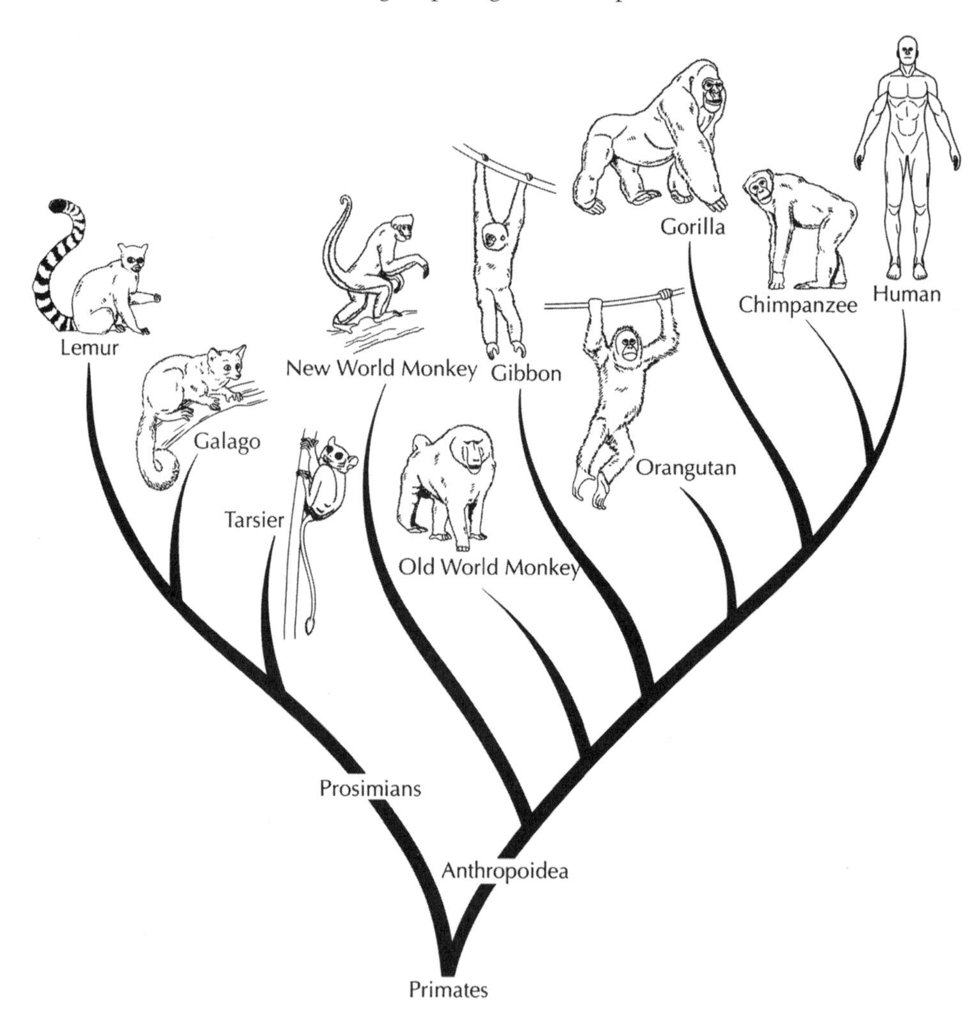

Figure 14.3. Family tree of the major groups of living primates, showing the close relationships between the great apes and humans and the more distant relationships of the Old World monkeys, New World monkeys, and lemurs, lorises, and bush babies.

(Drawing by Carl Buell. From Donald R. Prothero, *Evolution: What the Fossils Say and Why It Matters* [New York: Columbia University Press, 2007], fig. 15.4, p. 338. Reproduced with permission of the publisher.)

vacant by the massive extinction of the reptiles. It was at this time that the primates developed. The most important of these competitions, from the standpoint of human evolution, was that between rodents and primates.

The primates are a largely arboreal group of mammals and so, like the birds, are not well represented as fossils, because their remains do not undergo rapid burial. They have tended to remain rather unspecialized animals, but have developed complex brains and acute vision.

The oldest primates are lemurlike creatures found in Paleocene rocks, whose ancestors were probably arboreal insectivores, perhaps resembling living tree shrews. Living lemurs, aye-ayes, and bush babies are archaic remnants of this early group. These small creatures, though looking quite unlike the higher apes, are linked to them by the living Tarsioids of the East Indies: small, large-eyed arboreal creatures, intermediate in appearance between living lemurs and monkeys.

Early primates, known collectively as *prosimians*, probably evolved from a type of insect-eating shrew during the Paleocene (66–56 million years ago) or early Eocene (56–38 million years ago). By the late Eocene, competition with rodents and a cooling climate had vastly reduced the number of prosimians worldwide and had brought about their almost complete disappearance from North America and Europe. By the Oligocene (33–23 million years ago), primate evolution was centered in Africa and Asia.

The earliest known hominoids, the ancestors of modern apes and humans, appeared about 40 million years ago in Burma. They have been placed in the genera *Amphipithecus* and *Pondaungia.* These higher primates are thought to have spread out of Asia to other parts of the world, crossing into Africa through what was then the narrow, swamplike Tethys Sea, which at the time separated the Asian and African continents. Other populations crossed the proto-Atlantic Ocean, which was far narrower than it is today, probably by island hopping on volcanoes. These became the ancestors of the New World monkeys.

During the Miocene epoch (23–5 million years ago), great mountain ranges were being built and the continents that once had been Gondwanaland continued to drift apart. Although Africa and Eurasia were still attached, volcanoes were actively changing the face of Africa, and the Great Rift Valley was born. It appears likely that these events accelerated the pace of hominid evolution in Africa.

During the Miocene, a number of relatively unspecialized primates known as dryopithecines lived in Asia, Europe, and Africa. These included the African genus *Proconsul,* the fossil remains of which were first discovered in the 1930s on an island in Lake Victoria. The dryopithecines apparently were "pre-brachiators," being most at home in trees, but also capable of living on the ground. They went on to diversify during the Miocene and later Pliocene (5–2 million years ago), but they disappeared before the end of the Pliocene, probably evolving into modern great apes.

By some 14 million years ago, a more evolved primate, *Ramapithecus* (or what Louis Leakey called *Kenyapithecus,* based on a fragment of jaw and teeth he had found in western Kenya), was present in India and East Africa.[7] *Ramapithecus,* which was about the size of a chimpanzee, had an essentially human arrangement of teeth and may have walked upright, perhaps using the knuckles of its fingers for additional support, much as modern great apes do. Although *Ramapithecus* was once considered a direct human ancestor, more recent work has suggested that it was, in fact, ancestral to orangutans.

After *Ramapithecus,* there is a gap in the fossil record until the earliest australopithecines appeared. And the gap is substantial: from about 8 million to 4 million years ago in Africa. Other fossils from the Siwalik Hills of India and Pakistan (the so-called God-Apes) are hominoid fossils, ranging from 15 million to 7.5 million years in age.[8] They include the genus *Sivapithecus* (which may well be a form of *Ramapithecus* and, because it was named first, would become the senior synonym, replacing the latter name). *Sivapithecus* is now generally regarded as the ancestor of the modern orangutan.

While the outlines of our ancestry are clear, it is tantalizing that the fossil evidence still leaves questions about the precise paths by which we have arisen. The fossil record, as we have seen, is, by its nature, incomplete. In fact, even if all the gaps were somehow filled, the physical evidence that fossils provide is but one major clue in the far longer and broader quest, as old as humanity itself. And this physical evidence leads to larger questions. Who am I? Where did I come from? How did I come to be here? These questions confront us as children. Yet they never leave us. They confront us still in adulthood. Typically, in our busy lives, we tend to push them aside and deal with other pressing concerns. When we do consider them at all, we never fully answer them. Our knowledge grows, but our assumptions change, ambiguities emerge, conclusions differ. How did I, a pinpoint in a cold, dark, limitless universe, come to be here?

Ancient peoples wrestled with these same questions of origins, as the oldest creation narratives bear witness. Our most distinguished scientists and most notable scholars grapple with them still, even as our data accumulate and our knowledge increases.

One thing, perhaps, is clear. With the arrival of our own species, the long-established process of evolution by natural selection has been modified, though not suspended. A single species—ours—now has the capacity to influence or disrupt not only the natural rhythms, distribution, and future patterns of life on Earth, but even its survival. In the billions of years of life on the planet, that is a unique and sobering capacity.

Consider the combined effects of half a dozen or so areas of human impact on the life of our planet:[9]

- The projected "built-in" increase of human population from a present total of 7.2 billion to an estimated peak population of 9.6 billion in 2050, before a decline and leveling off to a projected 8.5 billion. Most of the present growth is in developing countries.
- An estimated 3.7 million human beings are thought to suffer already from malnutrition.
- Widespread soil erosion and deforestation.
- Severe depletion of the water table in many areas.
- Rising energy demands with constrained and dwindling conventional energy supplies.
- Growing industrialization, urbanization, and environmental transformation.
- Growing rates of extinction, estimated to have now reached an annual rate of 33,000 species.

There are, to be sure, some beneficial aspects of our stewardship—including habitat restoration and wildlife conservation, for example—but these scarcely outweigh the inevitable requirements and impact of these other trends.

All this means that any discussion of human origins must inevitably consider not only the continuing origin of new species—now clearly demonstrated—but also the significant threat which the reproductive vigor and global impact of our own species poses to the future of all the rest.

15

"ENDLESS FORMS, MOST BEAUTIFUL AND MOST WONDERFUL"

THE NONVERTEBRATES

Over the course of half a dozen of the earlier chapters, we followed the remarkable story of the evolution of the vertebrates. But it is all too easy to interpret the history of life as *their* history, as they slowly made their laborious way from their fishlike ancestors through their various stages of development—amphibians, reptiles, birds, mammals—to arrive, at last, at our own species, *Homo sapiens*. But, wonderful as are the vertebrates, they make up a very small percentage of all known animal species. There are, for example, some 5,700 named species of mammals, as well as 10,000 bird species, 9,000 reptile species, and 31,000 fish species, all recognized and named. That's a grand total of some 55,700 or so species of vertebrates. Against that, there are 1.3+ million named species of invertebrates, including about 1 million species of insects, not to mention some 281,000 species of flowering plants. Experts differ on the precise numbers of each group, but these round figures are close, and another 15,000 or so new species are recognized and named every year.

Fascinating, then, as our own clan is, and the other animals and plants of the terrestrial world, we should at least take a moment to acknowledge the ongoing richness of life in the oceans, whose inhabitants play a vital role in the overall life and ecological balance of our planet.

THE PROTISTS

Most of the oceans' most prolific members are invertebrates. The simplest and most abundant invertebrates are the living protists, almost all of microscopic size, including foraminifera, algae, and diatoms; between them, these form the foundation of the complex food chain that supports the richness of marine life.

Although we are describing these and a few other groups as "protists," by the way, they share only one essential property: they are unicellular (eukaryotic). But some, such as coccoliths and diatoms, are plantlike, meaning they photosynthesize, while others, such as foraminifera and radiolarians, behave as amoebalike animals. Between them, this huge group represents a host of ubiquitous, unicellular creatures, of such variety and simplicity that there is still debate as to the exact relationship between them. These "simple" organisms inhabit damp environments, from the oceans to hot springs, from soil to the dark interiors of other animals. Some are agents of disease; others are essential for health. Among this vast variety and abundance, we will discuss a few forms, such as diatoms; but only a few.

Although fossil *radiolarians*, with their elaborate "Christmas tree ornament" spherical and bell-shaped siliceous tests, are known from rocks of Cambrian and Ordovician age, they underwent a major expansion and diversification in Triassic times, and a further radiation in the Cenozoic. So common are they in some parts of the ocean that, as we have seen, their fossil remains form a distinctive deposit: radiolarian ooze, a siliceous mud that covers vast areas of the ocean floor and is preserved in places as radiolarian chert.

Foraminifera are protistans with a test that may be made of agglutinated particles or of calcium carbonate. We have already seen something of Paleozoic foraminifera, including the widespread fusulinids of Pennsylvanian-Permian times. The Triassic foraminifera marked the beginning of an expansion of their successors into a wide range of both benthic (bottom dwelling), pelagic, and planktonic (floating) marine environments. The oldest known planktonic forms are Jurassic. All pre-Jurassic forms were benthic, as are the vast majority of all species. Rapid diversification in Jurassic and Cretaceous times was marked by their worldwide distribution, but planktonic forms were severely depleted and almost wiped out in the K-T extinction event, though they rapidly recovered. A wide variety of Cenozoic groups—some the size and shape of small coins—flourished, though several smaller Cenozoic extinction events influenced the history of both planktonic and benthic groups.

Coccolithophores are unicellular, photosynthetic, flagellates that construct a test covered with coccoliths, ultraminute calcareous plates that are circular, disklike, or trumpet-shaped. They belong to the golden-brown algae; their fossil remains are so abundant that they form a large part of the phytoplankton. They are excellent biostratigraphic markers in rocks of Mesozoic and Cenozoic age, because of their rapid diversification. They originated in the Jurassic, about the same time as planktonic foraminifera. They underwent a sharp decline at the end of the Cretaceous but made a rapid recovery in the Paleocene.

Diatoms build siliceous tests that range from spindle-shaped to disk-like to triangular in form. They are photosynthetic planktonic algae and abundant in the upper waters of the world's oceans, where they form a significant portion of the base of the food chain, as well as occurring in freshwater and near surface terrestrial environments. They are abundant both in existing oceans and in fossil deposits, where their dense remains form diatomaceous earth, which is widely used in commercial filtration and as a filler or absorbent. Diatoms are found in rocks from Jurassic times onward, and they became increasingly abundant from Cretaceous times onward. They were little affected by the K-T extinction event, though several significant later Tertiary changes seem to reflect changing oceanic conditions.

THE INVERTEBRATES

Almost all the fossils we have described so far are of microscopic size. Among the more conspicuous metazoans there were also many changes. Mesozoic *bryozoans* were quite different in structure from those of the Paleozoic, and they underwent a great expansion during Cretaceous times. They remain a widespread, though minor, group today.

Perhaps the most striking difference between the animals of Paleozoic seas and those of the Mesozoic was the great decline of the *brachiopods*. Of the hordes of Paleozoic groups, only the rhynchonellids, the terebratulids, and the eternal *Lingula* persisted through the Mesozoic (fig. 15.1G, I). The creatures that seem largely to have displaced them were *pelecypods*, and later the *gastropods*. The early bivalves (pelecypods) were marine, but nonmarine forms became widespread in the coal-forming swamps of Pennsylvanian times. Although bivalves were somewhat reduced by the Permo-Triassic extinction, they flourished in later Triassic times, expanding in numbers and, in some cases, new ways of life, such as the free-swimming *Pecten* group.

Most Mesozoic pelecypods were not unlike living forms; oysters and pectens (scallops), for example, were common, Some, however, were quite different. One distinctive, hornlike group, the rudistids, reached a length of 2 feet and developed a coral-like shape, with one valve forming a long conical structure and the other a lidlike covering. So abundant were they in mid- and late Cretaceous times that they became the dominant reef builders of warm, tropical seas. Other distinctive Mesozoic bivalves included the curiously ridged and curved *Gryphaea*—sometimes referred to as "Devil's Toenails" by superstitious people of earlier times (fig. 15.1E)—while some giant species of *Inoceramus* reached a size of 4 feet.

Gastropods are relatively rare as fossils in most Paleozoic rocks. Although the oldest known terrestrial snails are of Pennsylvanian age, ancestors of the "modern" land snails appeared in the Cretaceous. Gastropods become more common in the Mesozoic, with freshwater forms becoming so abundant that their Jurassic remains formed the "Purbeck Marble," a rock widely used as an internal building stone in English churches and cathedrals. The gastropods were common in the Mesozoic, but they underwent their greatest expansion in Cenozoic times. With the pelecypods, they represent one of the dominant groups among living creatures of the sea.

But the most distinctive marine invertebrates of the Mesozoic were the cephalopods (literally "head-foot"), of which there were three distinct groups. In these groups, the foot has been modified into a head that tends to be well developed and have a crown of tentacles, with eyes, a parrotlike beak, and a well-developed brain. All are or were active marine predators.

The Coleoidea include the two-gilled (dibranchiate), extinct belemnites, which were the forerunners of the living octopuses, squid, and cuttlefish. Belemnites are usually represented as fossils by their cigar-shaped internal skeletons, which are abundant in many Mesozoic strata (fig. 15.2E–F). Several hundred species are known, the largest being *Megateuthis gigantean,* with an estimated length of 9 feet. Unusually well-preserved fossil specimens show that they were streamlined, tentacled creatures, very similar to their living descendants. These cephalopods, like the living squid and octopuses, rely for protection, not on a heavy shell into which they may withdraw, as did the ammonites, but on speed. The mantle is modified both into flaps which assist in locomotion, and into a funnel, through which water can be so forcibly ejected as to give the animal a type of jet propulsion. In both living and fossil forms, an ink sac is also present, from which a cloud of dark material may be emitted when the animal is attacked. Belemnites are known from rocks ranging from Upper

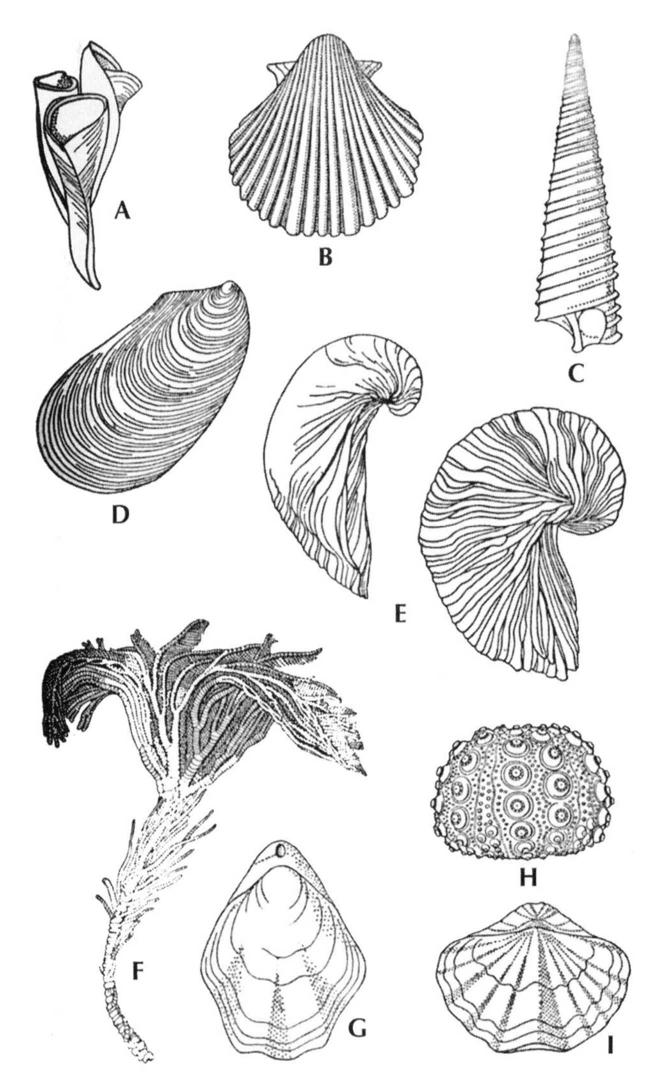

Figure 15.1. Mesozoic invertebrates. (A–D) Cretaceous: (A) *Monopleura*, a rudistid pelecypod, length 2 inches; (B) *Pecten*, a clam, width 1.5 inches; (C) *Turritella*, a gastropod, length 2 inches; (D) *Inoceramus*, a pelecypod, some species measured 4 feet across the shell; (E–I) Jurassic: (E) *Gryphaea*, two species, a pelecypod, length 2 inches; (F) *Pentacrimus,* a stalked crinoid, "armspan" about 3 inches; (G) *Goniothyris*, a terebratulid brachiopod, length 1.5 inches; (H) *Hemicidaris*, a regular echinoid, diameter about 1 inch; (I) *Spiriferina,* the last of the spiriferid brachiopods, width 1 inch. Mesozoic invertebrates.

(From Frank Rhodes, *Evolution of Life* [New York: Penguin, 1976], fig. 51, p. 271. Reproduced with permission.)

Mississippian to Cretaceous in age, often being abundant in Jurassic and Cretaceous rocks. Like their ammonite cousins, they did not survive the K-T extinction.

The nautiloids (Nautiloidea) include the living pearly *Nautilus* and range from Cambrian to Recent in age. Living forms include five species, all from the Indo-Pacific region. They have simple suture lines and are widespread as fossils. They were ancestral to ammonoids and coleoids.

The ammonite cephalopods (Ammonoidea), generally regarded as four-gilled (tetrabranchiate), underwent profound changes throughout the Mesozoic. The suture line of the shell became intensely crenulated in many forms, while the shape and ornament of the shell show almost every conceivable variation. These free-swimming creatures lived in countless numbers in the Mesozoic seas, and they are among the most useful index fossils known. Some forms reached gigantic proportions, with a shell diameter of 6 feet, but most were only a few inches in diameter.

Paleozoic forms had straight, conical (early Paleozoic), or coiled (later Paleozoic) shells (fig. 15.2) with internal divisions (septa) that became increasingly complex in form in Mesozoic times. They appeared in the Devonian and became relatively common in the Late Paleozoic, where they were represented by some thirty families. Only two genera survived the Permo-Triassic extinction, but they later exploded in numbers in Triassic times, with more than 500 genera described. By Jurassic and Cretaceous times they had expanded even further, developing into a myriad of forms (both coiled and uncoiled), though they vanished in the K-T extinction.

Sponges tend to show rather marked provincialism, presumably because of their sessile, benthic habit. Calcareous sponges underwent rapid expansion in the Permian, and some Mesozoic forms became so abundant that they are important components of reef communities, especially in Europe. In general, however, both calcareous and siliceous sponges have a patchy distribution and exhibit little substantial change in form over time.

The *corals* are the most conspicuous fossil members of the phylum Cnidaria. Paleozoic corals were, as we saw in chapter 5, made up of tabulate and rugose groups. These groups did not survive the Permian extinction. Scleractinian or stony corals consist of both solitary and colonial forms, with skeletons of aragonite, whose septa have a six-fold symmetry. They first appeared in the Triassic and at first formed reefs of modest size, but the group has undergone great and almost continuous expansion since then, building the vast coral reef systems that form such distinctive features of warm, shallow marine waters. A second group of scleractinian corals are solitary in habitat and widely distributed in the oceans in waters ranging from temperate to polar.

The *arthropods* showed dramatic changes in post-Paleozoic times. Though this chapter is concerned with the oceans, we should also say a word about the *insects*, the most prolific arthropods.

The huge insects that inhabited the coal-forming swamps of the Pennsylvanian did not survive, but Pennsylvanian and Permian fossils contain representatives of all the major groups of winged insects. It has been remarked of the Permian that during "no other geological period has such a diverse insect fauna existed," but the late Permian saw the extinction of six of the nine existing orders. They underwent explosive evolution in Cretaceous times; by late Cretaceous times, most of the modern insect superfamilies appeared, some developing with the rise of mammals and flowering plants. Fleas, for example, are first known from the Cretaceous, as are ants, while butterflies and several other forms appear in the early Cenozoic. Insects remain the most prolific of all groups, both in number of species and in number of individuals.

Other arthropod groups include the *chelicerates,* including the *arachnids*, which consist mainly of air-breathing terrestrial forms, such as spiders and scorpions, as well as the extinct eurypterids or "sea scorpions," an aquatic (Ordovician-Permian) group, some of whose members were more than 7 feet long. The *xiphosurids* are the horseshoe crabs, the earliest member of which appeared in the late Ordovician of Manitoba. The horseshoe crab is often called a "living fossil" because of its ancient ancestry, though it showed more variety of form than this name suggests. *Crustaceans* are a group that includes both marine and terrestrial forms. Crabs, shrimps, and lobsters (decapods) are crustaceans that have an outer covering of calcium carbonate and are well represented, not only as fossils in Mesozoic and Cenozoic strata, but also in living faunas. The group also includes the barnacles and the bivalved ostracods, which are widely distributed microfossils.

The *trilobites*, so dominant in the seas of early Paleozoic times, but in almost continuous decline since late Ordovician times, became extinct near the end of the Permian.

The other great invertebrate phylum, the *echinoderms*, is well-represented in rocks of Mesozoic and Cenozoic age. These "spiny-skinned" marine creatures first appeared in the Cambrian, represented by fixed, short-lived varieties of globular, disklike and stalked, cuplike forms,

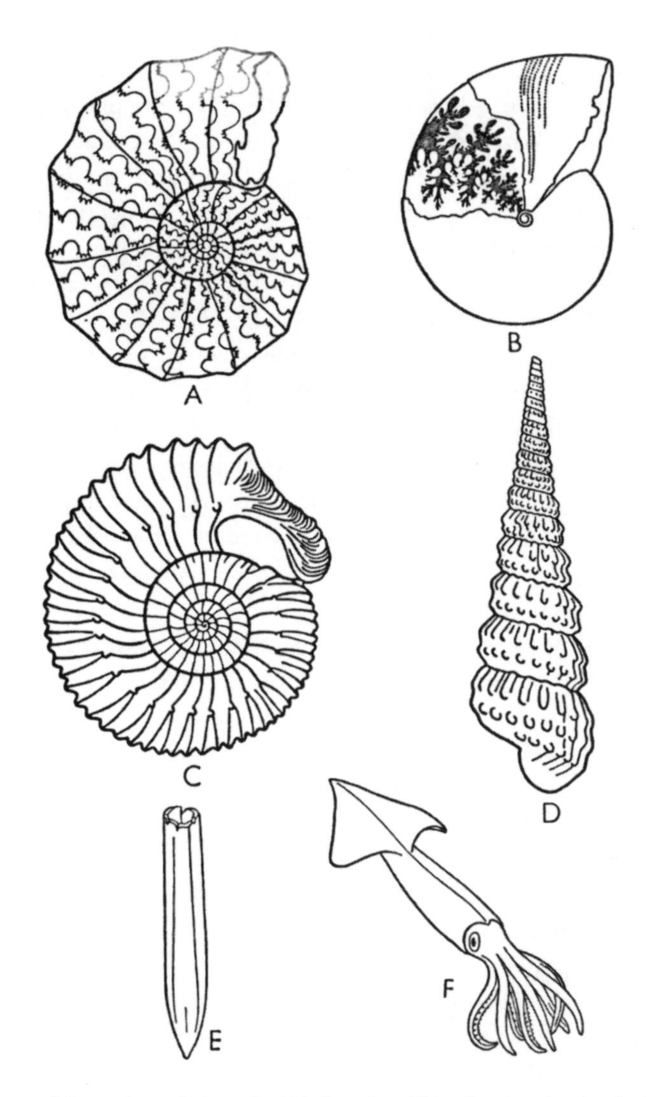

Figure 15.2. Mesozoic cephalopods. (A) *Ceratites*, Triassic, showing beginning of crenulation in the suture line, diameter about 2.5 inches; (B) *Phylloceras*, Jurassic, showing complexity of structure line, diameter about 2 inches; (C) *Stephanoceras*, Jurassic, showing complex ornamentation of the shell, diameter about 2 inches; (D) *Turrilites*, Cretaceous, showing complexity of coiling shell, length about 5 inches; (E) *Belemnitella*, Upper Cretaceous, one of the last survivors of the belemnites; (F) A reconstruction of a belemnite.

(After Hayes. From Frank Rhodes, *Evolution of Life* [New York: Penguin, 1976], fig. 52, p. 273. Reproduced with permission.)

including the abundant crinoids and blastoids. The other major group, the free-living eleutherozoans, includes starfish, sea cucumbers, brittle stars, and sea urchins. Almost all the fixed echinoderm groups, including the widespread blastoids, became extinct during the Paleozoic, but the crinoids, though greatly reduced by the $P\text{-}T_R$ extinction, and the echinoids expanded in Mesozoic and Cenozoic times. Many crinoids became free-living, as opposed to stalked, though some stalked forms (such as the surviving *Pentacrinus*—fig. 15.1F) had a stem 50 feet long. The sea urchins (fig. 15.1H) underwent a marked development, with many of the regular forms developing long spines, and a variety of irregular, heart-shaped and globular forms appearing in the Mesozoic and continuing throughout the Cenozoic.

All the major groups of living marine invertebrates were well established by Mesozoic times, and, apart from the extinction of the ammonites and belemnites, the changes that took place were of a comparatively minor kind. What did take place, however, was the Major Marine Revolution of Mesozoic times, involving a major restructuring of marine ecosystems, brought about by increased predation pressure. Most of the changes we have described in this chapter, except for the insects, involve marine creatures, many of which exist in literally countless numbers. And, as we emphasized at the beginning of the chapter, all are invertebrates. But we should also perhaps note the major changes in the other great group of marine creatures, the vertebrates.

THE VERTEBRATES

The vertebrate dwellers of the seas have in some ways, as we have seen, undergone more change than many of the invertebrate groups. The rise and extinction of the great marine reptiles, the evolution of sea birds and of ocean-going mammals are all events that have taken place within the span of Mesozoic and Cenozoic time. The fishes, too, have undergone considerable change.

The Mesozoic history of the sharks was one of increasing specialization and modernization. The primitive jaw suspension was replaced in most forms by the more advanced and efficient jaw attachment of modern forms, and many Mesozoic sharks were similar in their general streamlined body and strong fins and tail to living forms of the open oceans. Others became bottom dwellers, with flattened bodies, whiplike tails and teeth adapted to grinding the shellfish on which they fed. These were the ancestors of the skates and rays and they first appeared in early Mesozoic times.

But in spite of the persistence and the success of sharks, it is the bony fish that represent the peak of vertebrate adaptation to an aquatic life, not only in the seas but also in inland waters. Their Paleozoic ancestors included, as we have seen, the little, ray-finned palaeoniscids of the Devonian, as well as the lobe-finned coelacanths and lungfish. Although the living sturgeon and a few other fish are remnants of early ray-finned fish, most of the primitive forms were replaced in Mesozoic times by the holostean fishes.

These holosteans, of which the North American garpike and bowfish are the only survivors, had a more completely ossified skeleton, shorter jaws, more specialized skulls, a more symmetrical tail, and thinner scales than their predecessors, and all these were evolutionary trends that were carried even further in their descendants, the teleosts, which replaced them in Cretaceous times. From then until now the teleosts have undergone an expansion of unparalleled proportions, and they are now the dominant group of both the seas and inland waters, in both numbers of species and individuals.

Living teleosts display an astonishing variety of form, ranging from salmon, herring, carp, catfish, eel, sea horse, and pike to the flounder and angel fish. They exist in countless numbers in almost every aqueous environment, from the depths of the oceans to polar areas and tropical streams, and even within these environments they exhibit endlessly varied habits, habitats, and diets.

Yet it is the nonvertebrates, these "endless forms, most beautiful and most wonderful" of Charles Darwin, that collectively provide the basis of the food chain of the oceans, and it was on that food chain that our most remote ancestors depended.

We, creatures of the land, all too easily overlook the continuing significance of the extraordinary richness and abundance of the life of the oceans. Only an occasional glimpse of that unseen world reminds us of our evolutionary debt to it and our continuing dependence on it.

16

ON EXTINCTION

THE REALITY OF EXTINCTION

One of the most perplexing challenges for the founders of geology in the late eighteenth and early nineteenth centuries was to interpret the history of life as it was represented by the sequence of fossils that was gradually being pieced together. We saw something of this struggle—a struggle of both scholarship and religious conviction—in the case of William Buckland, as he wrestled with the so-called diluvium and the evidence and impact of recent glaciation. A closely related question involved the evidence for extinction—not only of particular species, but also of much larger groups, such as dinosaurs—of the causes that lay behind it, and its moral and spiritual implications. "Could a benevolent Creator" it was asked, "allow the possibility of extinction?"

But as field studies provided more and more material, the evidence for extinction became conclusive.

"No one can have marveled more than I have done at the extinction of species," wrote Charles Darwin. "When I found in La Plata the tooth of a horse embedded with the remains of Mastodon, Megatherium, Toxodon and other extinct monsters, which all co-existed with still living shells at a very late geological period, I was filled with astonishment."[1]

Any history of life, any account of origins, inevitably stresses first appearances, evolutionary relationships and continuity. But the fossil record also bears widespread evidence, not only of appearances, but also of disappearances; not only of continuity but also of a particular kind of discontinuity: extinction. In fact, it is widely estimated that of all the species that have ever lived, more that 99.9 percent are now extinct. We have, it is true, a handful of "living fossils": the coelacanth *Latimeria*, the horseshoe "crab" *Limulus*, the lizard *Sphenodon,* the small brachiopod *Lingula,* and a few more. Several of these have persisted for hundreds of millions of years. They are sometimes described as "Lazarus taxa." But the vast majority of fossil species had a relatively limited life span. Various estimates have been made of the "average life span" represented by fossil species (see table 16.1). The estimates vary widely from one group of organisms to another; for example, species of marine invertebrates are calculated to have had an average existence of 5–10 million years, Cenozoic mammals, 1–2 million years,[2] and Silurian graptolites, 2 million years. One fact is clear, however: species, like their members, are mortal.

Extinction represents the disappearance of a species or subspecies, and its fossil record helps us to pinpoint the end point of its existence. For recent forms, there may be some doubt about the exact time of disappearance or extinction, depending on the last sighting of a species member. The ivory-billed woodpecker, for example, long thought to be extinct, has been reported in a few recent unconfirmed sightings. In other cases there is no doubt, either of the fact of extinction, or its cause. The recent extinction of some species can be attributed to human predation: the passenger pigeon and the dodo, for example. In fact, human hunting, destruction of habitat, and introduction of non-native species are demonstrably increasing the rate of extinction of living species, and many biologists are concerned that such human influence could greatly increase the overall rate of extinction.

Table 16.1. Estimates of the duration of species, from origination to extinction

Taxon	Source of estimate	Species' average duration (millions of years)
All invertebrates	Raup 1978	11
Marine invertebrates	Valentine 1970	5–10
Marine animals	Raup 1991a	4
Marine animals	Sepkoski 1992	5
All fossil groups	Simpson 1952	0.5–5
Mammals	Martin 1993	1
Cenozoic mammals	Raup and Stanley 1978	1–2
Diatoms	Van Valen 1973	8
Dinoflagellates	Van Valen 1973	13
Planktonic foraminifera	Van Valen 1973	7
Cenozoic bivalves	Raup and Stanly 1978	10
Echinoderms	Durham 1970	6
Silurian graptolites	Rickards 1977	2

Source: John Lawton and Robert M. May, *Extinction Rates* (Oxford: Oxford University Press, 1995). Reprinted with the permission of Oxford University Press. Sources in the table may be found in the Related Reading section at the end of this book.

But except, perhaps, for the last examples, we can think of extinction as a kind of continuous process, by means of which existing species become extinguished, victims to such things as changing ecological conditions, dwindling food supply, climate change, and competition from other species. The fact that fossil species display life spans of only a few million years means that this "background extinction" is more or less a continuous process.

There was a time when the existence of extinction, or even its possibility, was strenuously resisted and the evidence for it strongly denied—and this was a position espoused by some of the leading naturalists of their day. Carl Linnaeus, for example, the founder of modern biological classification, thought it impossible for any species to become extinct: "Species tot sunt diversae quot diversas formas ab initio creavit infinitum Ens." ("There are as many species as the Infinite Being created in the beginning") he wrote. Jean Baptiste de Lamarck (1744–1829), the father of invertebrate zoology, is generally remembered not only as the advocate of a now-discredited theory of inheritance of acquired characteristics, but also as someone who denied the reality of extinction. Lamarck's view of extinction, like his view on evolution, however, was more subtle than this would imply. He recognized adaptive change in organisms (evolution) in response to environmental change, though he regarded use and disuse of a structure or organ as the cause.

Nor was his rejection of the possibility of extinction as absolute as his critics assert. Though he was doubtful that "species or races have been so inadequate that entire races are now extinct or lost," he continued, "If there really are lost species, it can be doubtless only among large animals that live . . . where man exercises absolute sway."[3] And he then went on to suggest these might include "*Palaeotherium, Anaplotherium, Megalonix, Megatherium, Mastodon.*"[4]

He went even further in reviewing his own charge—the invertebrates—suggesting that, since man cannot have been responsible for their destruction, "the fossils in question belonged to species still existing, but which have changed since that time and became converted into the similar species that we now find." Here was a man wrestling with a problem which he never satisfactorily resolved.

Lamarck's life and achievements, however, deserve more respect than he received in his lifetime—he became blind in his old age and died in poverty—or is accorded today. He is rightly regarded as the founder of invertebrate biology.

Perhaps the first to establish—rather than suggest—the reality of extinction was Georges Léopold Crútien Fredéric Dagobert Cuvier (1769–1832). Cuvier, or Baron Cuvier, as he later became, was one of the greatest scientific thinkers and leaders of the eighteenth century, as well as the founder of modern comparative anatomy and vertebrate paleontology. The degree to which Cuvier combined science with brilliant statesmanship was no less remarkable than his scientific achievements. He served in senior scientific and administrative appointments (including Inspector General of Public Education and State Councillor) under three successive but opposing French governments (the Revolutionary, the Napoleonic, and the restored Monarchy).

By 1795 Thomas Jefferson had published notes on the American mastodon—which he referred to as "the mammoth"—and discovered that two different species of this elephant had been identified, but he did not believe in extinction and thought the mammoth might still be living in the American west.

It was in 1806 that Cuvier first articulated not only the difference between living African and Indian elephants, but also the differences of both from the American mammoth and mastodon, which, he argued, must thus be presumed to be extinct. This revolutionary suggestion—flying in the face of traditional views of the permanence of species—he followed up by also claiming that several other Pleistocene fossil forms—the giant woolly rhinoceros, Irish elk, cave bear, and a much older Mesozoic mosasaur marine reptile—were far too large and conspicuous to have been living still undiscovered in remote refugia. They must indeed, he argued, be accepted as extinct. Just *how* they became extinct was less clear. Cuvier, himself the great functional anatomist, thought it unlikely that such perfectly adapted forms could succumb except to some devastating, catastrophic change. And that is what he proposed: that Earth had passed through a series of violent catastrophes, of which the flood of Noah was the most recent, and each of which destroyed Earth's inhabitants, which were subsequently replaced by a new creation. Though such a catastrophist theory later gave way to a more gradualistic understanding of Earth's processes and history, Cuvier's abiding insight of the *reality* of extinction endures, as does his wider contribution of creating the fields of comparative anatomy and vertebrate paleontology.

In more recent times, the reality of extinction has been forcefully made clear. The passenger pigeon, for example, a swift bird of rare beauty, was once so abundant that it is thought to have contributed up to 40 percent of the total bird population of the United States. Early observers describe flocks so large that they darkened the skies and took several hours to pass overhead. Yet by the early twentieth century, no birds were to be found in the wild, and the last captive bird, Martha, died in the Cincinnati Zoo in 1914. The passenger pigeon had suffered habitat loss by deforestation, but the main reason for its loss was that it was hunted to extinction. Boxcars of birds were shipped to eastern cities for human consumption.

The dodo, in contrast, was a flightless bird, about 3 feet tall and confined to the island of Mauritius in the Indian Ocean. Once abundant there, it was slaughtered by sailors for food and preyed on by introduced species. The last known survivor was recorded in 1662, less than a century after the first notice of its existence.

THE CAUSES OF EXTINCTION

This targeted human predation is but the most conspicuous example of the wider impact of our own species—intentional or not—on countless other species with whom we share the planet. For all species, survival is a precarious business, and, as we have seen, extinction even in historic times has overtaken some creatures that once were abundant. Can we learn from these examples how extinction comes about? It may arise from the *introduction,* whether natural or man-made, of *competitive pests*, such as rats or cats. It can also arise from the *loss of essential species*, such as the extinction of an essential pollinator or prey, for example.

Or it may arise from the destruction or slow *loss of habitat,* such as clear-cutting of tropical rainforests for agriculture, or the draining and filling of marshlands.

It can also arise by *competition*, as in Hawaii, where the arrival of external settlers led to the extinction of at least thirty-nine species of endemic land birds, and New Zealand, where colonization resulted in the extinction of moas and other flightless birds.

Small, isolated populations are especially vulnerable, because of the possibility of genetic drift (a random change in the genetic composition of a population produced by chance or natural events, rather than by natural selection) or inbreeding that produces deleterious effects.

Climate change can be another cause of extinction, as with, for example, the loss of large mammals—mammoths, woolly rhinoceroses, giant ground sloths, and others—after the Ice Age, and particular physical events such as major volcanism can also cause extinction, especially of small populations.

None of these causes is rare or isolated. Every species ultimately competes for space, food, and materials with others, both those species that are closely related and closely situated and those that are not. Leigh van Valen's Red Queen hypothesis —"it takes all the running you can do to stay in one place"—applies here. Every species, all the time, tries to keep up with change, and eventually all fail. "All species that ever lived," as David Raup has written, "are, to a first approximation, dead." Without such extinction, we—our species—might not be here. E. O. Wilson has estimated that if current rates of habitat destruction and other human influence continue, half of all living species will become extinct within a century. Wilson also predicts that "climate change alone, if left unabated, could be the primary cause of extinction of a quarter of the species of plants and animals on the land by midcentury."[5]

Such examples of extinction as those we have so far described, apart from those due to human influence, can be thought of as "background extinction": everyday instances of a natural, continuing process. It is the price paid for evolution: the price for life itself. And it involves not just individual species, but also larger groups; classes, for example, such as trilobites and placoderms. But the growing refinement of the geologic time scale and the steady, cumulative increase in the description of new fossil species and higher groups have given rise to attempts to quantify the *rate* and *pattern* of extinction. The pioneer work in this field was done by Raup and Sepkoski.[6]

MASS EXTINCTIONS

We have already noticed several episodes of major extinction and turnover: the disappearance of the "Ediacaran fauna" in early Cambrian times (540 million years ago), the disappearance of the early Cambrian trilobites (some 510 million years ago), the marked turnover of marine invertebrates in late Ordovician times. There are ten or a dozen of these extinction spikes that have been recognized by various authors.

But there have also been a few occasions in Earth's history when what have been identified as episodes of "Mass Extinction" have taken place. The definition of a mass extinction, as opposed to episodes of "major extinction" or the ongoing occurrence of background extinction, is not precise but is generally understood to involve the sudden worldwide disappearance of a significant proportion (say, +40%) of different biological groups—genera, families, and orders—occupying a range of ecological niches over a relatively brief interval of geologic time.

But that type of definition begs two questions: First, how rapid does a mass extinction have to be to qualify as "mass"? Here we run up against our inability to measure time with any great precision in older rocks. Does extinction over a period of, say, 10,000–100,000 years still qualify as mass extinction?

Second, how many species, genera, families or orders does it take to qualify disappearance as "mass extinction," as opposed to "background

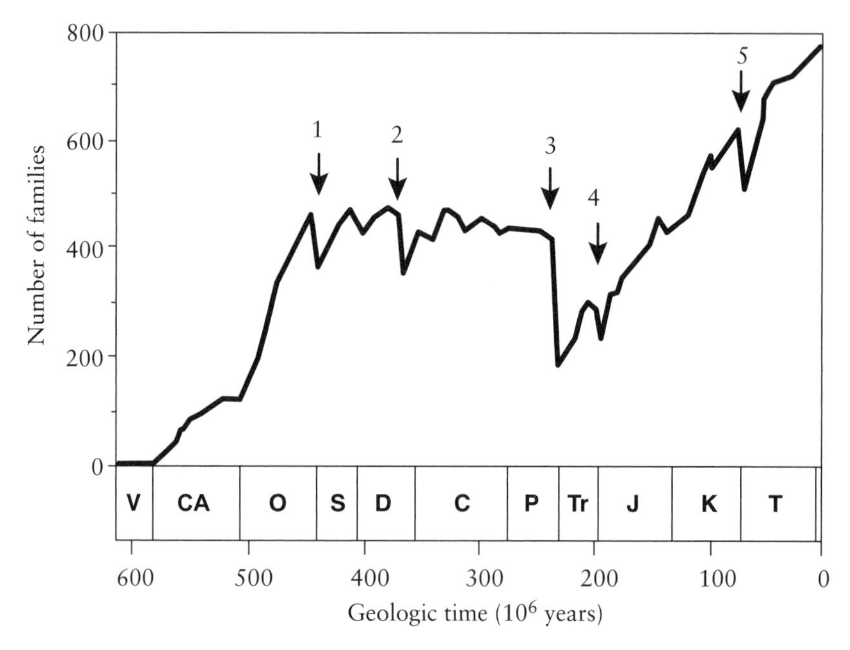

Figure 16.1. The history of the diversity of marine animal families throughout the Phanerozoic, as a function of time. The curve connects 77 discrete data points, each showing the total number of well-skeletonized families known from a particular stratigraphic stage. The arrows labeled 1 to 5 identify the five major events of mass extinction: 1, end Ordovician (O); 2, late Devonian (D); 3, end Permian (P); 4, end Triassic (Tr); 5, end Cretaceous (K). The durations of the various geological periods are indicated on the time axis. In addition to the five just given, the abbreviations for the remaining periods are V, Vendian; CA, Cambrian; S, Silurian; C, Carboniferous; J, Jurassic; T, Tertiary. Pattern of repeated extinction.

(After Sepkoski 1992. From John Lawton and Robert M. May, *Extinction Rates* [Oxford: Oxford University Press, 1995], fig. 1.1, p. 4. Reproduced with permission of Oxford University Press.)

extinction"? How do we quantify and compare extinction rates for, say, marine invertebrates—whose fossils are relatively common—and small terrestrial mammals, whose fossils are rare?

There are no clear-cut answers to these questions, but the work of Raup and Sepkoski, who plotted the number of extinctions of fossil marine invertebrates per million years, provided a striking illustration of five prominent peaks in rates of extinction, and it is these that have generally come to be regarded as examples of "mass" extinction.

The five mass extinctions generally recognized are the following:

- *End of Ordovician extinction*, some 443 million years ago. About 12 percent of families, and about 65 percent of known fossil marine invertebrate species became extinct. There were, at the time, no true vertebrates or land plants. There seem to have been two pulses of extinction, separated by a few hundred thousand years. Trilobites and some early groups of echinoderms and graptolites were all greatly reduced, as were some once-widespread brachiopods, bryozoans, and corals. This period was marked by extensive continental glaciation, with consequent lowering of sea level, both of which may have been factors, perhaps among others, in leading to this widespread extinction. In fact, the extinction "event" covered a span of some 2 million years.
- *Late Devonian extinction* events occurred from about 365 million years ago. As many as three-quarters of all species and perhaps half of all genera were involved in this event, with particular destruction in reefs and shallow water marine faunas. Cephalopods, brachiopods, ostracods, corals, and trilobites, as well as early fish, were affected, but early amphibians and land plants do not appear to have suffered. Two or more extinction pulses have been recognized within this overall peak, but even the "peak" itself seems to have involved a series of extinction "spikes," each occurring over a period that may have been as long as 3 or 4 million years.
- *End Permian extinction*, some 245 million years ago. This, the so-called Time of Great Dying, was the greatest of all extinctions and is estimated to have involved the loss of more than half of all families, more than 95 percent of all marine species and perhaps as many as 70–90 percent of amphibians, reptiles, insects, and some plants. The whole marine invertebrate biota was affected, with the long-dominant Paleozoic brachiopod-crinoid-bryozoan-coral

assemblage being overturned and replaced by the bivalve-gastropod-echinoid-assemblage, which became characteristic of the Mesozoic. Even groups that survived (such as the cephalopods) did so with new members. The species level losses of some of these groups may have reached 95 or more percent. The extinction itself spanned a period of some 8 million years. At this time, a single continent, Pangaea, included all Earth's land masses; extensive glaciation was occurring in the Southern Hemisphere as well as intense volcanism ("the Siberian Traps"). The extent to which climate change, acid rain, habitat loss, and extensive volcanism interacted and reinforced one another is difficult to determine. It has also been suggested that this combination of factors may have led to oceanic stagnation, which may also have been a toxic factor.

- *End Triassic extinction*, some 208 million years ago. During the last 18 million years of Triassic times, two or three phases of widespread extinction of land vertebrates occurred, including the disappearance of many large amphibians and most therapsid reptiles. In the oceans, some 90 percent of bivalve species, 80 percent of brachiopod species, many ammonites, all the conodonts, and many gastropods, crinoids, corals, and other reef dwellers all vanished. Potential causes are not easy to identify, though widespread volcanic eruptions, and their associated climatic and atmospheric impacts, may have been a factor.
- *End Cretaceous extinction*, some 66.4 million years ago. Though it was not the largest mass extinction of all time—the End Permian has that honor—the End Cretaceous extinction event is by far the best known. It marked the end of an era (literally, in a geologic sense) with the loss of the dinosaurs, but it also brought the extinction of flying reptiles (pterosaurs), many marine reptiles (mosasaurs and plesiosaurs), as well as many insects and flowering land plants. Ichthyosaurs became extinct about 30 million years before the end of Cretaceous times. This event also saw the extinction of a number of marine invertebrate groups, such as the ammonite cephalopods and some groups of marine plankton that had flourished throughout the Mesozoic Era. This extinction, then, was clearly global in extent, rapid in time, widespread in biological effects, and extensive in environmental impact, so any recognition of its cause must explain those four features of its impact.

THE SEARCH FOR A CAUSE

Let us use the End Cretaceous extinction event, the best documented of all, to explore in more detail the challenges of identifying cause and effect in the fossil record. There is, however, one other puzzling feature of this last great episode of mass extinction. Several major groups—insects, birds, mammals, and some reptiles (lizards, freshwater turtles, and land tortoises, for example)—not only survived, but also prospered, occupying some of the same niches and expanding rapidly into the very environments vacated by those groups that became extinct. Other groups, although temporarily reduced by this episode of extinction, later expanded and flourished. The angiosperms, for example, were reduced but rapidly recovered to blossom in the succeeding Tertiary period, and became the dominant group of land plants. Fishes, corals, and many mollusks survived and prospered. Some groups of microplankton were affected; others, such as diatoms, were not.

That poses a dilemma if we seek to involve only a single explanation for this and other great episodes of extinction. Can any *single* explanation—asteroid, volcanism, toxic emissions, changes in ocean chemistry, or any other currently popular mechanism—produce, not only profound global impact on some major biological groups, but also permit such puzzling exceptions by allowing other groups, in comparable environments, to survive?

Several kinds of physical events have been suggested as agents of doom. One suggestion is "Nemesis"—the "Death Star"—which, as has been suggested by David Raup, could be a binary companion of the Sun and might attract comets, which could bombard Earth. Others argue that the culprit might have been an asteroid impact. These, however, are rare, though clearly demonstrated, events. The giant meteor crater in Arizona represents a recent example. Older examples are known from Germany (the Ries crater) and Canada (Sudbury, Ontario). In these cases the exact form of the craters is less conspicuous because they have eroded over time, though they are clearly visible in aerial photographs.

Interest in the possibility of such an event at the K-T boundary was encouraged when two scientists from Berkeley, Louis and Walter Alvarez, and their colleagues discovered in 1980 a highly significant increase in iridium content in rocks at Gubbio, Italy, that span the K-T boundary. Iridium is rare on Earth, but abundant in meteorites. The rocks with this iridium "spike" also contained "shocked quartz," generally regarded as the result of high impact, together with tiny spherules formed from molten rock and molten glass, buttonlike structures (tektites). The authors of the paper argued that the impact of a meteorite with diameter of 10 km could have thrown up a dust cloud, reducing sunlight, disrupting photosynthesis in both land plants and ocean phytoplankton, and so disrupting the food chain.

Subsequent study has detected similar features in rocks of the same age (65 million years) from scores of localities, scattered throughout the world. Further evidence that this might be the "smoking gun" came from the later discovery of a giant submarine crater, located in the Chicxulub area of Mexico, that seemed to be of the same age. Subsequent detailed collecting has shown that sharp extinction in marine faunas is matched by sharp changes in pollen ratios of land plants, indicating a sudden replacement of angiosperms by ferns, before the angiosperms later recovered.

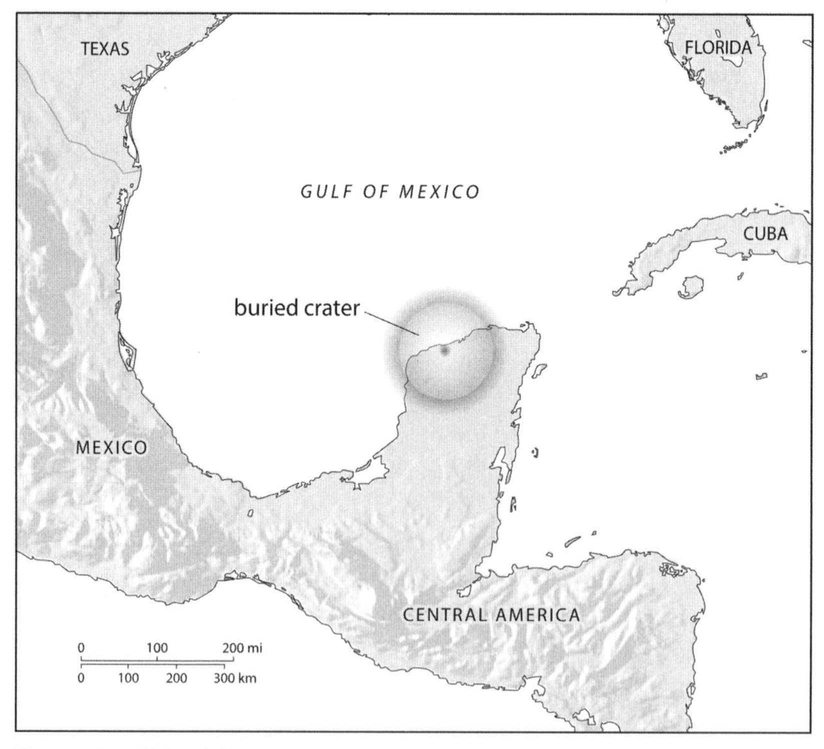

Figure 16.2. Chicxulub crater.

The Chicxulub asteroid was quickly embraced by many as the agent of K-T doom, but a minority of specialists—especially many paleontologists—have remained skeptical. The precise times of extinction, it was argued, differed significantly from one biological group to another, while the negligible impact on some biological groups made it unlikely that such a catastrophic event could account for the differences.

But was there a less catastrophic explanation? Oddly enough, the K-T extinction event also coincides with another major plume volcanic outburst in the Deccan Traps, whose lavas cover almost 250,000 square miles of western India and are a mile or more thick in some areas. They are thought to have erupted over an original area some half the size of modern India, with the release of gases, including sulfur dioxide, having a severe cooling effect on climate, very close to the K-T extinction. It is argued that the effects of this vast outpouring of gases would dwarf any possible effects from the impact that created the Chicxulub crater and that its timing is coincident with the recovery of marine plankton after the event. These two possible mechanisms—meteorite-asteroid and massive volcanism—for each of which there is strong evidence, may even have reinforced one another to produce the equivalent of a "nuclear winter" by injecting aerosols into the atmosphere.

The extent of these possible influences on the patterns of extinction is shown in table 1.1 (p. 11).

The K-T extinction has been studied in far more detail than any other, and the rock sequences that represent it have now been recorded from scores of localities, scattered all over the world. In the most complete sections of terrestrial deposits—those of the American west—the disappearance of dinosaurs is interpreted as abrupt. But just how abrupt is a different question; did extinction take place within a year or two, a few hundred thousand years, or over a million or more years?

There is still no unanimity in either interpreting or explaining these results. Those who interpret the decline of these various groups as gradual (though rapid) argue that climatic changes, competition, and changing geography and sea level are more plausible explanations than the "catastrophic" explanation provided by the asteroid impact.

While extraterrestrial agents could account for global-scale, extensive effects and sudden extinction, they cannot, it is argued, easily account for the striking differences in effect between different groups sharing common habitats. And these, as we have just seen, were conspicuous in the

K-T extinction event. One estimate is that 70–75 percent of all the families survived.

The "catastrophists"—in contrast, pointing to both the similarity of worldwide sequences of K-T rocks, including as they do, the presence of high iridium content, shocked quartz, and glassy spherules: all of them evidence of a high-impact asteroid event—argue that the occurrence of that event with such an episode of mass extinction cannot be a mere coincidence and must represent cause and effect.

What other kinds of terrestrial events, then, might have a global impact, sudden effect, and yet such strikingly differential results? There are at least four, which have been discussed: *widespread glaciation*, with its effects on overall climate, *changes in continental configuration*, with their effects on sea level and ocean circulation: *disease* or genetic malfunction, and *growing stress* in some biological groups attempting to adapt to ever-changing conditions. Furthermore, several of these could have impinged on one another. Cooling, for example, might have reduced availability of food for herbivorous dinosaurs, which in turn provided food for the carnivores. Each of these hypotheses has its own proponents, but none has yet provided a fully convincing explanation of such mass extinction. There may not, of course, be a simple, *single* explanation for these very different extinction events. Some observers, for example, who favor the widespread volcanism represented by the Indian Deccan Traps, tend to see this huge volcanic outburst as producing increasing environmental change in CO_2 content, acid rain, global anoxia, and atmospheric dust, thus invoking a mixture of interacting causes.

For all the precision provided by these sequences, there remains a degree of ambiguity, both about the exact times of extinction of the various groups involved, and about whether or not such an impact could produce such dire effects in groups inhabiting both terrestrial and aquatic environments.

The debate continues, with lively advocacy by proponents of both viewpoints. The evidence—inevitably—is fragmentary, the record incomplete. One "explanation" may not exclude another. The great contribution of the Alvarez team, father and son and their colleagues, is to have revived serious debate by giving what had once been a largely speculative discussion, a new quantitative basis.

There is one additional aspect of mass extinction that deserves our review. E. O. Wilson[7] and others have argued that another episode of mass extinction may be looming, describing present increases of "critically endangered" species as an "ongoing biological catastrophe." In several particular cases, human-induced habitat loss and pollution are identified as "manifestly the primary reason for this growing wave of extinction." Wilson concludes, "with certainty we are the giant meteorite of our time, having begun the sixth mass extinction of Phanerozoic history." In this sense, a study of the extinction events of the past may well be an important guide to preventing a looming mass extinction event in our own future. Far from being another example of the present being the key to the past, this could be a significant case of the past being a key to the future.

17

"HAVE BEEN AND ARE BEING EVOLVED": THE DEVELOPMENT OF LIFE

THE DISCOVERY OF EVOLUTION

July 1, 1858, was a fine summer day in London. The temperature in Hyde Park at 9:00 a.m. was 70 degrees. A drought and a water shortage had been occurring. The Globe Theater advertised a diorama of the Indian Mutiny. The clipper *Lonchiel* sailed for Sydney, with a voyage estimated at seventy-nine days. Parliament debated the Universities (Scotland) Bill and what W. S. Gilbert called "that annual blister, marriage with deceased wife's sister." Good progress was reported in laying the Atlantic Cable. Queen Victoria and the Prince Consort attended a performance of *Il Trovatore* by the Royal Italian Opera Company. And, in a cricket match at the Oval, the Gentlemen made 158 against the Players, who replied with 103 for 4 wickets, after a stand by Wisden and Lillywhite.

But one event that occurred that day passed unrecorded in the *Times* the following morning. At a meeting of the Linnean Society, Alfred Russel Wallace and Charles Robert Darwin presented a joint paper titled, "On the tendency of species to form varieties, and on the perpetuation of varieties and species by natural means of selection." The paper was calmly received, and few of those who heard it could have predicted its subsequent impact on scientific thought.

Yet this paper was of such major importance that it marked one of the great turning points in the history of human knowledge. In 1543, Copernicus had seen an order in the endless constructions of Aristotelian cosmology, and with a unified conception of the universe had brought unity to what had before been confusion. Almost a century later, Newton reduced the order of the universe to mechanism. In the years between 1628 and 1858, although the world had seen mechanism at work in the physical sphere, the realm of living things appeared to remain distinct.

As with the physical world, so with the organic; the existence of order had been demonstrated more than a century before. Linnaeus's *Systema*

Naturae had shown the orderliness of the organic world, but the orderliness of his classification had been almost universally interpreted as the result of the special creation of subsequently immutable species. It was the contribution of Darwin and Wallace to bring order and mechanism to an understanding of the organic world; a mechanism, they suggested, no less comprehensible and real than that of the inorganic world.

The association of the two authors was remarkable. In February 1858, Wallace lay stricken with fever at Ternate, in the jungles of the Moluccas. As his mind wandered over the problem of the development of species, a subject that has exercised his attention for a number of years, he suddenly recalled an *Essay on Population* by Rev. Robert Malthus that he had read twelve years before. Malthus argued that the human race would increase in geometric progression were it not for the fact that many of its members failed to survive and to reproduce. "In a sudden flash of insight," Wallace realized the applicability of this notion to the organic world as a whole and conceived the idea of natural selection in the development of species. Within a week he sent to Darwin a summary of his conclusions under the title "On the Tendency of Varieties to depart indefinitely from the Original Type." Wallace wrote that the idea expressed seemed new to him, and asked Darwin to show it to Sir Charles Lyell if he himself also thought it was new.

Darwin received the essay with astonishment and dismay, for Wallace's hypothesis was identical to what he himself had formulated. Darwin, who curiously enough had also been much influenced by Malthus's essay, had devoted the previous twenty years to the patient accumulation of evidence that he proposed to publish as a book. Year after year Darwin accumulated more and more data and slowly the enormous treatise took form, although he had, in fact, prepared an outline of his theory as far back as 1842, and a more lengthy account two years later. Of these, he wrote to Lyell, "I never saw a more striking coincidence; if Wallace had my MS sketch written out in 1842, he could not have made a better short abstract! Even his terms now stand as heads of my chapters." It was under such circumstances that Lyell and Sir Joseph Hooker suggested a joint presentation of papers announcing the theory. Darwin and Wallace readily agreed, and the joint publication included Wallace's essay and an extract from Darwin's manuscript of 1844, together with an extract from one of his letters to Asa Grey written in October 1857 "in which" (as Lyell and Hooker noted in their accompanying letter of presentation) "he repeats his views, and which shows that these remained unaltered from 1839 to 1857." The joint paper was calmly received; few of those who heard it could have predicted the way in which this new theory of evolution was soon to shatter the tranquility of Victorian thought.

The following year Darwin completed a brief abstract (as he called it) of his researches and on November 24, 1859, the most important book of the century was published—*On the Origin of Species by Means of Natural Selection, or the Preservation of Favoured Races in the Struggle for Life*. On the day of its publication the first edition of 1,250 copies was sold out.

The effect of *The Origin* on public opinion was immediate and profound, but its effect on the natural sciences was revolutionary. It provided the key that not only integrated and interpreted the maze of biological data but also gave new impetus and urgency to every avenue of research.

To Darwin, as to Wallace, the ultimate solution of the problem of the origin of species came suddenly, "In October 1838," he wrote, "I happened to read for amusement Malthus on *Population*, and being well prepared to appreciate the struggle for existence which everywhere goes on, from long-continued observation of the habits of animals and plants, it at once struck me that under these circumstances favorable variations would tend to be preserved, and unfavorable ones to be destroyed . . . I can remember the very spot in the road whilst in my carriage, when to my joy the solution came to me."

The new theory of evolution attracted little public attention until *On the Origin of Species* was published in 1859. The content of the book we will discuss later, but perhaps the most remarkable thing about the publication of *The Origin* was not the reception it was given by the scientific public, but the worldwide convulsion and outburst it produced among men and women of all interests and persuasions. Philosophers, politicians, theologians, literary critics, historians, classical scholars, and the man in the street—all alike took it upon themselves to assess its worth. And as so varied a group studied it, so their verdicts also varied—some accepted and respected Darwin's conclusions, others viewed them with suspicion, but most rejected them out of hand, and denounced both Darwinism and all its supposed implications. "As for the book, some treasured it, some burnt it, and some, undecided, like the Master of Trinity College, Cambridge, merely hid it!" Scientific theories, philosophies, political systems, ethical standards, revolutionary movements, social reforms, and economic laissez-faire—all these and more were established, modified, or justified on the basis of Darwinian premises. Indeed, Darwinism soon became all things to all men.

But the real debate focused on the theory itself. Was evolution true? Had living things evolved over unimaginable periods of time from quite distinct and different ancestral forms? These became the burning questions of the day.

One of the unique features of the study of fossils is the opportunity it provides to draw conclusions about the processes by which living things have developed. This does not mean, of course, that we can draw no such conclusions from the study of living things—far from it. Indeed, it is only to some extent on the basis of our knowledge of the physiology, genetics, and behavior of living organisms that we can interpret some features of fossils. But it is true that it is only by examining the broad history of life that we can hope to recognize any general patterns of development that may have taken place. We have already noted some of these features; it may be useful to summarize them.

THE EXPANSION OF LIFE

Throughout geological time, a more or less continuous expansion of living things has occurred, and this expansion has been reflected in a number of different ways. The total number of species, the degree of diversity, and the range of adaptations they represent have all increased throughout geological time. But within this broad expansion, no constant rate of increase, no common regularity of expansion, no overall trend has been observed.

Certain periods have been marked by quite spectacular expansion (the early Paleozoic and the early Mesozoic, for example); others (such as the Permo-Triassic boundary), by equally spectacular extinction. The same absence of regularity is seen in the increase in relative diversity. Certainly living creatures are more diverse than those of, say, Cambrian times, but the main patterns of animal radiation (apart from the vertebrates, insects, and plants) were established early in the fossil record. The reason for this seems clear: the seas were the first environment to be colonized, and the conditions of life there have remained broadly constant. It was the invasion of new environments on the land and in the air that brought new types of structure, new levels of complexity, and new ways of life, and it is these that are represented in the vertebrates and insects.

THE CONTINUITY OF LIFE

Our knowledge of living creatures makes it clear that any increase in the number of individuals can come about only by deriving them from preexisting parents. The fossil record makes it equally clear that the increase in diversity of living things has been brought about in the same way: that new species have developed over long periods from the cumulative modification of existing species. This is the process we saw in horses, and all the evidence suggests that the process is continuous. There are gaps in the fossil record, to be sure; more gaps than fossils, in fact, but they appear to be gaps of nonpreservation, and not of nonexistence. Furthermore, the continuity extends to major groups of animals, as well as to individual species: new classes, representing totally new structures adapted to new ways of life, are developed in just the same way and with just the same continuity. This was demonstrated in the origin of the amphibia, the reptiles, the birds, and the mammals. In fact, we can summarize the development of all new kinds of organisms as a process of "descent with modification"—a formula that is the equivalent of the word "evolution." We hear much of evolutionary theory, but if we accept this definition of evolution as a process of descent with modification, the fossil record makes it clear that evolution is a fact.

THE INTERRELATIONSHIPS OF LIFE

No man, no person, is an island, and there is a literary, a social, and a moral sense in which we are all aware of it. There is also an evolutionary sense in which it is equally true, for no individual, no species, no way of life can ever be independent of, or insulated from, others. This mutual interdependence extends to all levels and influences the whole of life. Competitors for food or space or sunlight, host and parasite, prey and predator, male and female, parent and offspring, all exist within an intricate and constantly changing equilibrium. We have seen something of its broader aspects in such things as the profound influence of the spread of mid-Tertiary grasslands on mammalian development.

THE ENVIRONMENT OF LIFE

The dependence of organisms extends to their environment no less than to their neighbors. Every organism, every community, is influenced by, and in turn influences, its environment, and a given type of life is bounded by the extent of a given environment. Fish are confined (with rare exceptions) to water, and the exploitation of the land involved a new type of life, just as the later return of reptiles and mammals to the water involved the return to a superficial fishlike body. Indeed, quite unrelated groups of animals living under similar environmental conditions tend to develop similar structures. This is the pattern of convergence or homeomorphy we have seen in so many groups.

Now, a given environment can support only a given total of living creatures. There are limits to the number of sheep that can be supported in a given field, the number of large trees in a small garden, and even the number of kittens in a particular household. Change in structure in a group of organisms can therefore be brought about in some environments only by migration or replacement, and the rise of many new groups involves the replacement of old ones, less well adapted to the same environment. It was this pattern of change that we saw in Mesozoic plants, Paleocene mammals, bony fish, and recent South American mammals, and such replacement is a frequent feature of the development of life.

THE PERSISTENCE OF LIFE

All living things are mortal, but life is continuous. We are linked to our ancestors by a frail physical thread (the reproductive cells) that is the legacy, the source, and the bond of life down the ages. One of the most striking features of life has been the persistence of all the major phyla of living things: none has become extinct. In a few cases groups, even genera of organisms have persisted for millions of years with no essential change, but this is the exception and not the rule. In the great majority of cases the taxonomic groups to which individuals belong are no less mortal than their members, and species, genera, families, orders, and classes have all shared the common fate of extinction.

We have seen the difficulty in interpreting the cause of this extinction, but there are a few groups from which some tentative conclusions might be drawn. The South American marsupials seem clearly to have owed their extinction to the sudden influx of placental competitors from the north, and it is difficult to escape the conclusion that the great Permo-Triassic wave of extinctions was related in some way to the unique combination of contemporary physical conditions. Certainly, in many, if not all cases, extinction seems to have involved organic competition or lack of adaptation to changing environmental conditions, or both.

The same is true of the closely related problem of survival. Some groups (*Lingula,* for example) seem to persist because they are eminently well adapted to a persistent environment.

Survival and extinction inevitably lead us to consider their corollary: rate of change. Survival of a group may involve no rate of change (it may remain static in form), considerable rate of change (it may survive by means of modification), or the ultimate cessation of change (it may become extinct). Rates of change might be expressed in various ways. Two of the most obvious are the consideration of rate of appearance of new groups per unit of time, and the rate of change of some selected body structure, either per unit of time or relative to some other structure. Numerous studies of this kind have been made and, although at present there are so many variables and imponderables that little agreement can be reached on their interpretation, such studies seem to offer great promise in the future.

THE EVOLUTIONARY PROCESS: THE DARWINIAN RECIPE

It would be inappropriate in this book to attempt a detailed consideration of the intricate and varied aspects of the evolutionary process. But we have seen that fossils demonstrate the fact of evolution, and it is difficult to escape the question, "How did it come about?"

The answer now generally accepted is that first suggested, as we have seen, by Charles Darwin and Alfred Russell Wallace in 1858, and later amplified and justified in Darwin's great book, *On the Origin of Species* (1859). The essentials of Darwin's argument are simple (and, from our retrospective viewpoint, obvious). Darwin claimed that living species have arisen from preexisting species by a process of descent with modification. Now, this suggestion was not new. Other naturalists (Lamarck, Buffon, and Darwin's grandfather, Erasmus Darwin, among them) had suggested descent with modification long before this, but the scientific world remained skeptical, and generally hostile. It was not until the eighteenth century that the true nature of fossils had been generally recognized, and its interpretation was governed by the widespread acceptance of Archbishop Ussher's chronology (chapter 2), which, as we have seen, reckoned creation to have taken place in 4004 BC at 9:00 a.m. on October 26.

Because of this general climate of opinion, the "orthodox" view of creation had come to be based on a theory of catastrophism, which maintained that the Earth had experienced a number of successive cataclysmic revolutions (of which the Noachian deluge was the most recent). Each of these catastrophes was thought to have completely destroyed all living things, so that after an interval of time a new creation took place, whose beings were in turn entombed in the strata of the next cataclysm.

It is easy to smile at what we now regard as such a naive concept, but it held undisputed sway in the scientific world for more than a century and found as its champions many of the greatest pioneers in the development of the natural sciences.

The importance of Darwin's work lies largely, therefore, not in the fact that he was the first to suggest the possibility of evolution, but in the fact that he convinced the great majority of scientists that evolution had taken place, and this he did by the presentation of a wealth of detailed data that supported his theory of the mechanism by which evolutionary changes had been effected.

The evolutionary hypothesis of Darwin and Wallace rested on three essential foundations. Two were observations, and the third an inference. They noted, first of all, that all organisms tended to overproduce. This proliferation of offspring is something that seems characteristic of the whole world of living things. A far greater number of young are produced than ever survive to maturity, because on the whole, the number of individuals within particular species remains more or less constant. Many calculations of this superabundance of nature have been made. The oyster, for example, produces something like 600 million eggs per season, and we are told that if the great-great-grandchildren of one such group all survived without mortality, and reproduced, their shells would number 66×10^{33} and would make a mountain eight times the size of the Earth.

Second, Darwin and Wallace observed that individuals within any species are not identical but show variation in certain characteristics, and that, although some variation is spontaneous, in many cases variations (however produced) may be inherited. The more obvious of these variations include such things as size, appearance, speed of movement, health, fertility, instincts, physiological efficiency, and so on. The mechanism of this inherited variation was unknown to Darwin and Wallace, but since their day genetic studies of inheritance and mutation have produced an understanding of the process. From these two observations, Darwin and Wallace concluded that some of these variant forms must inevitably stand a greater chance of survival than others; thus they would tend to produce relatively more offspring than those less-fitted for the environment in which they lived. Therefore, on the whole, reproduction would be nonrandom, the better-adapted producing relatively more offspring, and those would have characters that included those advantageous ones the parents possessed. This was natural selection, the mechanism by which Darwin and Wallace sought to explain the vast diversity of living things.

EVOLUTION IN ACTION: LIVING POPULATIONS

Although Darwin stressed the fitness of organisms to their environment as well as the drastic changes produced by domestic breeding, natural selection remained a hypothesis. It is now established as a fact and has been convincingly demonstrated in a number of living forms.

Darwin's Finches

Charles Darwin returned from the *Beagle* voyage on October 2, 1836, and headed home to Shrewsbury. On January 4, 1837, he was in London, presenting his findings and his specimens to the Zoological Society. During the five weeks that he had spent in the Galápagos Islands, from September 16 to October 20, 1835, Darwin had been struck by the similarity between a mockingbird he had collected on Chatham Island and those he had earlier seen on the mainland of Chile, and he later discovered that a third, distinctive mockingbird existed on Charles Island.

Darwin paid considerably less attention to the group of dark, sparrow-sized birds now known as "Darwin's finches." He noted them, but thought them to be related to warblers, wrens, blackbirds, and finches and failed to record the particular islands from which he collected each of his various specimens, though fortunately, Darwin's faithful servant, Syms Covington, as well as Captain Robert Fitzroy and his servant, who had also been collecting, did record the source of their various specimens.

A week after Darwin presented his material to the society, at the very next meeting, John Gould, the ornithologist, to whom the birds were given for identification, announced a startling discovery. The finches Darwin had collected were unique to the Galápagos Islands; they represented fifteen distinct species and all fifteen of them were new. In Darwin's *Journal of Researches*, later expanded as *The Voyage of the Beagle*, Darwin wrote,

Figure 17.1. Darwin's finches. Darwin theorized that the variety of finches on the Galápagos Islands arose from mainland species. He believed that many generations after their ancestors' arrival, finches had adapted to the islands' various environments and food sources and had become unique, reproductively isolated species. Similar examples of evolutionary diversification after separation from ancestral populations occur among plants.

(From Randy Moore, W. Dennis Clark, and Darrell S. Vodopich, *Botany*, 2nd ed. [McGraw-Hill, 1998], fig. 22.5, p. 529. Copyright McGraw-Hill. Reprinted with permission.)

"Seeing this gradation and diversity of structure in one small, intimately related group of birds, one might really fancy that, from an original paucity of birds in this archipelago, one species has been taken and modified for different ends."[1] And he later concluded, "Hence, both in space and time, we seem to be brought somewhere near to that great fact—that mystery of mysteries—the first appearance of new beings on this earth."[2]

The Galápagos finches provide a striking example of evolution in action. The Galápagos Islands lie near the equator in the Pacific, some 500 miles west of the coast of Ecuador, of which they form a political part. The islands, which have both English and Spanish names, are all of relatively recent volcanic origin. There are fifteen major islands, three minor islands, as well as a hundred or so islets and rocks. The islands rise steeply from the bottom of the Pacific, some summits reaching 9,000 feet above the ocean floor, and they display a variety of ecological niches.

Some of Darwin's finches—which he at first had lumped together—are illustrated in figures 17.1 and 17.2. Fifteen different species are now identified. They can be roughly divided in terms of diet, and this is reflected in their beaks. The warbler finch (*Certhidea olivacea*), with a sharp, pointed beak, is an insect eater. Another group of five finches are primarily insect eaters, though they also eat some plants. Their beaks are heavy (though less so than the plant eaters) and pointed. This group includes the large insectivorous tree finch (*Camarhynchus psittacula*), the medium insectivorous tree finch (*C. pauper*), the small insectivorous tree finch (*C. parvulus*), and the mangrove finch (*C. heliobates),* as well as the tool-using or woodpecker finch (*Camarhyncus or Cactospiza pallidus*), which uses cactus spines to probe for insects.

The diet of the vegetarian tree finch (*Camarhynchus crassirostris,* sometimes referred to as *Platyspiza crassirostris*) is limited to plants, whereas the remaining six species of ground finch, though primarily plant eaters, also eat some insects. The large, medium-sized, and small ground finches (*Geospiza magnirostris, G. fortis,* and *G. fuliginosa*) have massive curved beaks, in contrast to the more slender, sharp beaks of the sharp-beaked (*Geospiza difficilis*), common cactus (*G. scandens*), and large cactus (*G. conirostris*) finches.

Peter and Rosemary Grant of Princeton, who have spent a significant part of the last forty years camping on Daphne Major Island to observe the finches, have given a consummate account of the behavior, breeding, and feeding habits of finches under changing weather conditions.[3] They have recently described the development of reproductive isolation in a

Figure 17.2. Darwin's finches. The finches numbered 1–7 are ground finches. They seek their food on the ground or in low shrubs. Those numbered 8–13 are tree finches. They feed primarily on insects. (1) Large cactus finch (*Geospiza conirostris*); (2) Large ground finch (*Geospiza magnirostris*); (3) Medium ground finch (*Geospiza fortis*); (4) Cactus finch (*Geospiza scandens*); (5) Sharp-beaked ground finch (*Geospiza difficilis*); (6) Small ground finch (*Geospiza fuliginosa*); (7) Woodpecker finch (*Cactospiza pallida*); (8) Vegetarian tree finch (*Camarhynchus crassirostris*); (9) Medium tree finch (*Camarhynchus pauper*); (10) Large tree finch (*Camarhynchus psittacula*); (11) Small tree finch (*Camarhynchus parvulus*); (12) Warbler finch (*Certhidia olivacea*); (13) Mangrove finch (*Cactospiza heliobates*).

(From Biological Sciences Curriculum Study, *Biological Science: Molecules to Man*, 3rd ed. [Houghton Mifflin, 1973]).

lineage of finches (tagged 5110) and cautiously conclude that this may represent a new species in the process of development.

The Peppered Moth

Although Darwin stressed the fitness of organisms to their environment, and the drastic changes produced by domestic breeding, natural selection remained a hypothesis. It has now been established as a fact and has been convincingly demonstrated in a number of living groups. A particularly good example is the work of H. B. D. Kettlewell[4] on industrial melanism in moths. Until about the middle of the twentieth century, the British peppered moth, *Biston betularia,* existed commonly in its typical light form, having a gray background with pepperpot markings on the wings. A very rare, dark melanic variety was also known and described as *carbonaria.* This variety is controlled by a single dominant mendelian gene and is slightly more vigorous than the normal gray type. Because of its conspicuous color against the lichen-covered trees on which it rested, this dark form, wherever it appeared, was constantly eliminated, and it made up only about 1 percent of the populations in which it occurred. It persisted only because of the repetition of the same mutation. The spread of industrial pollution within the last century has brought about a change in the environment, darkening the bark of the trees on which the moths rest, and under these conditions the *carbonaria* variety has increased in numbers and has proved to be physiologically "hardier" than the peppered form. It now makes up about 99 percent of the total population in industrial areas, and it is now more black than it was a century ago.

Kettlewell showed that the rate of predation of this moth by various birds corresponds well with the relative frequency of the two forms in different environments. The light form still persists and is still dominant in areas unaffected by industrial pollution. This is an elegant example of natural selection in action; in a very short period, preadaptation in the form of the melanic variety has been encouraged by the spread of a polluted atmosphere. The mutant appeared considerably later in some parts of Germany. With the more recent reduction in industrial pollution, the light-colored variety has again become more common in once-polluted areas.

Although Kettlewell's conclusions and interpretation have been challenged by several subsequent writers, they have not been shown to be invalid, and recent similar repeated experiments have confirmed them. Industrial melanism has now been described in about sixty other species

Figure 17.3. Industrial melanism in the British peppered moth *Biston betularia.* (A) Dark and light moths on tree unaffected by industrial pollution. (B) Dark and light forms on a tree in an industrial area, affected by pollution.

of moths in western Europe and the United States, but it affects only those species resting on tree trunks or other exposed positions.

Land Snails

It would be wrong, however, to suggest that natural selection acts simply as a bludgeonlike extinguisher. The subtlety of its action has been well shown by the work of Cain and Sheppard,[5] for example, on the survival rate of land snails of the species *Cepaea memoralis* from southern England. They studied these snails from twenty-five areas and concluded that slightly different color and banding patterns on the shells varied with the background and color on which the snails lived. The more uniform habitats tended to have more unbanded shells, and the color of the background influenced the color of the shells. They also showed that in this case, selection depended on the season, one particular color being better suited to one season of the year than another, but being relatively less well suited at other times. This raises an important point with regard to the whole question of natural selection. It is, in fact, always a compromise: a compromise between different seasons, between different aspects of the habits of the organism, between different members of the population, between the individual and the community.

Although some subsequent writers have challenged some of Cain and Sheppard's conclusions, they are generally regarded as sound. Many other examples of natural selection might be quoted; a problematic one, so far as we are concerned, is the growing resistance of a number of bacteria to antibiotics. It has been shown, for example, that the bacterium *Escherichia coli* has become resistant to streptomycin, not because of any gradual increase in resistivity, but because of a mutation that is viable in a streptomycin medium. Mutations are already there, and once a favorable environment occurs, they are selected.

EVOLUTION IN ACTION: THE FOSSIL RECORD

The demonstration of this mechanism of natural selection presents problems in the fossil record, however, because fossils are generally only fragmentary remains. Can we also observe natural selection in the fossil record? It is, of course, quite impossible to demonstrate small changes in color in fossils, but we can find evidence of other comparable changes. The evidence at present is limited, but it exists. Let us consider a few examples.

Cave Bears

Björn Kurten has made a study of the European cave bear, *Ursus spelaeus*, from the Pleistocene of Odessa, which he compared with skulls of the living bear, *Ursus arctos*, from Finland.[6] In both species there is a positive correlation of height with length for the second molar, yet the average relation between the two dimensions is identical. The fact that the two species have a similar index of tooth height to length suggests that this shape is optimal for the tooth crown in question, although it must have been produced by a genetic change within the earlier history of one or both species. The two allometries can be traced back to the Lower Pleistocene common ancestor, *Ursus etruscos,* but the similarity could not have arisen without the influence of a genetic change, because continuation of existing trends would have produced dissimilar indices.

Kurten has been able to go further than this, however, and to show the way in which certain differences appear to be selected by environmental conditions. He studied a large fossil sample from caves near Odessa, and the results indicate both the type and amount of information that detailed fossil analysis may now provide. There is a marked separation of younger individual fossil bears into a series of discrete stages of development; this separation can be seen in samples from many localities, and the stages are virtually identical. This pattern is best interpreted as reflecting the fact that collections were made from caves that were inhabited only during hibernation, so that the growth stages are separated by periods of a year. The youngest specimens, equal in size to newborn cubs of the related living species, *Ursus arctos,* are rare, presumably because of the fragility of the bones. The predominant ontogenetic stage is one in which all permanent teeth are already formed, but in which only a few of them are in place. Comparison with living bears shows this stage to represent an age of four to five months. The cubs were evidently born during the period of hibernation, and the preponderance of specimens of this stage, with only very slight variation in both directions, indicates a mortality peak of short duration toward the end of hibernation. The succeeding growth stage represents an age of about 1.4 years, indicating the absence of bears from the cave until the next hibernation. Growth stages in successive years show that at about four years the bears are fully grown. Sexual dimorphism, although traceable in the canine teeth back to the first years of life, becomes pronounced in other structures in the third year, which is thus apparently the period of puberty.

The rate of mortality for each of the age groups can be calculated by dividing the total number of teeth in a given age group by the sum of these and all older homologous teeth. After careful allowance for bias and difficulties of interpretation in older specimens, Kurten constructed a life table. This table is of sigmoid form and shows a striking resemblance

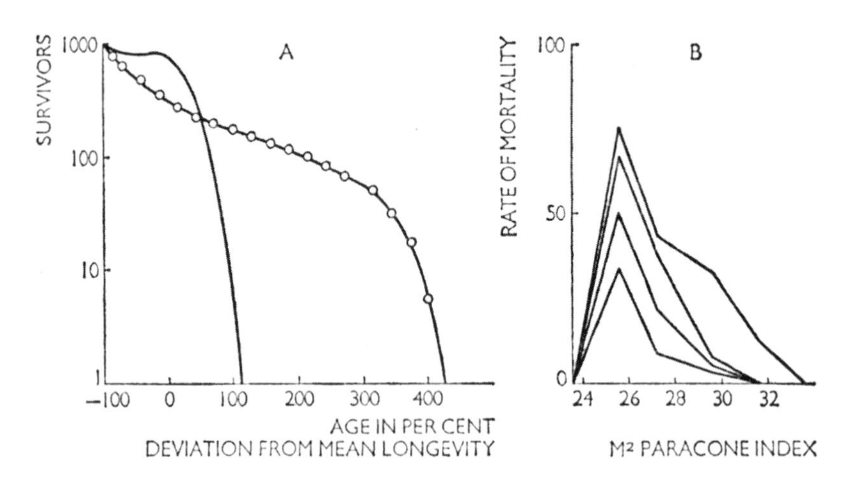

Figure 17.4. Natural selection in evolution of Pleistocene cave bears. (A) Survivorship curves from birth for *Ursus spelaeus* from Odessa and *Homo sapiens* (white males, continental United States); (B) Centripetal selection with a linear component simultaneously acting on the relative paracone length of the molar M^2 of *Ursus spelaeus* from Odessa.

(After Kurten, 1958; redrawn with permission of the author. From Frank Rhodes, *Evolution of Life* [New York: Penguin, 1976], fig. 54, p. 289. Reproduced with permission.)

to similar tables for other groups, including humans. It was further tested by comparison with other cave collections and with living bears. It includes decreasing mortality in young individuals, stable rates for those in their prime (five to ten years), and increasing rates for the old. Life expectation at birth was about three and a half years and the maximum age eighteen years.

Successful hibernation results mainly from adequate feeding during the preceding season, so that bear teeth are especially suited for studies of natural selection. Kurten studied the index, calculated as length of the largest cusp (the paracone) of the second molar divided by total length of the second molar. This index shows a marked change in mean and standard deviation with age, the standard deviation showing a slow but steady reduction in variation with age. The change in the mean index is highly significant, and its continuous reduction with increasing age is a clear indication of natural selection in favor of a smaller than average paracone, with an optimal index of 26 (a skewed point within the range). This selection started a very early stage, as soon as the teeth became functional; within two or three years the nonadapted had been eliminated, and the subsequent standard deviation of the sample reduced. Other tooth dimensions were tested and gave comparable results, which were explicable in terms of functional efficiency. Samples from other localities gave broadly similar, but not identical, results, suggesting differential selection in varying microenvironments. This intensive selection of an apparently trivial, but actually important character was especially strong in the years before breeding (the reduction in mean index during the first year was greater than in subsequent age groups). This study indicates the powerful effects of selection recorded in populations of fossils, just as in living populations.

Sea Urchins

Only one echinoid genus—the symmetrical Christmas tree ornament-like *Miocidaris*—survived the P-T_R extinction, and it was from this genus that the great variety of Mesozoic irregular, asymmetrical echinoids later arose. One of these, *Micraster* (the "heart urchin"), from the Cretaceous chalk of southern England (the "White Cliffs of Dover"), has been the object of both one of the earliest detailed studies of evolution, a landmark paper by A. W. Rowe in 1899,[7] and several more detailed studies in the last two decades. A number of changes in the form of *Micraster* species appear related to its burrowing habits. Thus, over Cretaceous times (fig. 17.5), the test became gradually broader, higher, and more wedge-shaped, with the mouth and its "lip" moving forward, its ambulacra (the fixed "arms" or petals) becoming elongated, the tubercles on its surface becoming more conspicuous, and the anal opening moving upward on the test. These changes, which were not uniformly present in the various related species and did not proceed at a uniform pace, seem to have been related to a variety of burrowing habits.

Although recent analyses have shown that the evolution of *Micraster* was less directional and more complex than Rowe suggested, a study by David Nichols[8] argues that continuous evolution of individual characters in successive fossil populations of *Micraster* can also be traced, in addition to changes regarded as adaptations to changing environmental conditions as reflected in the rocks in which they occur (particle size, depth, and so on). Nichols made an intensive study of seven living echinoid species and their habits and concluded, "these changes in the urchins probably reflect either a change in niche or an improvement in the adaptation of the animal in an effectively unchanged niche."

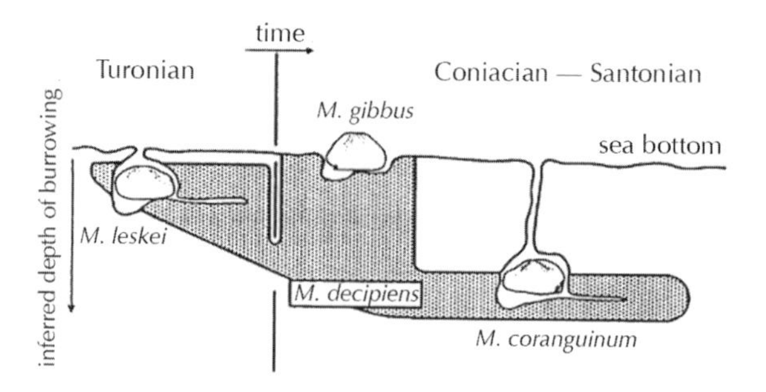

Figure 17.5. Evolution of the late Cretaceous heart urchin *Micraster*.
(After Rose and Cross. From Michael Benton and David Harper, *Introduction to Paleobiology and the Fossil Record* [Hoboken, NJ: Wiley-Blackwell, 2009], fig. 15.14, p. 405. Reproduced with permission of Wiley-Blackwell. Permission conveyed through the Copyright Clearance Center, Inc.)

Ammonoids

A worldwide study of Permian ceratitoid xenodiscoid ammonoids[9] is of particular significance, because these cephalopods were the ancestral stock of virtually all the hosts of subsequent Mesozoic ammonoids, which became the most distinctive fossils of Jurassic and Cretaceous times. The authors of this study, Claude Spinosa, William Furnish, and Brian Glenister, interpret the sequence and distribution of these fossils as representing two lineages, one of which became extinct relatively rapidly in Upper Permian times, while the other, *Cibolites waageni*, branched to provide the rootstock, from which all subsequent ammonoids evolved.

The diagram of this evolutionary sequence (fig. 17.6) illustrates the ranges and relationships of various genera involved and shows their changing form and profile, as well as the detailed pattern of their suture lines, which represent the trace of the inner chambers on the shell wall. It is noteworthy that this successive speciation represents a time interval of some 17 million years.

This broad pattern of branching evolutionary change, with slow, "progressive" change or stasis in some groups interspersed with sudden and rapid "bursts" of change in others, has been described as "punctuated equilibrium" by Stephen J. Gould and Niles Eldridge.[10] There has been debate as to whether this pattern or that of "phyletic graduation" best describes and reflects the change we can observe in the fossil record. In practice, the answer seems to be that both are present.

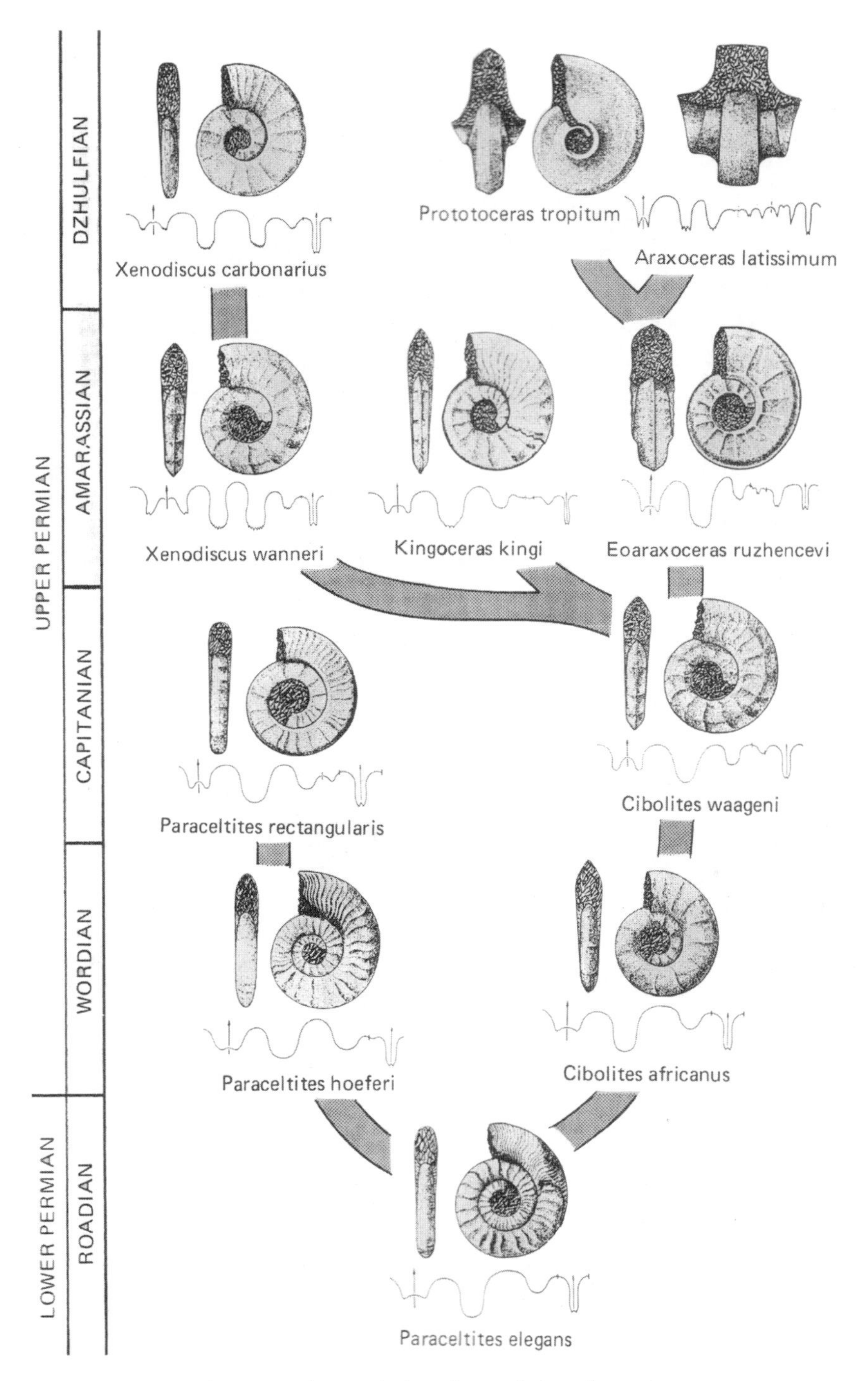

Figure 17.6. Evolutionary changes in Permian cephalopods.

(After Spinosa, Furnish, and Glenister 1975. From Harold L. Levin, *The Earth through Time*, 3rd ed. [Hoboken, NJ: Saunders College Publishing, 1988], fig. 4-18, p. 109. Reproduced with permission of Saunders College Publishing. Permission conveyed through the Copyright Clearance Center, Inc.)

Protists

Finally, evidence exists in a number of cases of continuity between species in protistans. In almost any species group studied in detail, there is evidence of gradual transition. Micropaleontologists tend to work with samples that are particularly well suited to demonstrating this. We might take one such example. Kennett has described evolution in the Upper Miocene and Lower Pliocene *Textularia miozea-kapitea* lineage (fig. 17.7)[11] His specimens—from the Cape Foulwind area, New Zealand—were collected over a stratigraphic interval of some 450 feet and show a progressive decrease in width-depth ratio. Taxonomically they are interpreted as representing both intra- and infraspecific evolution. Although the sizes of some of Kennett's samples are small, his general conclusions appear to be valid. Many comparable microfaunal examples could be quoted.

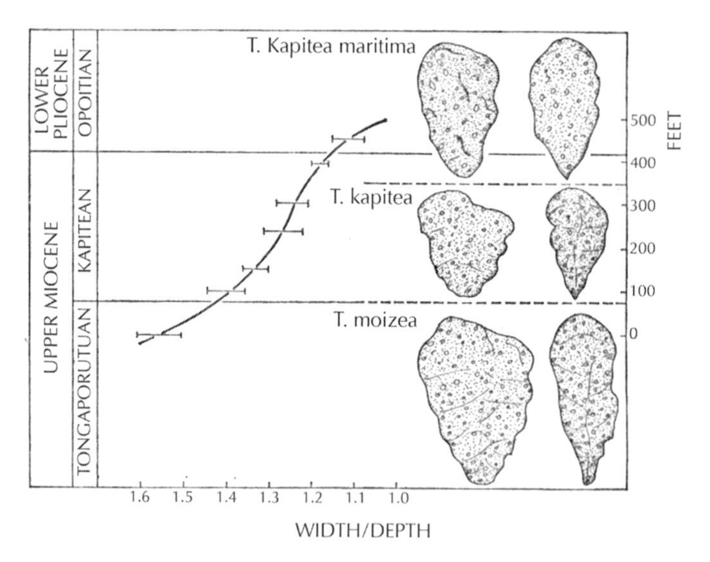

Figure 17.7. Evolution in Tertiary foraminifera. Curve showing average width-depth ratios for successive population samples of the *Textularia miozea-kapitea* lineage from the Upper Tertiary of Cape Foulwind, New Zealand. The standard deviation for each sample is shown, and representative individuals illustrated.

(From Frank Rhodes, *Evolution of Life* [New York: Penguin, 1976], fig. 55, p. 293. Reproduced with permission.)

CONCLUSION

These and similar studies of both living and fossil forms show natural selection to be an important mechanism of change, not so much as a fierce "nature red in tooth and claw" relationship, but rather as the product of differential survival and reproduction, which tend to involve systematic change in the gene pool of a population.

Was Darwin's explanation correct? Most (but not all) zoologists, botanists, geneticists, and paleontologists believe that it was, although it is now accepted with some minor modifications. The fact of evolution as such has been chiefly, but by no means exclusively, demonstrated by the study of fossils, but it is not surprising that it is the study of living rather than fossil forms that has provided most information about the process of evolution. Certainly, however, the fossil record confirms both the abundance of life in the past and the universality of infraspecific variation and the cumulative continuity of change. Furthermore, in the few cases where it is possible to draw any general conclusions (in problems of extinction, survival, competition, replacement, and so on) the fossil evidence seems to support Darwin's theory. What the fossil record has not yet provided, and never will, is an unbroken and detailed record of the history of life, a series of frame-by-frame stills of fossil organisms, so complete that they could be played back to provide a continuous film of evolution. This should not surprise us. Yet, imperfect as the fossil record inevitably is, it provides ringing endorsement of Darwin's theory.

Darwin was conscious that the fossil record represented the only opportunity to study the course of evolution and, as he viewed it, he acknowledged that, far from providing a continuous record of the development of life, it consisted of a series of gaps. The fossil record is known in far greater detail than it was in Darwin's day. How then are we to read the fossil record? Our biggest concern is still with the problem that faced Darwin. In brief, does the fossil record show evidence of continuity, or of discontinuity, in the rise of new species and higher groups? Although most paleontologists now regard it as indicating an imperfect sample of a once essentially continuous evolutionary sequence, a small number of distinguished workers regard the admitted "higher" gaps in the fossil record as inherent in the evolutionary process itself, rather than as geological in origin. It therefore seems worthwhile to examine this key question.

Let us first summarize the ways in which the fossil record shows evidence of continuity. We need review this only briefly, for paleontological literature provides many such traditional evolutionary "proofs." The first and most obvious way is that the fossil record shows a general progression from what we now regard as "simple and primitive" forms to more "complex." There is also throughout the history of life an apparent expansion in numbers of species, in diversity, in the total volume of the biosphere, and in the independence of some organisms from their environment. This gradational, successional, "progressive" character of the fossil record is so well known that it is often overlooked, but it is of major importance in evolutionary thinking. The expanding, diversifying, environment-expanding, replacing development of living things is just what one might predict a priori on the basis of evolutionary theory.

Second, at all taxonomic levels, there are now, in a growing number of cases, examples of continuity. Let us first take high taxonomic levels. We have already seen, especially in the vertebrates, remarkable transitional forms between various classes. Between the crossopterygian fish and the amphibia, we have the ichthyostegids, part fish, part amphibian, known from the Upper Devonian or Lower Mississippian of Greenland (p. 120). Between birds and reptiles, we have the renowned *Archaeopteryx* (see fig. 10.3). Between amphibia and reptiles, we have the seymouriamorphs (p. 129). Between reptiles and mammals, we have the therapsids (p. 184). We have, in the vertebrates at least, good evidence of continuity between classes. It may not be a matter of pure coincidence that in the vertebrates the classes were established much later in geological time than in other groups. The majority of invertebrate classes originated in either Precambrian or early Paleozoic times. Could the lateness of the vertebrates' establishment have favored their preservation?

Third, at a lower taxonomic level, between genera, for example, we also have a substantial number of transitional sequences. One of the best of all is the sequence of horses linking the whippet-sized, primitive, Eocene form, *Hyracotherium*, with the living horse. This was one of the first fossil sequences ever described. It was first described by Kovalevsky in 1874, later amplified by Marsh, and interpreted by Huxley. The beautiful gradational sequence shown by these fossils has been already described (fig. 1.5), so we need only summarize its major features. These involved the increase in body size, the increase in size and change in the shape of the skull, changes in teeth involving the premolarization of molars, and the deepening of the teeth from low crowned to high crowned, together with the infilling of depressions in the upper surfaces with cement. With these were associated changes in the limbs, with gradual reduction in the number of toes, and in construction of the limbs associated with the change in posture from pad-footed to spring-footed. Now this series provides clear evidence of the transition of one genus to another over a period of something like 70 million years, and it is a series that could be paralleled in many other groups: in the titanotheres, in the ceratopsian dinosaurs, in the proboscideans at the one end of the scale, and in the protozoans at the other, as well as in many other groups.

Here, then, both in fossil forms and in living forms, is evidence of the process of "descent with modification" that Darwin recognized and described. It can be demonstrated at all levels—from subspecies to classes, and in almost all groups, from protistans to vertebrates. And it provides evidence, not only of the existence of such descent and modification, but also of the influence of natural selection, both in fossil and in living populations. It is this mechanism that lies behind the endless diversity, extraordinary "fitness" and remarkable beauty of Earth's teeming creatures.

EPILOGUE

"There is grandeur in this view of life," wrote Charles Darwin, as he completed the sixth and final edition of *The Origin*, "with its several powers having been originally breathed by the Creator into a few forms or into one; and that, whilst this planet has gone cycling on according to the fixed law of gravity, from so simple a beginning endless forms, most beautiful and most wonderful have been, and are being evolved."[1]

This soaring conclusion reads far more eloquently than the rest of the book, clear and well written though that is. But what did "grandeur" imply; what was it, exactly, that was grand, lofty, or sublime about "this view of life"? Darwin in his later life was not, as we know, a devout religious man, much as his wife, Emma, regretted and lamented it. Darwin did, however, seek to demonstrate that there was no ultimate inconsistency between his advocacy of the mechanism of natural selection and a belief in a "Divine Power." In the second and later editions, he not only inserted "by the Creator" into the final sentence of *The Origin,* but also included epigraphs by William Whewell, Joseph Butler, and Francis Bacon, each of which justified the complementarity of mechanism and larger purpose.

We have caught a glimpse of this grandeur as we have traced the steady development and growth of life from its hazy origin and early manifestations, down the long corridors of time that lead to the present. It is a wonderful story, a history of millions upon millions of millions of individuals, of millions of different species, through millions of years. There is nothing more breathtaking in the whole of human experience than the contemplation of this ceaseless cavalcade of living things, hovering between birth and death on the surface of our tiny planet. In an endless procession, animals and plants have spread and multiplied and vanished, each for a fleeting moment a part of the continuing process we call life.

But this outer view is only one facet of the grandeur and mystery and beauty of it all, for within and through the bodies of each of these countless individuals there has pulsed the breath of life—at a score of levels of complexity, a bewildering maze of intricate chemical and physical and biological changes have interacted to produce and maintain this frail thread of life that binds us all. Yet "he who stops at the fact misses the glory," and if in the search for the pattern and process of its development we miss the grandeur and the wonder of life and fail to grasp its deep significance for our thinking, then we have missed the glory.

For the man or woman who has caught a glimpse of life in this perspective can see it only as a thing of reverence and wonder. How life evolved we now begin to understand; that it evolved remains a source of wonder; why it evolved is not a question with which science as such is concerned, but it is a question that links the other two, and gives to the scientific quest a purpose and a harmony within the broader and deeper unity of human contemplation. And it is ultimately upon our answer to this question that we build the fabric of our own portion of the history of life.

NOTES

Preface

1. Peter Dodson, "Origin of Birds: The Final Solution?" *American Zoologist* 40, no. 4 (2000): 504–12.

1. Defrosting the Mammoth

1. "Mammoths," *National Geographic* online, http://ngm.nationalgeographic.com/2009/05/mammoths/mueller-text/2.
2. Austin Clark, "Thomas Jefferson and Science," *Journal of the Washington Academy of Sciences* 33, no. 7 (1943): 1–93.
3. Simon J. Knell, *The Great Fossil Enigma: The Search for the Conodont Animal* (Bloomington: Indiana University Press, 2013).
4. F. Burkhardt, D. M. Porter, J. Harvey, and J. R. Topham, eds., *The Correspondence of Charles Darwin*, vol. 10 (Cambridge: Cambridge University Press, 1997), xxvi, 183, 188–89.

2. Terrestrial Timepieces

1. Simon Winchester, *The Map That Changed the World: William Smith and the Birth of Modern Geology* (New York: Harper Collins, 2002).
2. Keith Sircombe, "Rutherford's Time Bomb," *New Zealand Herald*, May 15, 2004, http://www.nzherald.co.nz/nz/news/article.cfm?c_id=1&objectid=3566551.

3. "From So Simple a Beginning"

1. Frances Darwin, 1887, Life and Letters of Charles Darwin, Darwin Letter 1871, http://darwin-online.org.uk/EditorialIntroductions/Freeman_LifeandLettersandAutobiography.html.
2. Charles Darwin, *The Origin of Species by Means of Natural Selection*, 6th ed. (London: John Murray, 1888), 285–87.
3. Freeman Dyson, *Origins of Life* (Cambridge: Cambridge University Press, 1985).
4. James Lovelock, *GAIA—A New Look at Life on Earth* (Oxford: Oxford University Press, 1979).

5. A. G. Cairns-Smith, *Seven Clues to the Origin of Life: A Scientific Detective Story.* (Cambridge: Cambridge University Press, 1985).
6. Andrew H. Knoll, *Life on a Young Planet* (Princeton, NJ: Princeton University Press, 2003).
7. Dyson, *Origins of Life.*
8. S. M. Miller and L.E. Orgel, *The Origins of Life on the Earth* (Englewood Cliffs, NJ: Prentice Hall, 1974).
9. J. William Schopf and Anatoliy B. Kudryavstev, "Biogenicity of Earth's Earliest Fossils: A Resolution of the Controversy in Gondwana Research," *Gondwana Research* 22, no. 3–4 (2012): 761–71.

4. Classification

1. Illinois Natural History Survey Reports, May–June 1998.
2. http://www.glacierbayalaska.com/rainforest-of-southeastalaska.
3. http://www.naturesfootprint.com/community/..../earthworms-introduction.
4. Juliette Jowett, "Scientists Prune List of the World's Plants," *Guardian*, Sept. 19, 2010, http://www.guardian.co.uk/science/2010/sep/19/scientists-prune-world-plant-list.

5. Spineless Wonders

1. Colin Tudge, *The Variety of Life: A Survey and a Celebration of All the Creatures that Have Ever Lived* (Oxford: Oxford University Press, 2000), 108. This work also provides an excellent introduction both to the classification of living things and to their general characteristics.
2. See also chapter 3.
3. Andrew Knoll, *Life on a Young Planet* (Princeton, NJ: Princeton University Press, 2003), 7.
4. Stephen J. Gould, *Wonderful Life: The Burgess Shale and the Nature of History* (W.W. Norton and Company, 1989).
5. Knoll, *Life on a Young Planet,* 6–11.

6. Bone, Scales, and Fins

1. G. Graffin, "A New Locality of Fossiliferous Harding Sandstone: Evidence for Freshwater Ordovician Vertebrates," *Journal of Vertebrate Paleontology* 12, no. 1 (1992): 1–10.
2. California Academy of Sciences, Catalog of Fishes (continuously updated). http://www.calacademy.org/scientists/projects/catalog-of-fishes.

7. The Greening of the Land

1. David Beerling, *The Emerald Planet: How Plants Changed Earth's History* (Oxford: Oxford University Press, 2007).

2. Beerling, *Emerald Planet*, 18.
3. Charles Darwin, *Autobiography*, ed. Nora Barlow (New York: W.W. Norton, 1958), 52.
4. Gar W. Rothwell, William L. Crepet, and Ruth Stockey, "Is the Anthophyte Hypothesis Alive and Well? New Evidence from the Reproductive Structures of Bennettitales," *American Journal of Botany* 96 (2009): 296–322.

9. The Reign of the Reptiles

1. Darwin letter to Marsh (Aug. 31, 1880) http://www.pbs.org/wgbh/americanexperience/features/primary-resources/dinosaur-letter/.
2. Benton, pp. 98–100 in Stephen J. Gould, ed., *The Book of Life: An Illustrated History of the Evolution of Life on Earth* (New York: W. W. Norton, 1993).

11. The Blossoming Earth

1. Elizabeth Pennisi, "On the Origin of Flowering Plants," *Science* 324 (2009): 28–31.
2. William E. Friedman, "The Meaning of Darwin's 'Abominable Mystery'," *American Journal of Botany* 96 (2009): 5–21.
3. William L. Crepet and K. J. Niklas, "Darwin's Second 'Abominable Mystery': Why Are There So Many Angiosperm Species?" *American Journal of Botany* 96 (2009).
4. William L. Crepet and G. D. Feldman, "Earliest Fossil Evidence of Grasses in the Fossil Record," *American Journal of Botany* 78 (1991): 1010–14.

12. The Rise of the Mammals

1. Colin Tudge, *The Variety of Life: A Survey and a Celebration of All the Creatures That Have Ever Lived* (Oxford: Oxford University Press, 2000).
2. For a thoughtful discussion of the complexities of cladistics, see John Long and Peter Schouten "Origin of Birds: The Final Solution?" *American Zoologist* 40, no. 4 (2000): 504–12.
3. John V. Straaten, "Fossil Info—The Pleistocene Era, The Mammoth," Henskens Fossils, http://www.henskensfossils.nl.
4. Sid Perkins, "Fossils Throw Mammalian Family Tree into Disarray," *Nature News*, 2013, http://www.nature.com/news/fossils-throw-mammalian-family-tree-into-disarray.

13. The Mammalian Explosion

1. George Gaylord Simpson, Anne Roe, and Richard C. Lewontin, *Quantitative Zoology* (New York: Dover, 2003).
2. George Gaylord Simpson, *Tempo and Mode in Evolution* (New York: Columbia University Press, 1944).

3. George Gaylord Simpson, *Horses: The Story of the Horse Family in the Modern World and through Sixty Million Years of History* (Oxford: Oxford University Press, 1951).
4. George Gaylord Simpson, "The Principles of Classification and a Classification of Mammals," *Bulletin of the American Museum of Natural History* 85 (1945): 1–350.

14. The Leakeys' Legacy

1. Virginia Morrell, *Ancestral Passions: The Leakey Family and the Quest for Human Origins* (New York: Simon and Schuster, 1995), 473ff.
2. Morrell, *Ancestral Passions*, 485.
3. P. V. Tobias, "Louis Leakey, Self-styled White African: Appreciation and Some Personal Recollections," *Transactions of the Royal Society of South Africa* 58 (2003): 41.
4. Mary Leakey, Archaeologist and Anthropologist, http://pin.primate.wisc.edu/edu/careers/leakey.html.
5. Poem used by permission of Punch Limited, London, UK.
6. Tobias, "Louis Leakey," 41.
7. E. Simons, "Man's Immediate Forerunners," *Philosophical Transactions of the Royal Society of London* 292 (1981): 21–41.
8. Kenneth A. R. Kennedy, *God-Apes and Fossil Men: Paleoanthropology of South Asia* (Ann Arbor, MI: University of Michigan Press, 2000).
9. Frank H.T. Rhodes, *Earth: A Tenant's Manual* (Ithaca, NY: Cornell University Press, 2012).

16. On Extinction

1. Charles Darwin, *On the Origin of Species by Means of Natural Selection* (London: John Murray, 1888), 294.
2. John H. Lawton and Robert M. May, *Extinction Rates* (Oxford: Oxford University Press, 1995).
3. Jean-Baptiste Lamarck, *Zoological Philosophy* (London: Macmillan, 1914), 44.
4. Lamarck, *Zoological Philosophy*, 45.
5. E. O. Wilson, *The Creation: An Appeal to Save Life on Earth* (New York: W. W. Norton, 2006), 74.
6. David Raup and John J. Sepkoski, "Mass Extinctions in the Marine Fossil Record," *Science* 215, no. 4539 (1982): 1501–3.
7. Wilson, *Creation*, 80.

17. "Have Been and Are Being Evolved"

1. Charles Darwin, *Journal of Researches into the Natural History and Geology of the Countries Visited During the Voyage of the Beagle Round the World* (London: John Murray. 1890), 364.

2. Darwin, *Journal of Researches*, 362.
3. Jonathan Weiner, *Beak of the Finch: A Story of Evolution in Our Time* (New York: Alfred A. Knopf, 1994).
4. H. B. D. Kettlewell, "Selection Experiments on Industrial Melanism in the Lepidoptera," *Heredity* 9 (1955): 323–42
5. A. J. Cain and P.M. Sheppard, "Natural Selection in *Cepaea*," *Genetics* 39 (1950): 89–116.
6. Bjorn Kusten, *The Cave Bear Story: Life and Death of a Vanished Animal* (New York: Columbia University Press, 1955).
7. A. W. Rowe, "An Analysis of the Genus *Micraster* as Determined by Rigid Zonal Collecting from the Zone of *Rhyconella cuvieri* to That of *Micraster coranguinum*," *Quarterly Journal of the Geological Society of London* 55 (1899): 494–547.
8. D. Nichols, "Changes in the Chalk Heart Urchin *Micraster* Interpreted in Relation to Living Forms," *Philosophical Transactions of the Royal Society of London. Series B, Biological Sciences* 242, no. 693 (1959): 347–437.
9. Claude Spinosa, W. M. Furnish, and Brian Glenister, "The Xenodiscidae, Permian Ceratitoid Ammonoids," *Journal of Paleontology* 49, no. 2 (1975): 239–83.
10. Stephen J. Gould and N. Eldridge, "Punctuated Equilibria: The Tempo and Mode of Evolution Reconsidered," *Paleobiology* 3, no. 2 (1977): 115–51.
11. J. P. Kennett, "The Kapitean Stage (Upper Miocene) at Cape Foulwind, West Coast," *New Zealand Journal of Geology and Geophysics* 5, no. 4 (1962): 620–25.

Epilogue

1. Charles Darwin, *The Origin of Species by Means of Natural Selection, or the Preservation of Favoured Races in the Struggle for Life*, 6th ed. (London: John Murray, 1888), 429.

GLOSSARY

Acanthodians. Extinct group of small, enigmatic Paleozoic fish, often known as "spiny sharks" but were not sharks but placoderms. Bony spines projected from the front of their fins and the body was covered in diamond-shaped scales. Many of them bore a series of paired spiny fins along the ventral part of the body. Fragmentary material from Late Ordovician, but mostly Upper Silurian–Permian.

Actinopterygii (actinopterygians). Ray-finned fishes. Part of class Osteichthyes with webbed fins supported by bony spines as opposed to fleshy, lobed fins. Outnumber all other living vertebrates. Upper Silurian–Recent.

Agnatha. Class of vertebrates, represented by living lampreys and hagfish and including the earliest fossil vertebrates (ostracoderms, anapsids, and coelolepids). They lack jaws and paired fins. Ordovician–Recent.

Amnion. Membrane that surrounds the embryo in the amniotic egg of reptiles, birds, and mammals.

Ammonites. Group of extinct fossil cephalopod mollusks, related to the living *Nautilus.* Upper Silurian–Upper Cretaceous.

Amniote. A vertebrate that produces amniotic eggs; i.e., a reptile, bird, or mammal.

Amniotic egg. Land egg containing not only an embryo, but also several membranes for protection.

Amphibia. Major group of cold-blooded tetrapod vertebrates whose members typically live on land but reproduce in water. Includes living newts, salamanders, frogs, toads, and caecilians, as well as many fossil forms. Upper Devonian–Recent.

Anapsids. Group of reptiles based on skull type. Anapsid skulls have no hole behind the eye socket.

Angiosperms. Flowering plants. Defined by their evolution of highly protected seeds, enclosed in a fruit, grain, pod, or capsule. Far outnumber all other living plants. Cretaceous–Recent.

Animalia. Animals, one of five major kingdoms of living things.

Ankylosaurs. Small group of medium-sized, heavily armored, herbivorous ornithischian dinosaurs. Cretaceous.

Annelida. Phylum that includes ringed or segmented worms (bristle worms, earthworms, leeches). Precambrian–Recent.

Anthracosaurs. Group of early amphibians (embolomeres), probably close to the ancestry of reptiles. Pennsylvanian.

Antiarchs. Group of placoderms, extinct primitive armored fish: small with heavy, boxlike, armored head shield, strong pectoral fins, and body scales. Middle Devonian–Upper Permian.

Archaean. Era of geologic time, spanning roughly 4,000–2,500 million years ago.

Archaeobacteria. Group of single-celled organisms (prokaryotes) that may be the oldest, most primitive living cells. Include methanogens, often existing under extreme conditions. Precambrian–Recent.

Archosaurs. Major group of diapsid vertebrates that includes dinosaurs, pterosaurs, crocodilians, and birds.

Arthrodires. Group of large, powerfully built placoderm fishes having heavy, bony head shield, large eyes, and widely articulated jaws. Some grew to be 30 feet long. Devonian.

Arthropoda. A major invertebrate phylum that includes insects, crustaceans, spiders, trilobites, and other groups. All have a segmented body, jointed appendages, and external skeleton. Precambrian–Recent.

Arthrophyta. Group of "jointed plants" belonging to the Tracheophyta and including the living horsetails (*Equisetum).* Also known as sphenopsids. Devonian–Recent.

Artiodactyla. A large group of terrestrial, herbivorous, hoofed mammals with an even number of toes, including cattle, sheep, deer, antelope, pigs, goats, camels, and other forms. Eocene–Recent.

Bacteria. Ubiquitous, unicellular, microscopic organisms lacking a nucleus. Precambrian–Recent.

Banded iron formations (BIF). Major rock type formed dominantly between 2,500 million and 1,800 million years ago, probably by the oxidation of iron in seawater.

Barnacles. Marine arthropods belonging to the class Cirripedia with calcareous shells that encrust rocks and ships. Cambrian–Recent.

Belemnites. Extinct group of dominantly Mesozoic cephalopod mollusks. The internal fossil shells are cigarlike in shape. Carboniferous–Cretaceous.

Bivalves (pelecypods). Major class of marine and freshwater mollusks. Includes familiar groups such as oysters, clams, mussels, and scallops. Cambrian–Recent.

Blastoidea. Class within phylum Echinodermata. An extinct group of marine, stalked animals. Ordovician-Permian.

Brachiopoda. Phylum of marine animals, with a two-valved shell: "lamp shells." Represented by comparatively few living members but many fossil forms, especially in the Paleozoic. Cambrian–Recent.

Brontotheres. See Titanotheres.

Bryophyta. Division of the Plant Kingdom, including liverworts, hornworts, and mosses. A small and rather primitive group of land plants. Carboniferous–Recent.

Bryozoa. Phylum of small, aquatic, usually fixed and colonial animals. Sea mats and "corallines" (superficially resembling hydroids, but considerably more complex). Often called Polyzoa. Ordovician–Recent.

Cambrian. Period of geological time that began about 540 million years ago and ended about 485 million years ago. The oldest system of the Paleozoic Era. The oldest group of rocks to contain abundant fossils.

Canidae. Family of carnivorous and omnivorous mammals, including dogs, wolves, foxes, jackals, and coyotes. Eocene–Recent.

Caniforms. "Doglike" carnivores, including dogs, wolves, foxes, bears, badgers, weasels, otters, raccoons, and skunks; cf. Feliforms.

Carboniferous. Period of geological time that extended from 358 million years ago to about 280 million years ago, containing most of the world's major coal deposits. Often divided into lower (Mississippian) and upper (Pennsylvanian) periods.

Carnivora. Order of mammals that includes bears, cats, and badgers, as well as some aquatic forms. Eocene–Recent.

Carnosaurs. Informal name for a group of very large carnivorous dinosaurs.

Catastrophism. A once widely held theory that Earth's history had involved a series of catastrophes by which all living things were destroyed, to be followed by a new creation. It is in contrast with uniformitarianism, which suggests that present processes are sufficient to account for the change observed in the geologic record; often summed up as "the present is the key to the past," uniformitarianism argues for the adequacy of existing causes.

Cenozoic. The most recent geological era. The era of recent life, including the Tertiary and Quaternary periods. The "age of mammals." The last 68 million years of geologic time.

Cephalochordates. Group of small, marine, soft-bodied chordates defined by a notochord that persists through life. Includes the lancelets (*Amphioxus*). Cambrian–Recent.

Cephalopoda. Class of marine mollusks, characterized by the evolution of a well-developed head; typically swimming carnivores. Includes squid, *Nautilus,* and extinct belemnites and ammonites. Cambrian–Recent.

Cetacea. Whales, porpoises, and dolphins. An aquatic order of placental mammals. Eocene–Recent.

Chalicotheres. Group of herbivorous, odd-toed (perissodactyl) ungulate mammals, typically horse-sized, with elongated forelimbs and long clawed fingers. Spread throughout North America, Europe, Asia, and Africa during the Early Eocene to Early Pleistocene.

Chelicerata. One of the major subdivisions of the phylum Arthropoda; includes horseshoe crabs, scorpions, spiders, mites, harvestmen, ticks.

Chiroptera. Bats. Mammals whose forelimbs form webbed wings; only mammals capable of true powered flight. Almost 1,000 living species. Two major groups: fruit bats and flying foxes. Eocene–Recent.

Chondrichthyes. The cartilaginous fishes. Sharks and rays. Devonian–Recent.

Chordata. Phylum of animals with a notochord, hollow dorsal nerve cord, and gill slits. Includes vertebrates the Tunicata (sea squirts) and Cephalochordata (lancelets). Cambrian–Recent.

Clade. A set of organisms grouped together because they (and only they) are thought to have evolved from a common ancestor.

Cladistic classification. A biological classification that seeks to reflect the evolutionary steps that occurred during the history of a selected group of organisms.

Cladogram. A branching, treelike diagram that portrays relationships among three or more taxa, with no implication of time sequence.

Club mosses. An ancient plant group with few living species but many extinct forms—especially diverse during the Carboniferous Period when they grew to the size of large trees and were the chief contributors to the great coal deposits. Also known as lycopods, lycophytes, or lycopsids.

Cnidaria (Coelenterata, coelenterates). Major metazoan phylum that contains jellyfish, corals, and sea pens. Have very simple sheetlike structure of tissue; also characterized by tentacles and nematocysts or stinging cells. Include both medusoid (swimming) and polyploid (fixed) forms. Aquatic, mostly marine. Precambrian–Recent.

Coccolithophores. Unicellular, eukaryotic phytoplanktonic algae that create a covering of calcareous plates (coccoliths). Exist in countless numbers in the oceans.

Coccospheres. Spherical microorganisms composed of coccoliths.

Coelacanth. Group of sarcopterygian (lobe-finned) fishes that evolved in Devonian times, including two living species. Middle Devonian–Recent.

Coniferophyta (conifers). Group of gymnosperm plants. Includes pine, spruce, cedar, yew, and larch. Usually tall evergreen forest trees. Pennsylvanian–Recent.

Conodonts. Group of extinct, marine, dominantly Paleozoic, phosphatic, toothlike fossils of eel-like, chordate affinities. Cambrian–Upper Triassic.

Corals. Members of the phylum Cnidaria (coelenterates).

Cordaites. Extinct group of conifer-like trees belonging to the gymnosperms. Largely confined to the Upper Paleozoic.

Creodonts. Large group of early carnivores, archaic in structure, generally small and slender but including some wolf-sized forms. Early Tertiary.

Cretaceous. Period of geological time that began about 145 million years ago and ended about 66 million years ago. The last system of the Mesozoic Era.

Crinoidea. Class within phylum Echinodermata, including feather stars and sea lilies. Usually fixed by a long stalk, but some forms are free-swimming. Ordovician–Recent.

Crocodilians. Major group of archosaurs, including the living crocodiles and alligators.

Crossopterygians. Group of lobe-finned bony fish (sarcopterygians). Suborder of the Choanichthyes. "Coelacanths," represented by only one living genus (*Latimeria*). Fossil forms represent ancestors of the land vertebrates. Lower Devonian–Recent.

Crustaceans. Major group of arthropods, typically aquatic; includes living crabs, crayfish, krill, barnacles, ostracods, and lobsters, as well as many fossils. Middle Cambrian–Recent.

Cyanobacteria. Blue-green bacteria or algae; major group of photosynthetic bacteria. Archaean–Recent.

Cycadeoids (Bennettiales). Group of extinct Mesozoic gymnosperm plants, widespread in distribution, differing from living cycads in their method of reproduction.

Cycads. Group of tropical to subtropical gymnosperms, the most primitive living seed plants. Permian–Recent.

Cynodonts ("dog-teeth"). Large group of therapsids, both omnivorous and carnivorous in habit, including the immediate ancestors of mammals. Wide geographic distribution, especially in Southern Hemisphere. Permian–Triassic.

Devonian. Period of geological time, began 418 million years ago and ended about 350 million years ago. Often spoken of as the "age of fishes."

Diapsids. Huge group of vertebrates having two holes in the skull behind the eye. Includes all living reptiles and birds.

Diatoms. Group of algae. Microscopic, unicellular plants, solitary or colonial. Siliceous "skeletons" form diatomaceous earth deposits. Jurassic–Recent.

Dinosaurs. Very large group of diverse archosaurs. Became the dominant Mesozoic terrestrial reptilian group. Triassic–Cretaceous.

Dipnoi. Lungfish. Group of lobe-finned (sarcopterygian) fishes that have evolved the capacity to breathe air. Represented by three living freshwater genera. Devonian–Recent.

Echinodermata. Phylum of animals, including sea urchins, sea cucumbers, brittle stars, starfish, sea lilies, and feather stars. Marine, usually radially symmetrical, with water vascular system. Include extinct fossil blastoids, cystoids, eocrinoids, edrioasteroids. (Cambrian–Recent).

Echinoids. Sea urchins and sand dollars. Class within phylum Echinodermata. Globular test of calcareous plates, containing five ambulacral arms. Ordovician–Recent.

Edaphosaurs. Group of large, vegetarian, "sail-back" pelycosaurs. Late Carboniferous–Early Permian.

Edentata (edentates or Xenarthra). Group of placental mammals that includes armadillos, anteaters, and sloths. Paleocene–Recent.

Eleutherozoans. Group of free-living Echinodermata. Includes sea urchins, sea cucumbers, brittle stars, and starfish.

Eubacteria. "True bacteria." Includes all bacteria except Archaeobacteria.

Eukaryotes. Organisms with cells that have organelles and DNA enclosed within nuclei. Capable of mitotic cell division. All plants and animals are eukaryotes.

Eurasia. The combined continental land masses of Europe and Asia.

Eurypterids. Extinct group of Paleozoic Arthropoda. Aquatic scorpions, up to 9 feet in length. Ordovician–Permian, but especially characteristic of Silurian and Devonian.

Eutheria. One of two major groups of mammals (cf. Metatheria), includes placental mammals. Late Jurassic–Recent.

Feliforms. Broad division of catlike carnivorous mammals (cf. Caniforms).

Ferns (Pteridophytes). Major group of vascular land plants; reproduce by means of spores rather than seeds. Includes also club mosses and liverworts. Carboniferous–Recent.

Foraminifera. Group of protozoans, mostly marine, many of which developed shells, usually calcareous. Most are microscopic. Ordovician–Recent.

Fungi. Subdivision of Thallophyta. Mushrooms, molds, yeasts, and others. Simply organized plants, lacking chlorophyll.

Gastroliths. Stones that are deliberately ingested into the digestive tract by some living and fossil vertebrates. They act as grindstones to help macerate food, or, it was once thought, possibly as ballast in some aquatic air-breathers.

Gastropoda. Class within phylum Mollusca. Includes snails, slugs, and pteropods. Marine, freshwater, and terrestrial. Distinct head. Often a single coiled shell. Cambrian–Recent.

Ginkgoes. Ancient lineage of gymnosperm plants, represented by only a single living species. Permian–Recent.

Glyptodonts. Very large armored mammals, characteristic members of South American Cenozoic faunas.

Gondwanaland. A southern supercontinent made up of the present southern continents plus India; the southern part of Pangaea.

Graptolites. Extinct colonial marine organisms, apparently related to hemichordates. Cambrian–Carboniferous.

Great Oxygenation Event (GOE). Introduction of free oxygen into Earth's atmosphere by cyanobacteria about 2.4 billion years ago.

Gymnosperms. Conifers and their allies (cycads, ginkgoes, cycadeoids, cordaites). Primitive seed plants, differing from angiosperms by having naked, unprotected seeds. Devonian–Recent.

Hadrosaurs. The duck-billed dinosaurs, a major group of ornithischian dinosaurs.

Holosteans. Group of primitive bony fishes that constitute one of three major groups of the ray-finned fishes (Actinopterygii). Include living bow-fins and gars. Devonian–Recent.

Holothuroidea (holothuroids). Sea cucumbers. Class of Echinodermata. Soft cylindrical body, bearing microscopic "plates." Cambrian–Recent.

Horsetails. Small group of living vascular plants ("scouring rushes"). Stems, with whorls of narrow leaves. There are some twenty-five living species but fossil forms were numerous and grew to almost 100 feet high, forming major constituents of coal. Carboniferous–Recent.

Ichthyosaurs ("fish lizards"). Major group of marine Mesozoic reptiles. Middle Triassic–Late Cretaceous.

Ichthyostegids. Early group of amphibia, exhibiting many fishlike characteristics. Late Devonian.

Jurassic. Period of geological time that began 201 million years ago and ended 145 million years ago. The middle system of the Mesozoic.

Labyrinthodonts ("maze-toothed"). Early heterogeneous group of squat amphibia. The dominant amphibians of late Paleozoic and early Mesozoic times. Ancestral to later land-living vertebrates. Devonian–Triassic.

Lagomorphs. Members of the order Lagomorpha, which includes rabbits, hares, and pikas. Paleocene–Recent.

Lepospondyls (minosaurs). Early tetrapods, including both aquatic and terrestrial forms. Include eel-like, snakelike, and lizardlike forms, as well as the wedge-headed *Diplocaulus*. All have simple, spool-like vertebrae. Mississippian–Early Permian.

Liverworts. Simplest true land plants. Bryophytes, comprising about 9,000 living species of mostly small spore-producing plants. Ordovician–Recent.

Lycopsids. Oldest group of vascular plants; major contributors to the coal forests of the Carboniferous. Living forms include club mosses, spike mosses, and quillworts. Devonian–Recent.

Mammals. Class of tetrapod vertebrates. Humans, dogs, whales, many more. Characterized by hair, milk secretion, diaphragm used in respiration. Jurassic–Recent.

Mammoths. Group of extinct elephants (*Mammuthus*), notable for their portrayals in cave paintings in Europe, and for preserved frozen specimens in Siberia and Alaska. Teeth are flat and ridged (cf. Mastodons), suited to the grazing diet of mammoths. Pleistocene–Holocene.

Marsupials. Group of nonplacental Mammalia, including kangaroo and opossum. From Australia and North and South America. Young are born in very undeveloped state and sheltered in mother's pouch. Cretaceous–Recent.

Mastodons. Group of extinct elephants ("mastodon" = "breast tooth") inhabiting North and Central America from Late Pliocene to end Pleistocene. Primarily forest-dwelling browsers with conical, cusped teeth reflecting their diet.

Mesozoic. Era of geological history from 252 to 66 million years ago. Followed the Paleozoic Era and preceded the Cenozoic Era.

Metatheria. Marsupial mammals, including about 270 living species.

Mollusca. A large phylum of invertebrate animals that includes snails, mussels, clams, cephalopods, octopuses, and slugs. Mostly aquatic, soft-bodied, with a hard shell, unsegmented, and with a head and muscular foot. Cambrian–Recent.

Monotremata (monotremes). Subclass of mammals. Duck-billed platypus and two genera of spiny anteater. Australia and New Guinea. Very primitive group. Lay eggs and have some reptilian characteristics. Cretaceous–Recent.

Mosasaurs. An extinct group of large, carnivorous marine reptiles also known as "sea lizards." Cretaceous.

Multituberculates. Extinct group of mammals, characterized by teeth with many cusps. Late Jurassic–Oligocene.

Nautiloids. Shelled cephalopods having an external chambered shell either straight or variously curved or coiled, with simple septa forming sutures that are simple lines without marked flexures. Late Cambrian–Recent.

Nectridians. Small group of early aquatic amphibians, often classified with lepospondyls. Early Carboniferous–Permian.

Nothosaurs. Group of early Mesozoic long-necked, fish-eating marine reptiles, up to 12 feet in length, included in the sauropterygians. Triassic.

Ornithischians. Order of herbivorous, "bird-hipped" dinosaurs, the pelvis resembling that of birds. Great variety of forms, including duck-billed ornithopods, ceratopsians, ankylosaurs, and stegosaurs. Triassic–Cretaceous.

Ornithopods. The largest group of ornithischian dinosaurs, including the iguanodonts and the hadrosaurs (duck-bills).

Osteichthyes (bony fish). Class of phylum Chordata. Bony fish having a skeleton of bone, as well as cartilage, including *Actinopterygii* ("ray fins") and *Sarcopterygii* ("lobe-fins"). Late Silurian–Recent.

Osteolepid. Lobe-finned fishes that included the immediate ancestors of the tetrapods. Late Devonian.

Ostracoderms. Early jawless fishes, so called because their body surfaces were covered with phosphatic plates. Silurian–Devonian.

Pakicetids ("Pakistani whales"). Carnivorous terrestrial cetaceans. Eocene.

Paleozoic. The oldest of the three geological eras of life. The era of ancient life; the "age of invertebrates." Includes all systems from the Cambrian to the Permian. Began about 542 million years ago and lasted until 251 million years ago.

Pangaea ("entire Earth"). The supercontinent that formed by coalescence of all the individual continents during the late Paleozoic and then fragmented during the Early Mesozoic.

Pantodonts. Extinct group of early herbivorous eutherian mammals. Paleocene–Eocene.

Pelecypoda. Class within phylum Mollusca. Lamellibranchia. Aquatic, bivalved mollusks including mussels, clams, and oysters. Ordovician–Recent.

Pelycosaurs. Artificial group of extinct, "primitive," synapsid, egg-laying, mammal-like reptiles. Includes the sail-back lizards and others from which the mammals probably arose. Carboniferous–Permian.

Pennsylvanian. Period of geological time comprising the upper Carboniferous, from 323 to 298 million years ago.

Perissodactyla. Group of herbivorous mammals with an odd number of toes; includes horses, rhinos, and tapirs. Eocene–Recent.

Permian. Period of geological time. The last system of the Paleozoic era, which began about 298 million years ago and ended about 252 million years ago.

Photosynthesis. Conversion of light energy to chemical energy. Synthesis by green plants of carbohydrates from water and carbon dioxide, with the aid of energy absorbed from sunlight.

Phylogeny. The evolutionary development and history of a group of organisms.

Pinnipeds ("feather-footed"). Semiaquatic marine carnivorous mammals. Includes seals, sea lions, walruses, and sea elephants. Oligocene—Recent.

Placental mammals. Diverse group of mammals belonging to the subclass Placentalia, containing most living mammals. The embryo develops in the uterus, attached by a highly organized placenta. Include rodents, bats, whales, elephants, and many other living species. Cretaceous–Recent.

Placoderms. Extinct class of vertebrates (sometimes called Aphetohyoidea). Primitive armored fish, with archaic jaw suspension. Dominant fish of Devonian times. Lower Silurian–Permian.

Placodonts. Group of heavily plated Triassic marine reptiles.

Plantae. Kingdom including all land plants, such as mosses, conifers, ferns, flowering plants, and others. Ordovician–Recent.

Pleistocene. Subdivision of Quaternary period; began 2.6 million years ago and ended 10,000 years ago.

Plesiosaurs. Group of Mesozoic marine reptiles. More primitive forms had short necks and large heads, whereas later forms had long necks and short heads. Both had paddlelike limbs. Lower Jurassic–Upper Cretaceous.

Pliocene. Subdivision of the Tertiary Period of geological time. The most recent division of the Tertiary. Began 5 million years ago and ended 2.6 million years ago.

Polyplacophorans. Chitons and other marine mollusks (class Polyplacophora). Devonian–Recent.

Porifera. Sponges; phylum of multicellular but very primitive animals. No nervous system. "Collar" cells. Often develop spicules. Cambrian–Recent.

Precambrian. Vast interval of geological time, preceding the Cambrian. The first 4,000 million or so years of Earth history.

Proboscidea. Taxonomic order comprising elephants and their extinct relatives, such as mammoths and mastodons. Paleocene–Recent.

Prokaryotes. Single-celled organisms, without nuclei, that do not undergo mitotic cell division. Includes bacteria and archaea.

Proterozoic. Major division of Precambian Earth history from about 2,500 to 542 million years ago. Follows the Archaean Era and precedes the Paleozoic Era.

Protista. Single-celled eukaryotes.

Pterodactyloids. See Pterosaurs.

Pteridosperms. See Pteridophytes.

Pteridophytes. Division of plant kingdom comprising spore-bearing terrestrial plants, including ferns, horsetails, and club mosses. Well-developed sporophytes, alternation of generations, well-developed roots, stem, and leaves. Ordovician–Recent.

Pterosaurs ("winged lizards"). Fossil order of flying reptiles. Wings membranous, supported by greatly elongated fourth finger. Late Triassic–Cretaceous.

Radiolarians. Group of marine planktonic protozoans, having delicate siliceous skeletons. Precambrian–Recent.

Reptiles. Class of vertebrates. Includes turtles, lizards, snakes, crocodiles, and many extinct forms, such as dinosaurs. Cold-blooded tetrapods, dominant in the Mesozoic. Amniote egg. Early Carboniferous–Recent.

Rhamphorhyncoids. The stem group of pterosaurs; the Mesozoic flying reptiles. Late Triassic–Late Jurassic.

Rodents. Small mammals, including rats, mice, and squirrels having two constantly growing incisors in each jaw. Make up 40 percent of all living mammal species. Paleocene–Recent.

Saurischians. Order of fossil reptiles, to which many of the dinosaurs are referred (the other is Ornithischia). "Lizard-hip" structure. Triassic–Cretaceous.

Sauropods. Group of generally large, quadrupedal, herbivorous dinosaurs. Triassic–Cretaceous.

Scaphopoda. Elephant tusk shells. Small class of burrowing marine mollusks with tubular shell. Silurian–Recent.

Seed ferns. See Pteridophytes.

Silurian. A period of geological time that began 443 million years ago and ended 416 million years ago.

Sirenia. Manatees, dugongs, sea cows. An aquatic order of placental mammals having front flippers and vestigial hind legs. Related most closely to elephants and conies. Paleocene–Recent.

Snakes. Elongated, legless, scale-covered, carnivorous reptiles. Cretaceous–Recent.

Sphenacodonts. A group of carnivorous pelycosaurs. Pennsylvanian–Permian.

Sphenopsids. See Arthrophyta.

Sponges. Major invertebrate phylum (Porifera). Pore-bearing invertebrates, mostly marine but some freshwater. All adults are sessile and lack organs and tissues. Cambrian–Recent.

Stegosaurs. Group of armored ornithischian dinosaurs characterized by spikes and plates on the back and tail. Jurassic–Cretaceous.

Stromatolites. Trace fossils of mats of bacteria, often of cyanobacteria. The stromatolite itself typically consists of sediment caught up in the bacterial slime and deposited in layers, but cells are occasionally found. Precambrian–Recent.

Synapsid. Major group of reptiles, defined by the presence of one hole in the skull behind the eye. Includes living mammals and their ancestors.

Teleosts. Group of ray-finned fishes, including most living forms. Triassic–Recent.

Thallophyta. Large, heterogeneous group of plants, including algae, fungi, and lichens.

Theropod dinosaurs ("beast-footed"). Major group of bipedal, saurischian dinosaurs, ancestral to birds. Most were carnivores. Includes *T. rex*. Triassic–Cretaceous.

Therapsids. The "mammal-like reptiles." The group of reptiles descended from pelycosaurs and includes the ancestors of mammals.

Theria. Group of living mammals, such as marsupials and placentals, defined by presence of a specialized molar tooth type.

Titanotheres (Brontotheres). Early Cenozoic group of large, horned, herbivorous mammals related to living horses from the Northern Hemisphere. Massive in structure. Early Eocene–Oligocene.

Tracheophyta. The largest and most advanced division of the plant kingdom. The vascular plants. Silurian–Recent.

Triassic. Period of geologic time. The oldest period of the Mesozoic Era. Began 252 million years ago and ended 200 million years ago.

Triconodont. Group of very early mammals, defined by the three large cusps on their molar teeth. Late Triassic–Late Cretaceous.

Trilobites. Extinct group of Paleozoic marine arthropods. Early Cambrian–Late Permian.

Tyrannosaurs ("tyrant lizards"). Small group of very large predatory theropod dinosaurs. Group includes *Allosaurus, Tyrannosaurus,* and other similar genera. Jurassic–Cretaceous.

Uintatheres. Group of early Cenozoic North American vegetarian, horned, hoofed mammals that evolved rapidly to large size. Mid Eocene–Late Pliocene.

Ungulates. Large "informal" group of hoofed mammals that are typically vegetarian, including cattle, sheep, goats, and antelope among living members. Includes both artiodactyls (cloven- or even-hoofed) and perissodactyls (odd-toed) mammals, as well as aardvarks, hyraxes, manatees, and elephants. Taxonomy of group in state of flux. Paleocene–Recent.

Urochordates (sea squirts). Group of living marine chordates that have a notochord but not a backbone. Also known as "tunicates."

Vascular plants. Tracheophytes, or higher plants with tissue forming a continuous system throughout that functions in conduction of water, mineral salts, and synthesized food materials, and in mechanical support. Include club mosses, horsetails, ferns, conifers, and flowering plants. Early Devonian–Recent.

Xylem. Connective tissue in vascular plants that carries the products of photosynthesis around the plant and provides support.

RELATED READING

Chapter 1. Defrosting the Mammoth

Coleman, William. 1962. *Georges Cuvier, Zoologist: A Study in the History of Evolution Theory.* Cambridge, MA: Harvard University Press.
Fortey, Richard A. 1991. *Fossils: The Keys to the Past.* Cambridge, MA: Harvard University Press.
Gillespie, Charles Coulston. 1996. *Genesis and Geology.* Cambridge, MA: Harvard University Press.
Guthrie, R. Dale. 1990. *Frozen Fauna of the Mammoth Steppe.* Chicago, IL: University of Chicago Press.
Knell, Simon J. 2012. *The Great Fossil Enigma: The Search for the Conodont Animal.* Bloomington, IN: Indiana University Press.
Lister, Adrian, and Paul G. Bahn. 1994. *Mammoths.* New York: Macmillan.
Lovejoy, Arthur O. 1936. *The Great Chain of Being.* Cambridge, MA: Harvard University Press.
Rudwick, Martin J. S. 1976. *The Meaning of Fossils: Episodes in the History of Paleontology.* 2nd ed. New York: Science History Publications.
—. 1997. *Georges Cuvier, Fossil Bones, and Geological Catastrophes: New Translations and Interpretations of the Primary Texts.* Chicago, IL: University of Chicago Press.
—. 2007. *Bursting the Limits of Time: The Reconstruction of Geohistory in the Age of Revolution.* Chicago, IL: University of Chicago Press.
—. 2010. *Worlds before Adam: The Reconstruction of Geohistory in the Age of Reform.* Chicago, IL: University of Chicago Press.
Shoshani, Jeheskel, and Pascal Tassy. 1996. *The Proboscidea: Evolution and Paleoecology of Elephants and Their Relatives.* New York: Oxford University Press.
Ward, Peter D. 1997. *The Call of the Dinosaur Mammoths: Why the Ice Age Mammals Disappeared.* New York: Springer Verlag.

Chapter 2. Terrestrial Timepieces

Albritton, Claude C. Jr. 2002. *The Abyss of Time: Changing Conceptions of Earth's Antiquity after the Sixteenth Century.* New York: Dover Publications.

Benchfield, Joe D. 1998. "The Age of the Earth and the Invention of Geological Time." *Geological Society, London, Special Publications* 143:137–43.
Berry, William B. N. 1968. *Growth of a Prehistoric Time Scale Based on Organic Evolution.* San Francisco: W. H. Freeman.
Brush, Stephen G. 1996. *Transmuted Past: The Age of the Earth and the Evolution of the Elements from Lyell to Patterson.* New York: Cambridge University Press.
Calder, Nigel. 1983. *Timescale: An Atlas of the Fourth Dimension.* New York: The Viking Press.
Dalrymple, G. Brent. 1991. *The Age of the Earth.* Stanford, CA: Stanford University Press.
—. 2001. "The Age of the Earth in the Twentieth Century: A Problem (Mostly) Solved." *Geological Society, London, Special Publications* 190:205–21.
Gillespie, Charles C. 1996. *Genesis and Geology: A Study of the Relations of Scientific Thought, Natural Technology, and Social Opinion in Great Britain, 1790–1850.* Cambridge, MA: Harvard University Press.
Gould, Stephen J. 2001. *Time's Arrow, Time's Cycle: Myth and Metaphor in the Discovery of Geologic Time.* Cambridge, MA: Harvard University Press.
Gradstein, Felix M., James Ogg, and Alan Smith, eds. 2005. *A Geologic Time Scale 2004.* Cambridge: Cambridge University Press.
Grost, Martin. 2001. *Measuring Eternity.* New York: Broadway Books.
Haber, Francis C. 1959. *The Age of the World: Moses to Darwin.* Baltimore: Johns Hopkins University Press.
Hallam, A. 1989. *Great Geological Controversies.* 2nd ed. Oxford: Oxford University Press.
Holland, Charles Hepworth. 1999. *The Idea of Time.* Chichester: John Wiley Books.
Holmes, Arthur. 1913. *The Age of the Earth.* London: Harper and Brothers.
Lewis, Cherry. 2000. *The Dating Game: One Man's Search for the Age of the Earth.* Cambridge: Cambridge University Press.
Lewis, Cherry, and Simon J. Knell, eds. 2001. *The Age of the Earth: From 4004 BC to 2002 AD.* London: Geological Society of London.
Macdougall, Doug. 2008. *Nature's Clocks: How Scientists Measure the Age of Almost Everything.* Berkeley: University of California Press.
Ogg, James G., Gabbi Ogg, and Felix M. Gradstein. 2008. *The Concise Geologic Time Scale.* Cambridge: Cambridge University Press.
Toulmin, Stephen, and June Goodfield. 1965. *The Discovery of Time.* New York: Harper and Row.
York, Derek, and Ronald M. Farquhar. 1972. *The Earth's Age and Geochronology.* Oxford: Pergamon Press.

Chapter 3. "From So Simple a Beginning"

Cairns-Smith, A. Graham. 1984. *Genetic Takeover and the Mineral Origins of Life.* New York: Cambridge University Press.

—. 1985. *Seven Clues to the Origin of Life: A Scientific Detective Story.* Cambridge: Cambridge University Press.

Clarkson, Euan N. K. 1998. *Invertebrate Paleontology and Evolution.* 4th ed. Malden, MA: Blackwell Science.

Conway-Morris, Simon. 1998. *The Crucible of Creation: The Burgess Shale and the Rise of Animals.* Oxford: Oxford University Press.

Davies, Paul. 2003. *The Origin of Life.* New York, NY: Penguin Books.

Dyson, Freeman J. 1999. *Origins of Life.* 2nd ed. Cambridge: Cambridge University Press.

Fortey, Richard. 1998. *Life: A Natural History of the First Four Billion Years of Life on Earth.* New York: Alfred Knopf.

Fry, Iris. 2000. *The Emergence of Life on Earth: A Historical and Scientific Overview.* New Brunswick, NJ: Rutgers University Press.

Gould, Stephen Jay. 1990. *Wonderful Life: The Burgess Shale and the Nature of the Fossil Record.* New York: W. W. Norton.

Knoll, Andrew H. 2003. *Life on a Young Planet: The First Three Billion Years of Evolution on Earth.* Princeton, NJ: Princeton University Press.

Margulis, Lynn. 1982. *Early Life.* Boston: Scientific Books.

—. 2000. *Symbiotic Planet: A New Look at Evolution.* New York: Basic Books.

Margulis, Lynn, Clifford Matthews, and Aaron Haselton, eds. 2000. *Environmental Evolution: Effects of the Origin and Evolution of Life on Planet Earth.* 2nd ed. Cambridge, MA: MIT Press.

Schopf, J. William, ed. 1983. *Earth's Earliest Biosphere.* Princeton, NJ: Princeton University Press.

—. 2001. *Cradle of Life: The Discovery of Earth's Earliest Fossils.* Princeton, NJ: Princeton University Press.

—, ed. 2002. *Life's Origin: The Beginnings of Biological Evolution.* Berkeley: University of California Press.

—. 2006. "Fossil Evidence of Archaean Life." *Philosophical Transactions of the Royal Society of London* 361:869–85.

Schopf, J. William, and D. J. Bottjer, eds. 2009. "World Summit on Ancient Microscopic Fossils." Special issue, *Precambrian Research* 173 (1–4), 222 pp.

Schopf, J. William, and C. Klein, eds. 1992. *The Proterozoic Biosphere: A Multidisciplinary Study.* Cambridge: Cambridge University Press.

Walter, M. R., ed. 1976. *Stromatolites.* Amsterdam: Elsevier.

Wills, Christopher, and Jeffrey Bada. 2000. *The Spark of Life: Darwin and the Primeval Soup.* Cambridge, MA: Basic Books.

Wood, Rachel. 1999. *Reef Evolution.* Oxford: Oxford University Press.

Chapter 4. Classification: The Diversity of Life

Cain, A. J. 1960. *Animal Species and Their Evolution.* New York: Harper and Brothers.

Howard, Daniel J., and Stephen H. Berlocher, eds. 1998. *Endless Forms: Species and Speciation.* Oxford: Oxford University Press.
Kunz, Werner. 2012. *Do Species Exist?: Principles of Taxonomic Classification.* Weinheim, Germany: Wiley-Blackwell.
Larson, James L. 1971. *Reason and Experience: The Representation of Natural Order in the Work of Carl von Linne.* Burbank, CA: University of California Press.
Lerwill, C. J. 1971. *An Introduction to the Classification of Animals.* London: Constable & Robinson.
Margulis, Lynn, and Karlene V. Schwartz. 1997. *Five Kingdoms: An Illustrated Guide to the Phyla of Life on Earth.* 3rd ed. San Francisco: W. H. Freeman.
Mayr, Ernst. 1942. *Systematics and the Origin of Species.* Cambridge, MA: Harvard University Press.
Mayr, Ernst, and Peter D. Ashlock. 1991. *Principles of Systematic Zoology.* New York: McGraw-Hill.
Phillips, Adam. 2001. *Darwin's Worms: On Life Stories and Death Stories.* New York: Basic Books.
Ritvo, Harriet. 1998. *The Platypus and the Mermaid, and Other Figments of the Classifying Imagination.* Cambridge, MA: Harvard University Press.
Rudwick, Martin J. S. 1985. *The Meaning of Fossils: Episodes in the History of Paleontology.* Chicago, IL: University of Chicago Press.
Savory, Theodore H. 1970. *Animal Taxonomy.* Portsmouth, NH: Heinemann.
Schuh, Randall T., and Andrew V. Z. Brower. 2009. *Biological Systematics: Principles and Applications.* 2nd ed. Ithaca, NY: Cornell University Press.
Simpson, George Gaylord. 1961. *Principles of Animal Taxonomy.* New York: Columbia University Press.
Smith, Andrew B. 1994. *Systematics and the Fossil Record.* Oxford: Wiley-Blackwell.
Stephenson, Robert, Roger Browne, and Marilyn Clay. 1992. *Exploring the Variety of Life.* Chicago, IL: Raintree Steck-Vaughn.
Stewart, Amy. 2005. *The Earth Moved: On the Remarkable Achievements of Earthworms.* Chapel Hill, NC: Algonquin Books.
Tudge, Colin T. 2000. *The Variety of Life: A Survey and a Celebration of All the Creatures That Have Ever Lived.* Oxford: Oxford University Press.

Chapter 5. Spineless Wonders

Boardman, Richard S., Alan H. Cheetam, and Albert J. Rowell, eds. 1987. *Fossil Invertebrates.* Palo Alto, CA: Blackwell Scientific.
Briggs, Derek E. G., Douglas H. Erwin, Frederick J. Collier, and Chip Clark. 1995. *Fossils of the Burgess Shale.* Washington, DC: Smithsonian Institution Press.
Clarkson, Euan N. K. 1998. *Invertebrate Paleontology and Evolution.* 4th ed. London: Chapman & Hall.

Conway Morris, Simon. 1998. *Crucible of Creation: The Burgess Shale and the Rise of Animals.* Oxford: Oxford University Press.

——. 2003. *Life's Solution: Inevitable Humans in a Lonely Universe.* Cambridge: Cambridge University Press.

Coppold, Murray, and Wayne Powell. 2000. *A Geoscience Guide to the Burgess Shale: Geology and Paleontology in Yoho National Park.* Field, BC: The Yoho-Burgess Shale Foundation.

Eldredge, Niles, and Steven M. Stanley, eds. 1984. *Living Fossils.* New York: Springer Verlag.

Fortey, Richard. 2001. *Trilobite: Eyewitness to Evolution.* New York: Vintage Books.

——. 2012. *Horseshoe Crabs and Velvet Worms: The Story of Animals and Plants That Time Has Left Behind.* New York: Alfred A. Knopf.

Glaessner, Martin F. 1984. *The Dawn of Animal Life: A Biohistorical Study.* New York: Cambridge University Press.

Gould, Stephen Jay. 1989. *Wonderful Life: The Burgess Shale and the Nature of History.* New York: W. W. Norton.

Live-Setti, Ricardo. 1963. *Trilobites.* 2nd ed. Chicago, IL: University of Chicago Press.

McMenamin, Mark. 1998. *The Garden of Ediacara.* New York: Columbia University Press.

McMenamin, Mark A., and Dianna L. McMenamin. 1990. *The Emergence of Animals, the Cambrian Breakthrough.* New York: Columbia University Press.

Murray, J. W., ed. 1985. *Atlas of Invertebrate Macrofossils.* London: Longman.

Nielsen, Claus. 1995. *Animal Evolution: Interrelationships of the Living Phyla.* Oxford: Oxford University Press.

Parker, Andrew. 2003. *In the Blink of an Eye.* New York: Basic Books.

Rudwick, M. J. S. 1970. *Living and Fossil Brachiopods.* London: Hutchinson.

Ryland, J. S. 1970. *Bryozoans.* London: Hutchinson.

Schopf, J. William. 1999. *Cradle of Life: The Discovery of the Earth's Earliest Fossils.* Princeton, NJ: Princeton University Press.

——. 2002. *Life's Origin: The Beginnings of Biological Evolution.* Berkeley: University of California Press.

Schwartz, Jeffrey H. 1999. *Sudden Origins: Fossils, Genes, and the Emergence of Species.* New York: John Wiley.

Simonetta, Alberto M., and Simon Conway Morris, eds. 2009. *The Early Evolution of Metazoa and the Significance of Problematic Taxa.* New York: Cambridge University Press.

Tudge, Colin. 2000. *The Variety of Life.* Oxford: Oxford University Press.

Valentine, James W. 2004. *On the Origin of Phyla.* Chicago, IL: University of Chicago Press.

Walker, Gabrielle. 2003. *Snowball Earth: The Story of the Great Global Catastrophe That Spawned Life As We Know It.* New York: Crown/Random House.

Whittington, H. B. 1985. *The Burgess Shale*. New Haven, CT: Yale University Press.
—. 1992. *Fossils Illustrated: Trilobites*. Woodbridge: The Boydell Press.
Willmer, Pat. 1990. *Invertebrate Relationships*. Cambridge: Cambridge University Press.

Chapter 6. Bones, Scales, and Fins

Benton, Michael J. 2005. *Vertebrate Paleontology*. 3rd ed. Oxford: Blackwell.
Carroll, Robert L. 1988. *Vertebrate Paleontology and Evolution*. New York: W. H. Freeman.
Colbert, Edwin H., Michael Morales, and Eli C. Minkoff. 2001. *Colbert's Evolution of the Vertebrates: A History of the Background Animals through Time*. 5th ed. New York: Wiley-Liss.
Forey, P. L. 1998. *History of the Coelacanth Fishes*. London: Chapman and Hall.
Gee, Henry. 1997. *Before the Backbone: Views on the Origin of Vertebrates*. New York: Chapman & Hall.
Janvier, P. 1996. *Early Vertebrates*. Oxford: Clarendon Press.
Jarvik, Erik. 1981. *Basic Structure and Evolution of Vertebrates*. 2 vols. London: Academic Press.
Long, John A. 1995. *The Rise of Fishes: 500 Million Years of Evolution*. Baltimore, MD: Johns Hopkins University Press.
Maisey, J. G. 1996. *Discovering Fossil Fishes*. New York: Henry Holt.
Moy-Thomas, J. A., and R. S. Miles. 1971. *Paleozoic Fishes*. 2nd ed. London: Chapman & Hall.
Schultze, Hans-Peter, and Linda Trueb eds. 1992. *Origins of the Higher Groups of Tetrapods: Controversy and Consensus*. Ithaca, NY: Cornell University Press.
Shubin, Neil H. 2008. *Your Inner Fish*. New York: Pantheon.
Smith, J. L. B. 1956. *Old Fourlegs: The Story of the Coelacanth*. London: Longmark.
Thomson, Keith S. 1991. *Living Fossil: The Story of the Coelacanth*. New York: W. W. Norton.
Weinberg, Samantha. 2000. *A Fish Caught in Time: The Search for the Coelacanth*. New York: Harper Collins.
Zimmer, Carl. 1998. *At the Water's Edge: Macroevolution and the Transformation of Life*. New York: Free Press.

Chapter 7. The Greening of the Land

Beerling, David. 2007. *The Emerald Planet: How Plants Changed Earth's History*. Oxford: Oxford University Press.
Cleal, Christopher J., and Barry A. Thomas. 2009. *An Introduction to Plant Fossils*. Cambridge: Cambridge University Press.

Graham, Linda. 1993. *Origins of Land Plants.* New York: John Wiley & Sons.
Iwatsuki, Kunio, and Peter H. Raven, eds. 1997. *Evolution and Diversification of Land Plants.* London: Springer Verlag.
Kennick, Paul, and Paul Davis. 2004. *Fossil Plants.* Washington, DC: Smithsonian Books.
Kennick, Paul, and Peter R. Crane. 1997. *The Origin and Early Diversification of Land Plants: A Cladistic Study.* Washington, DC: Smithsonian Books.
Lovelock, J. E. 1979. *Gaia: A New Look at Life on Earth.* Oxford: Oxford University Press.
Niklas, Karl J. 1997. *The Evolutionary Biology of Plants.* Chicago, IL: University of Chicago Press.
Stewart, Wilson N., and G. W. Rothwell. 1993. *Paleobotany and the Evolution of Plants.* 2nd ed. Cambridge: Cambridge University Press.
Taylor, Thomas N., and Edith L. Taylor. 2009. *The Biology and Evolution of Fossil Plants.* 2nd ed. Burlington, MA: Academic Press.
Thomas, Barry A., and Robert Spicer. 2002. *Evolution of Paleobiology of Land Plants.* 2nd ed. London: Springer.
Willis, K. J., and J. C. McElwain. 2002. *The Evolution of Plants.* Oxford: Oxford University Press.

Chapter 8. The Amphibian Foothold

Anderson, Jason, and Hans-Dieter Sues, eds. 2007. *Major Transitions in Vertebrate Evolution.* Bloomington: Indiana University Press.
Carroll, Robert L. 1988. *Vertebrate Paleontology and Evolution.* New York: W. H. Freeman.
———. 2009. *The Rise of the Amphibians: 365 Million Years of Evolution.* Baltimore, MD: Johns Hopkins University Press.
Clack, J. A. 2012. *Gaining Ground: The Origin and Early Evolution of Tetrapods.* 2nd ed. Bloomington: Indiana University Press.
Colbert, Edwin H., Michael Morales, and Eli C. Minkoff. 2001. *Colbert's Evolution of the Vertebrates: A History of the Backboned Animals through Time.* 5th ed. New York: Wiley-Liss.
Cracraft, Joel, and Michael J. Donoghue. 2004. *Assembling the Tree of Life.* New York: Oxford University Press.
Duellman, William E., and Linda Trueb. 1994. *Biology of Amphibians.* Baltimore, MD: Johns Hopkins University Press.
Gould, S. J. 1993. *Eight Little Piggies.* London: Jonathan Cape.
Heatwole, Harold, and Robert L. Carroll, eds. 2000. *Paleontology: The Evolutionary History of Amphibia.* Vol. 4 of *Amphibian Biology.* London: Beatty and Sons.
Laurin, Michel. 2010. *How Vertebrates Left the Water.* Berkeley: University of California Press.
Panchen, Alec L., ed. 1980. *The Terrestrial Environment and the Origin of Land Vertebrates.* London: Academic Press.

Shubin, Neil. 2008. *Your Inner Fish: A Journey into the 3.5 Billion Year History of the Human Body.* New York: Random House.
Steyer, Sebastian, and Alain Benetau. 2012. *Earth before the Dinosaurs.* Bloomington: Indiana University Press.
Thomson, Keith S. 1991. *Living Fossil: The Story of the Coelcanth.* New York: W. W. Norton.
Zimmer, Carl. 1999. *At the Water's Edge.* New York: Touchstone.

Chapter 9. The Reign of the Reptiles

Bakker, Robert T. 1986. *The Dinosaur Heresies.* New York: Zebra.
—. 1996. *Raptor Red.* New York: Bantam Books.
Benton, Michael J. 1991. *The Reign of the Reptiles.* New York: Crescent Books.
—. 2003. *When Life Nearly Died.* New York: Norton.
Brett-Suman, Michael K., Thomas R. Holtz, and James O. Farlow, eds. 1997. *The Complete Dinosaur.* 2nd ed. Bloomington: Indiana University Press.
Brusatte, Stephen L. 2012. *Dinosaur Paleobiology.* London: Wiley-Blackwell.
Cadbury, Deborah. 2001. *The Dinosaur Hunters.* London: Fourth Estate.
Callaway, Jack M., and Elizabeth L. Nicholls, eds. 1997. *Ancient Marine Reptiles.* San Diego, CA: Academic Press.
Charig, Alan. 1979. *A New Look at the Dinosaurs.* London: British Museum of Natural History.
Colbert, Edwin H. 1984. *The Great Dinosaur Hunters and Their Discoveries.* New York: Courier Dover Publications.
Colbert, Edwin H., Michael Morales, and Eli C. Minkoff. 2001. *Colbert's Evolution of the Vertebrates: A History of the Backboned Animals through Time.* 5th ed. New York: Wiley-Liss.
Currie, Philip J., and Kevin Padian. 1997. *Encyclopedia of Dinosaurs.* San Diego, CA: Academic Press.
Dingus, Lowell, and Timothy Rowe. 1997. *The Mistaken Extinction: Dinosaur Evolution and the Origin of Birds.* New York: W. H. Freeman.
Fastovsky, David E., and David B. Weishampel. 2005. *The Evolution and Extinction of the Dinosaurs.* 2nd ed. New York: Cambridge University Press.
Gillette, David G. 1994. *Seismosaurus: The Earth Shaker.* Princeton, NJ: Princeton University Press.
Gould, Stephen Jay. 1991. *Bully for Brontosaurus.* New York: W. W. Norton.
Horner, John R., and Dob Lessem. 1993. *The Complete T. Rex.* New York: Simon & Schuster.
Horner, John R., and James Gorman. 1988. *Digging Dinosaurs.* New York: Workman.
Jaffe, Mark. 2000. *The Gilded Dinosaur: The Fossil War between E. D. Cope and O. C. Marsh and the Rise of American Science.* New York: Crown.
McGowen, Christopher. 1983. *The Successful Dragons.* Toronto: Samuel Stevens.

———. 1991. *Dinosaurs, Spitfires, and Sea Dragons.* Cambridge, MA: Harvard University Press.

Norman, David B. 1985. *The Illustrated Encyclopedia of Dinosaurs.* New York: Crescent Books.

Novacek, Michael J. 1996. *Dinosaurs of the Flaming Cliffs.* New York: Anchor Books.

Padian, Kevin, ed. 1986. *The Beginning of the Age of Dinosaurs.* Cambridge: Cambridge University Press.

Paul, Gregory S. 1988. *Predatory Dinosaurs of the World.* New York: Simon & Schuster.

———. 2012. *The Princeton Field Guide to Dinosaurs.* Princeton, NJ: Princeton University Press.

Sues, Hans-Dieter, and Nicholas C. Fraser. 2010. *Triassic Life on Land: The Great Transition.* New York: Columbia University Press.

Sumida, Stuart S., and Karen L. Martin, eds. 1997. *Amniote Origins: Completing the Transition to Land.* London: Academic Press.

Wallace, David Rains. 1999. *The Bonehunters' Revenge: Dinosaurs, Greed, and the Greatest Scientific Feud of the Gilded Age.* New York: Houghton Mifflin.

Weishampel, David B., Peter Dodson, and Halszaka Osmalska, eds. 2004. *The Dinosauria.* 2nd ed. Berkeley: University of California Press

Chapter 10. The Air

Ackerman, Jennifer. 1998. "Dinosaurs Take Wing: The Origin of Birds." *National Geographic* 194, no. 1 (July): 74–99.

Chang, Mee-Mann, Pei-Ji Chen, Yuan-Ging Wang, and Yuan Wang, eds. 2008. *The Jehol Fossils: The Emergence of Feathered Dinosaurs, Beaked Birds, and Flowering Plants.* Oxford: Elsevier.

Chatterjee, Sankar. 1997. *The Rise of Birds: 225 Million Years of Evolution.* Baltimore, MD: Johns Hopkins University Press.

Chiappe, Luis M. 2007. *Glorified Dinosaurs: The Origin and Early Evolution of Birds.* New York: Wiley-Liss.

Chiappe, Luis M. and Lawrence M. Witmer, eds. 2002. *Mesozoic Birds: Above the Heads of Dinosaurs.* Berkeley, CA: University of California Press.

Colbert, Edwin H., Michael Morales, and Eli C. Minkoff. 2001. *Colbert's Evolution of the Vertebrates: A History of the Backboned Animals through Time.* 5th ed. New York: Wiley-Liss.

Currie, Philip J., Eva B. Koppelhus, Martin A. Shugan, and Joanna L. Wright, eds. 2004. *Feathered Dragons: Studies in the Transition from Dinosaurs to Birds.* Bloomington: Indiana University Press.

Dyke, Gareth, and Gary Kaiser, eds. 2011. *Living Dinosaurs: The Evolutionary History of Modern Birds.* New Jersey: John Wiley & Sons.

Feduccia, Alan. 1999. *The Origin and Evolution of Birds.* 2nd ed. New Haven, CT: Yale University Press.

Hanson, Thor. 2011. *Feathers: The Evolution of a Natural Miracle.* New York: Basic Books.
Kaiser, Gary. 2008. *The Inner Bird: Anatomy and Evolution.* Seattle: University of Washington Press.
Long, John. 2008. *Feathered Dinosaurs: The Origin of Birds.* New York: Oxford University Press.
Martynink, Matthew. 2012. *A Field Guide to Mesozoic Birds and Other Winged Dinosaurs.* Vernon, NJ: Pan Aves.
Mayr, Gerald. 2009. *Paleogene Fossil Birds.* Frankfurt, Germany: Springer-Verlag.
Norell, Mark, and Mick Ellison. 2005. *Unearthing the Dragon: The Great Feathered Dinosaur Discovery.* New York: Pi Press.
Shipman, Pat. 1999. *Taking Wing: Archaeopteryx and the Evolution of Bird Flight.* New York: Simon and Schuster.
Wellnhofer, Peter. 1991. *The Illustrated Encyclopedia of Pterosaurs.* New York: Crescent Books.
Witten, Mark P. 2013. *Pterosaurs.* Princeton, NJ: Princeton University Press.

Chapter 11. The Blossoming Earth

Behrensmeyer, Anna K., John D. Damuth, William A. DiMichele, Richard Potts, Hans-Dieter Sues, and Scott. L. Wing. 1992. *Terrestrial Ecosystems through Time: Evolutionary Paleontology of Terrestrial Plants and Animals.* Chicago: University of Chicago Press.
Friis, Else M., Peter Crane, and Kaj Ramsguard Pedersen. 2011. *Early Flowers and Angiosperm Evolution.* Cambridge: Cambridge University Press.
Friis, Else M., William G. Chaloner, and Peter R. Crane. 1987. *The Origins of Angiosperms and Their Biological Consequences.* Cambridge: Cambridge University Press.
Good, Ronald. 1974. *The Geography of Flowering Plants.* London: Longman Publishing Group.
Kenrick, Paul, and Paul Davis. 2004. *Fossil Plants.* London: Natural History Museum.
Miller, Frederic P., Agnes F. Vandome, and John McBrewster. 2009. *Evolution of Plants.* New York: Alphascript Publishing.
Niklas, Karl J. 1997. *The Evolutionary Biology of Plants.* Chicago, IL: University of Chicago Press.
Soltis, Douglas E., Pamela S. Soltis, Peter K. Endress, and Mark W. Chase. 2005. *Phylogeny and Evolution of Angiosperms.* Sunderland, MA: Sinauer.
Stebbins, G. Ledyard. 1974. *Flowering Plants: Evolution above the Species Level.* Cambridge, MA: Harvard University Press.
Stewart, Wilson N., and Gar W. Rothwell. 1993. *Paleobotany and the Evolution of Plants.* 2nd ed. Cambridge: Cambridge University Press.

Thomas, Barry A., and Robert A. Spicer. 1995. *Evolution and Paleobiology of Land Plants.* 2nd ed. London: Chapman & Hall.
Willis, K. J. and J. C. McElwain. 2002. *The Evolution of Plants.* Oxford: Oxford University Press.

Chapter 12. The Rise of the Mammals

Colbert, Edwin H., Michael Morales, and Eli C. Minkoff. 2001. *Colbert's Evolution of the Vertebrates: A History of the Backboned Animals through Time.* 5th ed. New York: Wiley-Liss
Kemp, T. S. 1982. *Mammal-like Reptiles and the Origin of Mammals.* London: Academic Press.
—. 2005. *The Origin and Evolution of Mammals.* Oxford: Oxford University Press.
Lillegraven, Jason A., Zofia Kielan-Jaworowska, and William A. Clemens, eds. 1980. *Mesozoic Mammals: The First Two Thirds of Mammalian History.* Berkeley: University of California Press.
Rose, Kenneth D. 2006. *The Beginning of the Age of Mammals.* Baltimore, MD: Johns Hopkins University Press.
Rose, Kenneth D., and J. David. Archibald, eds. 2005. *The Rise of Placental Mammals: Origins and Relationships of the Major Extant Clades.* Baltimore, MD: Johns Hopkins University Press.
Ross, D. 1992. *Elephant: The Animal and Its Ivory in African Culture.* Los Angeles, CA: Fowler Museum of Cultural History.

Chapter 13. The Mammalian Explosion

Chinsamy-Turan, Anusuya, ed. 2011. *Forerunners of Mammals: Radiation, Histology, Biology.* Bloomington: Indiana University Press.
Colbert, Edwin H., Michael Morales, and Eli C. Minkoff. 2001. *Colbert's Evolution of the Vertebrates: A History of the Backboned Animals through Time.* 5th ed. New York: Wiley-Liss.
Goswani, Anjali, and Anthony Friscia, eds. 2010. *Carnivore Evolution: New Views on Phylogeny, Form and Function.* Cambridge: Cambridge University Press.
Kemp, T. S. 2005. *The Origin and Evolution of Mammals.* Oxford: Oxford University Press.
Kurten, Bjorn. 1972. *The Age of Mammals.* New York: Columbia University Press.
MacFadden, Bruce J. 1992. *Fossil Horses: Systematics, Paleobiology and Evolution of the Family Equidae.* Cambridge: Cambridge University Press.
Prothero, Donald R. 2005. *After the Dinosaurs: The Age of Mammals.* Bloomington: Indiana University Press.
Prothero, Donald R. 2009. *Greenhouse of the Dinosaurs: Evolution, Extinction, and the Future of Our Planet.* New York: Columbia University Press.

Prothero, Donald R., and Robert M. Schock. 2002. *Horns, Tusks, and Flippers: The Evolution of Hoofed Mammals.* Baltimore, MD: Johns Hopkins University Press.
Prothero, Donald R., and Scott E. Foss. 2007. *The Evolution of Artiodactyls.* Baltimore, MD: Johns Hopkins University Press.
Savage, R. J. G. 1986. *Mammal Evolution: An Illustrated Guide.* New York: Facts-on-File and The British Museum.
Szalay, Frederick S., Michael J. Novacek, and Malcolm C. McKenna, eds. 1993. *Mammal Phylogeny.* New York: Springer-Verlag.
van der Geer, Alexandra, George Lyras, John deVos, and Michael Derunitzakis. 2010. *Evolution of Island Mammals: Adaptation and Extinction of Placental Mammals on Islands.* Hoboken, NJ: Wiley-Blackwell.
Werdelin, Lars, ed. 2010. *Cenozoic Mammals of Africa.* Berkeley: University of California Press.

Chapter 14. The Leakeys' Legacy

DeSalle, Bob, and Ian Tattersall. 2008. *Human Origins: What Bones and Genomes Tell Us about Ourselves.* College Station: Texas A&M University Press.
Fleagle, John G. 1998. *The Primate Adaptation and Evolution.* 2nd ed. New York: Academic Press.
Hartwig, Walter C., ed. 2002. *The Primate Fossil Record.* Cambridge: Cambridge University Press.
Issac, Glynn L., and Elizabeth R. McCown. 1976. *Human Origins: Louis Leakey and the East African Evidence.* Menlo Park, CA:W. A. Benjamin.
Johanson, Donald, and Blake Edgar. 1996. *From Lucy to Language.* New York: Simon & Schuster.
Johanson, Donald, and Kate Wong. 2010. *Lucy's Legacy: The Quest for Human Origins.* New York: Broadway Books.
Jones, Stephen, Robert D. Martin, David Pilbeam, and Sarah Bunney, eds. 1992. *Cambridge Encyclopedia of Human Evolution.* Cambridge: Cambridge University Press.
Larsen, Clark S., Robert M. Matter, and Daniel L. Gebo. 1998. *Human Origins: The Fossil Record.* 3rd ed. Prospects Heights, IL: Waveland Press.
Leakey, Richard. 1996. *The Origin of Humankind.* New York: Basic Books.
Lewin, Roger. 1987. *Bones of Contention: Controversies in the Search for Human Origins.* New York: Simon & Schuster.
———. 2004. *Human Evolution: An Illustrated Introduction.* 5th ed. Cambridge, MA: Blackwell Scientific.
Lewin, Roger, and Robert A. Foley. 2003. *Principles of Human Evolution.* 2nd ed. Oxford: Blackwell.
Mai, Larry L., Marcus Y. Owl, and M. Patricia Kershing. 2005. *The Cambridge Dictionary of Human Biology and Evolution.* Cambridge: Cambridge University Press.

Martin, Robert D. 1990. *Primate Origins and Evolution: A Phylogenetic Approach.* London: Chapman & Hall.
McKee, Jeffrey K., Frank E. Poirier, and W. Scott McGraw. 2004. *Understanding Human Evolution.* 5th ed. Upper Saddle River, NJ: Pearson.
Mortel, Virginia. 1995. *Ancestral Passions: The Leakey Family and the Quest for Human Origins.* New York: Simon & Schuster.
Olson, Steve. 2003. *Mapping Human History: Unravelling the Mystery of Adam and Eve.* London: Bloomsbury Publishing.
Palmer, D. 2010. *Origins: Human Evolution Revealed.* London: Mitchell Beazley, Octopus Publishing Group.
Poirier, F. E. 1990. *Understanding Human Evolution.* 2nd ed. Englewood Cliffs, NJ: Prentice Hall.
Savage, R. J. G., and M. R. Long. 1986. *Human Evolution.* London: British Museum.
Simons, E. L. 1981. "Man's Immediate Forerunners." *Philosophical Transactions of the Royal Society of London* 292:21–41.
Stringer, Christopher B., and Clive Gamble. 1993. *In Search of the Neanderthals.* London: Thames & Hudson.
Stringer, Christopher, and Peter Andrews. 2012. *The Complete World of Human Evolution.* 2nd ed. London: Thames & Hudson.
Stringer, Christopher, and Robin McKie. 1997. *African Exodus: The Origins of Modern Humanity.* New York: Henry Holt.
Tattersall, Ian. 1993. *The Human Odyssey: Four Million Years of Human Evolution.* Upper Saddle River, NJ: Prentice Hall.
Tattersall, Ian, and J. Schwartz. 2000. *Extinct Humans.* New York: Westview.
Tudge, Colin. 1995. *The Day before Yesterday.* London: Cape Publishing.
Wood, Bernard. 2006. *Human Evolution: A Very Short Introduction.* Oxford: Oxford University Press.
Wood, Bernard, and Brian G. Richmond. 2007. *Human Evolution: A Guide to Fossil Evidence.* Boulder, CO: Westview Press.
Zimmer, Carl. 2007. *Smithsonian Intimate Guide to Human Origins.* New York: Harper Paperbacks.

Chapter 15. "Endless Forms, Most Beautiful and Most Wonderful"

Carroll, Sean B. 2005. *Endless Forms Most Beautiful: The New Science of Evo Devo.* New York: W. W. Norton.
Clarkson, Euan N. K. 1998. *Invertebrate Paleontology and Evolution.* 4th ed. Hoboken, NJ: Wiley-Blackwell.
Cowen, Richard. 2013. *History of Life.* 5th ed. Hoboken, NJ: Wiley-Blackwell.
Erwin, Douglas, and Jim Valentine. 2013. *The Cambrian Explosion.* Greenwood Village, CO: Roberts and Co.
Fedonkin, Mikhail A., James G. Gehling, Kathleen Grey, Guy M. Narbonne, and Patricia Vickers-Rich. 2008. *The Rise of Animals: Evolution and*

Diversification of the Kingdom Animalia. Baltimore, MD: Johns Hopkins University Press.
Lehman, Ulrich T. 1981. *The Ammonites: Their Life and Their World.* Cambridge: Cambridge University Press.
Maynard Smith, John, and Eors Szathmary. 1998. *The Major Transitions in Evolution.* Oxford: Oxford University Press.
Monks, Neale, and Philip Palmer. 2002. *Ammonites.* Washington, DC: Smithsonian Books.
Morton, J. E. 1967. *Molluscs.* London: Hutchinson.
Scales, Helen. *Spirals in Time: The Secret Life and Curious Afterlife of Seashells.* New York: Bloomsbury USA.
Taylor, Paul D., and David N. Lewis. 2007. *Fossil Invertebrates.* Cambridge, MA: Harvard University Press.
Vermeij, Gelrat J. 1987. *Evolution and Escalation: An Ecological History of Life.* Princeton, NJ: Princeton University Press.

Chapter 16. On Extinction

Alvarez, Luis W., Frank Alvarez, Frank Asaro, and Helen V. Michel. 1980. "Extraterrestrial Cause for the Cretaceous-Tertiary Extinction." *Science* 208:1095–1008.
Alvarez, Walter. 1997. *T. Rex and the Crater of Doom.* Princeton, NJ: Princeton University Press.
Archibald, J. David. 1996. *Dinosaur Extinction and the End of an Era: What the Fossils Say.* New York: Columbia University Press.
Benton, Michael J. 2003. *When Life Nearly Died.* New York: W. W. Norton.
Benton, Michael J., and Richard J. Twitchett. 2003. "How to Kill (Almost) All Life: The End-Permian Extinction Event." *Trends in Ecology and Evolution* 18:358–65.
Courtillot, Vincent. 1999. *Evolutionary Catastrophies: The Science of Mass Extinctions.* Cambridge: Cambridge University Press.
Cuppy, Will. 1941. *How to Become Extinct.* Boston, MA: David R. Godine.
Donovan, Stephen K., ed. 1979. *Mass Extinctions: Processes and Evidence.* New York: Columbia University Press.
Durham, J. W. 1970. "The Fossil Record and the Origin of the *Deuterostomata.*" *Proceedings of the North American Paleontology Convention*, Chicago, 1969, section H, pp 1104–32. Chicago, IL: University of Chicago Press.
Ehrlich, Paul, and Anne Ehrlich. 1981. Extinction: The Causes and Consequences of the Disappearance of Species. New York: Random House.
Eldridge, Niles. 1986. *Time Frames: Rethinking of Darwinian Evolution and the Theory of Punctuated Equilibria.* New York: Simon & Schuster.
Erwin, Douglas H. 1993. *The Great Paleozoic Crisis: Life and Death in the Permian.* New York: Columbia University Press.
—. 2006. *Extinction: How Life on Earth Nearly Ended 250 Million Years Ago.* Princeton, NJ: Princeton University Press.

Glen, William, ed. 1994. *Mass-Extinction Debates: How Science Works in a Crisis.* Stanford, CA: Stanford University Press.

Goldsmith, Donald. 1986. *Nemesis: The Death Star and Other Theories of Mass Extinction.* New York: Walker.

Hallam, Anthony. 1989. *Great Geological Controversies.* 2nd ed. Oxford: Oxford University Press.

Hallam, Anthony, and P. B. Wignall. 1997. *Mass Extinctions and Their Aftermath.* Oxford: Oxford University Press.

Jablonski, David. 2005. "Mass Extinctions and Macroevolution." *Paleobiology* 31:192–210.

Jablonski, David, and David M. Raup. 1995. "Selectivity of End-Cretaceous Marine Bivalve Extinctions." *Science* 268:389–91.

Koeberl, Christian, and Kenneth G. MacLeod. 2002. *Catastrophic Events and Mass Extinctions: Impacts and Beyond.* Special paper 356. Boulder, CO: Geological Society of America.

Kolbert, Elizabeth. 2014. *The Sixth Extinction: An Unnatural History.* New York: Henry Holt and Company.

Lawton, John H., and Robert M. May. 1995. *Extinction Rates.* Oxford: Oxford University Press.

Leakey, Richard, and Roger Lewin. 1995. *The Sixth Extinction: Patterns of Life and the Future of Humankind.* New York: Doubleday.

MacLeod, Norman, and Gerta Keller, eds. 1995. *Cretaceous-Tertiary Mass Extinctions: Biotic and Environmental Changes.* New York: W. W. Norton.

Martin, Paul S., and Richard G. Klein, eds. 1984. *Quaternary Extinctions: A Prehistoric Revolution.* Tucson: University of Arizona Press.

Martin, R. D. 1993. "Primate Origins: Plugging the Gaps." *Nature* 363:223–34.

McGhee, George R., Jr. 1995. *The Late Devonian Mass Extinction: The Frasnian/Famennian Crisis.* New York: Columbia University Press.

Newell, Norman D. 1967. *Revolutions in the History of Life.* Special paper 89. Boulder, CO: Geological Society of America

Nitecki, M. H., ed. 1984. *Extinctions.* Chicago, IL: University of Chicago Press.

Novacek, Michael J., and Quentin D. Wheeler, eds. 1992. *Extinction and Phylogeny.* New York: Columbia University Press.

Officer, Charles, and Jake Page. 1996. *The Great Dinosaur Extinction Controversy.* New York: Addison Wesley.

Powell, James L. 1998. *Night Comes to the Cretaceous: Comets, Craters, Controversy, and the Last Days of the Dinosaurs.* New York: W. H. Freeman.

Prothero, Donald R. 1994. *The Eocene-Oligocene Transition: Paradise Lost.* New York: Columbia University Press.

Raup, David M. 1978. "Cohort Analysis of Generic Survivorship." *Paleobiology* 4:1–15.

——. 1986. *The Nemesis Affair.* New York: W. W. Norton.

——. 1991a. "A Kill Curve for Phanerozoic Marine Species." *Paleobiology* 17:37–48.

—. 1991b. *Extinction: Bad Luck or Bad Genes.* New York: W. W. Norton.
Raup, D. M., and S. M. Stanley. 1978. *Principles of Paleontology.* 2nd. ed. San Francisco, CA: Freeman.
Rees, Martin. 2003. *Our Final Hour.* New York: Basic Books.
Rickards, R. B. 1977. "Patterns of Evolution in the Graptolites." In *Patterns of Evolution*, edited by A. Hallam, 333–58. Amsterdam: Elsevier.
Sepkoski, J. J., Jr. 1992. "Phylogenetic and Ecologic Patterns in the Phanerozoic History of Marine Biodiversity." In *Systematics, Ecology, and the Biodiversity Crisis,* edited by N. Eldridge, 77–100. New York: Columbia University Press.
Sharpton, Virgil, and Peter D. Ward, eds. 1990. *Global Catastrophes in Earth History: An Interdisciplinary Conference on Impacts, Volcanism, and Mass Mortality.* Special paper 247. Boulder, CO: Geological Society of America
Silver, Leon, ed. 1983. *Geological Implications of Impacts of Large Asteroids and Comets on the Earth.* Special paper 190. Boulder, CO: Geological Society of America
Simpson, G. G. 1952. "How Many Species?" *Evolution* 6:342–62.
Stanley, Steven M. 1987. *Extinction.* New York: Scientific American Books.
Taylor, Paul D, ed. 2004. *Extinctions in the History of Life.* Cambridge: Cambridge University Press.
Valentine, J. W. 1970. "How Many Marine Invertebrate Species?" *Journal of Paleontology* 44:410–15.
Van Valen, L. 1973. "A New Evolutionary Law." *Evolutionary Theory* 1:1–30.
Ward, Peter D. 1994. *The End of Evolution: On Mass Extinctions and the Preservation of Biodiversity.* New York: Bantam Books.
—. 2004. *Gorgon: Paleontology, Obsession, and the Greatest Catastrophe in Earth's History.* New York: Viking Press.

Chapter 17. "Have Been and Are Being Evolved"

Barton, Nicholas H., Derek E. G. Briggs, Jonathan A. Elsen, David B. Goldstein, and Nipam H. Patel. 2007. *Evolution.* Cold Spring Harbor, NY: Cold Spring Harbor Laboratory Press.
Carroll, Sean B. 2006. *Endless Forms Most Beautiful: The New Science of Evo Devo.* New York: W. W. Norton.
Carroll, Sean B., Jennifer K. Grenier, and Scott. D. Weatherbee. 2004. *From DNA to Diversity: Molecular Genetics and the Evolution of Animal Design.* Malden, MA: Wiley-Blackwell.
Dawkins, Richard. 1989. *The Selfish Gene.* 2nd ed. Oxford: Oxford University Press.
Eldredge, Niles, and Joel Cracraft. 1980. *Phylogenetic Patterns and the Evolutionary Process.* New York: Columbia University Press.
Futuyma, Douglas J. 2009. *Evolution.* 2nd ed. Sunderland, MA: Sirauer Associates.

Gould, Stephen Jay. 1977. *Ontogeny and Phylogeny.* Cambridge, MA: Harvard University Press.
—. 2002. *The Structure of Evolutionary Theory.* Cambridge, MA: Belknap Press of Harvard University Press.
—. 2007. *Punctuated Equilibrium.* Cambridge, MA: Harvard University Press.
Grant, Verne. 1985. *The Evolutionary Process: A Critical Review of Evolutionary Theory.* New York: Columbia University Press.
Kemp, T. S. 1999. *Fossils and Evolution.* Oxford: Oxford University Press.
Kinschner, Mark W., and John C. Gerhart. *The Plausibility of Life: Resolving Darwin's Dilemma.* New Haven, CT: Yale University Press.
Sepkoski, David, and Michael Reese, eds. 2009. *The Paleobiological Revolution: Essays on the Growth of Modern Paleontology.* Chicago, IL: University of Chicago Press.
Weiner, Jonathan. 1995. *The Beak of the Finch.* New York: Random House.

Epilogue

Ayala, Francisco J. 2007. *Darwin's Gift to Science and Religion.* Washington, DC: Joseph Henry Press.
Barrow, John D., Simon Conway Morris, Stephen J. Freeland, and Charles L. Harper Jr. 2008. *Fitness of the Cosmos for Life: Biochemistry and Fine-Tuning.* Cambridge: Cambridge University Press.
Conway Morris, Simon. 1998. *The Crucible of Creation: The Burgess Shale and the Rise of Animals.* Oxford: Oxford University Press.
—. 2003. *Life's Solution: Inevitable Humans in a Lonely Universe.* Cambridge: Cambridge University Press.
—, ed. 2008. *The Deep Structure of Biology: Is Convergence Sufficiently Ubiquitous to Give a Directional Signal?* West Conshohocken, PA: Templeton Foundation Press.
Dawkins, Richard. 1986. *The Blind Watchmaker.* New York: W. W. Norton.
—. 2004. *The Ancestor's Tale: A Pilgrimage to the Dawn of Evolution.* Boston, MA: Houghton Mifflin.
Dennett, Daniel C. 1995. *Darwin's Dangerous Idea: Evolution and the Meanings of Life.* New York: Touchstone.
Gould, Stephen Jay. 1999. *Rocks of Ages: Science and Religion in the Fullness of Life.* New York: Ballantine.
Gribbin, John. 2011. *Why Our Planet Is Unique.* Hoboken, NJ: John Wiley.
Larson, Edward J. 2002. *Theory of Evolution: A History of Controversy.* Chantilly, VA: Teaching Company.
Miller, Kenneth R. 1999. *Finding Darwin's God: A Scientist's Search for Common Ground between God and Evolution.* New York: Harper Collins.
Prothero, Donald R. 2007. *Evolution: What the Fossils Say and Why It Matters.* New York: Columbia University Press.

Quammen, David. 2006. *The Reluctant Mr. Darwin: An Intimate Portrait of Charles Darwin and the Making of His Theory of Evolution.* New York: Atlas Books.

Rees, Martin. 2003. *Our Final Hour.* New York: Basic Books.

Ruse, Michael. 2001. *Can a Darwinian Be a Christian:? The Relationship between Science and Religion.* Cambridge: Cambridge University Press.

Ward, Peter D., and Donald Brownlee. 2000. *Rare Earth: Why Complex Life Is Uncommon in the Universe.* New York: Copernicus Books.

INDEX

Page numbers followed by letters *f* and *t* refer to figures and tables, respectively.